Neurodegenerative Diseases

Daniela Galimberti • Elio Scarpini
Editors

Neurodegenerative Diseases

Clinical Aspects, Molecular Genetics and Biomarkers

Springer

Editors
Daniela Galimberti, PhD
Neurology Unit
Alzheimer Center
University of Milan
Fondazione Ca' Granda
IRCCS Ospedale Policlinico
Milan
Italy

Elio Scarpini, MD
Neurology Unit
Alzheimer Center
University of Milan
Fondazione Ca' Granda
IRCCS Ospedale Policlinico
Milan
Italy

ISBN 978-1-4471-7242-0 ISBN 978-1-4471-6380-0 (eBook)
DOI 10.1007/978-1-4471-6380-0
Springer London Heidelberg New York Dordrecht

Printed on acid-free paper

Springer is part of Springer Science+Business Media (www.springer.com)

Preface

More than a century ago, Alois Alzheimer and Gaetano Perusini described the intriguing case of a 51-year-old severely demented woman. They argued that specific changes in the brain (senile plaques and neurofibrillary tangles) were responsible for neuronal loss and dementia. Nevertheless, in the following decades, the study of this disorder, named Alzheimer's disease (AD), has been largely limited to descriptive neuropathology and mere psychological assessment, with poor understanding of the biological and genetic bases of neurodegeneration. Over the last three decades instead, thanks also to the availability of new technologies, the field of neurodegeneration and dementia has developed significantly. First of all, it has become clearer that dementia cannot be equated with AD. Different neurodegenerative disorders, each with specific pathologic hallmarks, can lead to symptoms that partially overlap with AD. Together with a better understanding of the disease, significant developments have been done also in Frontotemporal Lobar Degeneration (FTLD), Lewy body disease, Parkinson dementia, and vascular dementia. The second important discovery in this field is that a number of cases are monogenic, thus caused by autosomal dominant mutations in specific genes, whose penetrance is almost complete. Besides, there are sporadic cases in which genetic variants contribute to the susceptibility to each disease, but are not causal themselves. The presence of a mutation allows to predict the development of a pathological condition before symptom onset. This concept opens the way to the identification of biomarkers for anticipating the diagnosis of dementia even in sporadic cases, allowing, ideally, a disease-modifying treatment that could block the aberrant mechanisms at the basis of neuronal death before the appearance of overt dementia.

In this scenario, this book is aimed to cover different aspects of neurodegeneration, including basic findings, genetics, clinical issues, diagnosis, therapy, biomarkers, and their use for enrichment of cohorts to be included in trials with novel disease-modifying compounds.

In the first part of the volume, Laura Ghezzi describes the clinical aspects of AD and the current symptomatic therapy with anticholinesterase inhibitors and memantine. Chiara Fenoglio introduces the topic of genetics and epigenetics with general notions on monogenic and polygenic forms of neurodegenerative disorders, whereas Mahdi Ghani and Ekaterina Rogaeva go further into the description of causal genes for AD, and Onofre Combarros gives a description of the genetic risk factors involved in sporadic forms, focusing particularly on recent wide-genome association studies. Then, Ana Verdelho and Francesca Clerici consider additional risk factors that contribute to the development of dementia. In particular, Ana Verdelho describes the influence of vascular factors on the development of dementia, whereas Francesca Clerici analyzes the environmental and lifestyle factors, such as smoke, alcohol consumption, schooling, etc, that influence the development of dementia.

In the second part of the book, additional neurodegenerative conditions are described, and an overview of biomarkers and tools supporting the diagnosis and useful to monitor the course of the disease or the response to treatment is given. In detail, Maria Serpente describes the current knowledge about FTLD, focusing mainly on the causal genes discovered in the last few years, with particular attention to the heterogeneity of clinical phenotypes associated with such genetic defects. Janet van Eersel, Fabien Delerue, Lars Ittner, and Yazi Ke give a comprehensive description of animal models currently available for a better understanding of both AD and FTLD. Right after, the topic of biomarkers is discussed by Niklas Mattsson and Henrik Zetterberg, who go into details of cerebrospinal fluid biomarkers, and Marco Bozzali and Laura Serra, who describe the potential usefulness of structural and functional imaging for diagnosis and monitoring of AD and FTLD. In the last part of the book, additional conditions related to neurodegeneration are presented and discussed. Nicola Ticozzi and Vincenzo Silani address amyotrophic lateral sclerosis (ALS), which leads to the selective death of motor neurons. Intriguingly, in the last few years, it has been demonstrated that the same mutations that lead to FTLD can also be responsible for ALS, or both, thus leading to the hypothesis that a continuum between these two neurodegenerative disorders exists. Camilla Ferrari, Benedetta Nacmias, and Sandro Sorbi give an elegant overview of uncommon dementias that, despite being rare, should be considered in differential diagnosis workup.

Lastly, in the very last few years, it has been demonstrated that there are common features between neurodegenerative diseases and other conditions that are not properly "neurodegenerative" but may present a neurodegenerative component. In this regard, Bernardo Dell'Osso, Gregorio Spagnolin, Neva Suardi, and Carlo Altamura describe common features between psychiatric disorders, particularly schizophrenia and FTLD. The volume closes with a chapter by Axel Petzold, focused instead on neurodegenerative aspects of multiple sclerosis, a demyelinating disease in which, over time, neuronal death occurs, leading to disability accumulation and a progressive course.

As the last few years have witnessed remarkable progress in multiple areas of neurodegeneration, including diagnosis, therapy, basic knowledge, genetics, and biomarkers, we hope that these chapters provide a broad, comprehensive, and up-to-date overview of the current knowledge in this rapidly evolving field. We do think that efforts put in this area will help to speed up the process of drug discovery for neurodegenerative diseases, where there still remains an unmet medical need.

Milan, Italy

Daniela Galimberti
Elio Scarpini

Abbreviations

AChEI	Anticholinesterase inhibitors
AD	Alzheimer's disease *or* axial diffusivity
ADGC	Alzheimer's Disease Genetics Consortium
AI	Ambulation index
ALS	Amyotrophic lateral sclerosis
ALS-FTD	Amyotrophic lateral sclerosis with associated frontotemporal dementia
AM	Anatomical connectivity mapping
AOA2	Ataxia-ocular apraxia type 2
APOE	Apolipoprotein E
APOE-ε4	Apolipoprotein E type 4 allele
APP	Amyloid precursor protein
APP	Amyloid precursor protein
ARHSP-TCC	Autosomal recessive hereditary spastic paraplegia with thin corpus callosum
BACE	Beta-site APP-cleaving enzyme
BMI	Body mass index
BOLD	Blood oxygenation level dependent
bp	Base pair
BPSD	Behavioral and psychological symptoms of dementia
bvFTD	Behavioral variant frontotemporal dementia
C9ORF	Chromosome 9 open reading frame
CADASIL	Cerebral autosomal dominant arteriopathy subcortical infarcts and leukoencephalopathy
CANTAB	Cambridge Neuropsychological Test Automated Battery
CBD	Corticobasal degeneration
CBS	Corticobasal syndrome
CHARGE	Cohorts for Heart and Aging in Genomic Epidemiology
ChIP	Chromatin immunoprecipitation
CHMP2B	Multivesicular body protein 2B
CJD	Creutzfeldt-Jakob disease

CLN6	Ceroid-lipofuscinosis neuronal 6
CNS	Central nervous system
CNV	Copy-number variation
CSF	Cerebrospinal fluid
CT	Computed tomography
DAN	Dorsal attention network
DAP12	DNAX-activating protein
DAT	Dopamine transporter
DIAN	Dominantly Inherited Alzheimer Network
DIR	Double inversion recovery
DIS	Dissemination in space
DIT	Dissemination in time
DMN	Default-mode network
DNMT	DNA methyltransferase
DRPLA	Dentatorubral-pallidoluysian atrophy
DSM-IV-TR	*Diagnostic and Statistical Manual of Mental Disorders, 4th Edition, Text Revision*
EADF	European Alzheimer's Disease Initiative
EAE	Experimental autoimmune encephalomyelitis
ECF	Extracellular fluid
EDSS	Extended disability status scale
EFNS	European Federation of Neurological Society
EIF2B	Eukaryotic translation initiation factor 2B
ELISA	Enzyme-linked immunosorbent assay
EMA	European Medicines Agency
EOAD	Early-onset alzheimer's disease
eQTL	Expression quantitative trait loci
FA	Fractional anisotropy
fALS	Familial amyotrophic lateral sclerosis
FDG-PET	Fluorodeoxyglucose positron emission tomography
FMR1	Fragile X mental retardation 1
fMRI	Functional magnetic resonance imaging
FTD	Frontotemporal dementia
FTLD	Frontotemporal lobar degeneration
FUS	Tumor-associated protein fused in sarcoma *or* ubiquitinated protein fused in sarcoma
FXTAS	Fragile X-associated tremor/ataxia syndrome
GABA	Gamma-aminobutyric acid
GB	Glucocerebrosidase
GBE1	Glycogen branching enzyme gene
GD	Gaucher's disease
GERARD	Genetic and Environmental Risk in Alzheimer's Disease
GFAP	Glial fibrillary acidic protein gene
GM	Gray matter
GRN	Progranulin

gSAP	γ(gamma)-secretase-activating protein
GWAS	Genome-wide association study
HAAS	Honolulu-Asia Aging Study
HAD	National Hospital Anxiety and Depression scale
HCHWA-D	Hereditary cerebral hemorrhage with amyloidosis – Dutch type
HEXA	β (Beta)-hexosaminidase A enzyme
HIV	Human immunodeficiency virus
HPRT	Hypoxanthine-guanine phosphoribosyltransferase enzyme
9HPT	9-hole peg test
HR	Hazard ratio
IBM	Inclusion body myopathy
IBMPFD	Inclusion body myopathy associated with Paget disease of the bone and frontotemporal dementia
IGAP	International Genomics of Alzheimer's Project
IL	Interleukin
iPCS	Induced pluripotent stem cells
IQ	Intelligence quotient
LA	Logopenic aphasia
LADIS	Leukoaraiosis and disability
LBD	Lewy body dementia
LD	Linkage disequilibrium
LGN	Lateral geniculate nucleus
LOAD	Late-onset alzheimer's disease
MAF	Minor allele frequency
MAPT	Microtubule-associated protein tau
MCI	Mild cognitive impairment
MD	Mean diffusivity
MELAS	Mitochondrial encephalopathy, lactic acidosis, and stroke
MEP	Motor evoked potentials
MIR	Microtubule-interacting region
miRNA	Micro RNA
MIT	Microtubule-interacting transport
MMSE	Mini-mental state exam
MND	Motor neuron disease
MRI	Magnetic resonance imaging
mRNA	messenger RNA
MRS	Magnetic resonance spectroscopy
MS	Multiple sclerosis
MSFC	Multiple Sclerosis Functional Composite
MSSS	Multiple Sclerosis Severity Score
NA	Neuroacanthocytoses
NART	National Adult Reading Test
Nf	Neurofilament
NfH	Neurofilament heavy

NfL	Neurofilament light
NFT	Neurofibrillary tangle
NGS	Next-generation sequencing
NIA-AA	National Institute on Aging-Alzheimer's Association
NINCDS-ADRDA	National Institute of Neurological and Communicative Diseases and Stroke/Alzheimer's Disease and Related Disorders Association
NMDA	*N*-methyl-D-aspartate
OCB	Oligoclonal bands
OCT	Optical coherence tomography
OR	Odds ratio
PANK2	Pantothenate kinase 2
PASAT	Paced auditory serial addition test
PCA	Posterior cortical atrophy
PCR	Polymerase chain reaction
PD	Parkinson's disease *or* proton density
PEN2	Presenilin enhancer 2
PET	Positron emission tomography
PGRN	Progranulin
PIB	[^{11}C] Pittsburgh Compound-B
PKAN	Pantothenate kinase-associated neurodegeneration
PLP	Proteolipid gene
PNFA	Primary nonfluent aphasia
PPA	Primary progressive aphasia
PPMS	Primary progressive multiple sclerosis
PSP	Progressive supranuclear palsy
P-tau	Phosphorylated tau
RCT	Randomized clinical trial
RD	Radial diffusivity
REST	Repressor element-1 silencing transcription factor
RFLP	Restriction fragment length polymorphism
RGC	Retinal ganglion cell
RISC	RNA-induced silencing complex
RNFL	Retinal nerve fiber layer
RR	Relative risk
RRMS	Relapsing-remitting multiple sclerosis
SALS	Sporadic amyotrophic lateral sclerosis
SCA	Spinocerebellar ataxia
SD	Semantic dementia
SILK	Stable isotope labeling kinetics
SLC20A2	Sodium dependent phosphate transporter
SNP	Single-nucleotide polymorphism
SP	Senile plaques *or* spastic paraparesis
SPECT	Single-photon emission computed tomography
SPG	Spastic paraplegia

SPG2	Spastic paraplegia type 2
SPP	Signal peptide peptidase
STR	Short tandem repeat or microsatellite
SYST-EUR	Systolic Hypertension in Europe
TALEN	Transcription activator-like effector nucleases
TBSS	Tract-based spatial statistics
TDP-43	Transactive response DNA-binding protein of 43 kDa
TDT	Transmission disequilibrium test
TM	Transmembrane
TREM2	Triggering receptor expressed on myeloid cells 2
T-tau	Total tau
TWT	Timed walk test
VaD	Vascular dementia
vAD	Variant AD
VBM	Voxel-based morphometry
VCP	Valosin-containing protein
VCT	Valosin-containing protein
VEP	Visual evoked potentials
VLCFA	Very long chain fatty acid
VNTR	Variable number of tandem repeat
VPS13A	Vacuolar protein sorting 13 homolog A
WHO	World Health Organization
WM	White matter
WMC	White matter changes
WML	White matter lesion
ZfN	Zinc finger nucleases

Contents

Contributors

A. Carlo Altamura, MD Department of Psychiatry, University of Milan, Fondazione IRCCS Cà Granda, Ospedale Maggiore Policlinico, Milan, Italy

Marco Bozzali, MD Neuroimaging Laboratory, Santa Lucia Foundation IRCCS, Rome, Italy

Francesca Clerici, MD, PhD Department of Biomedical and Clinical Sciences, Centre for Research and Treatment on Cognitive Dysfunctions, Institute of Clinical Neurology, "Luigi Sacco" Hospital, University of Milan, Milan, Italy

Onofre Combarros, MD, PhD Neurology Service and School of Medicine, University Hospital "Marqués de Valdecilla", University of Cantabria, Santander, Spain

Fabien Delerue Transgenic Animal Unit, School of Medical Sciences, University of New South Wales, Sydney, NSW, Australia

Bernardo Dell'Osso, MD Department of Psychiatry, University of Milan, Fondazione IRCCS Cà Granda, Ospedale Maggiore Policlinico, Milan, Italy

Janet Van Eersel, PhD Dementia Research Unit, School of Medical Sciences, University of New South Wales, Sydney, NSW, Australia

Chiara Fenoglio, PhD Department of Pathophysiology and Transplantation, University of Milan, Centro Dino Ferrari, Fondazione Cà Granda, IRCCS Ospedale Policlinico, Milan, Italy

Camilla Ferrari, MD Department of Neuroscience, Psychology, Drug Research and Child Health (NEUROFARBA), University of Florence, Florence, Italy

Daniela Galimberti, PhD Neurology Unit, Alzheimer Center, University of Milan, Fondazione Ca' Granda, IRCCS Ospedale Policlinico, Milan, Italy

Mahdi Ghani, MD, PhD Tanz Centre for Research in Neurodegenerative Diseases, University of Toronto, Toronto, ON, Canada

Laura Ghezzi, MD Neurology Unit, Department of Pathophysiology and Transplantation, University of Milan, Fondazione Cà Granda, IRCCS Ospedale Policlinico, Milan, Italy

Lars M. Ittner, MD Dementia Research Unit and Transgenic Animal Unit, School of Medical Sciences, University of New South Wales, Sydney, NSW, Australia

Yazi D. Ke, PhD Dementia Research Unit, School of Medical Sciences, University of New South Wales, Sydney, NSW, Australia

Niklas Mattsson, MD, PhD Department of Veterans Affairs Medical Center, Center for Imaging of Neurodegenerative Diseases, San Francisco, CA, USA

Clinical Neurochemistry Laboratory, Institute of Neuroscience and Physiology, Sahlgren's University Hospital/Mölndal, Mölndal, Sweden

Benedetta Nacmias, PhD Department of Neuroscience, Psychology, Drug Research and Child Health (NEUROFARBA), University of Florence, Florence, Italy

Axel Petzold, PhD Department of Neurology, VU University Medical Center, Amsterdam, The Netherlands

Ekaterina Rogaeva, PhD Faculty of Medicine, Neurology Division, Tanz Centre for Research in Neurodegenerative Diseases, University of Toronto, Toronto, ON, Canada

Elio Scarpini, MD Neurology Unit, Alzheimer Center, University of Milan, Fondazione Ca' Granda, IRCCS Ospedale Policlinico, Milan, Italy

Maria Serpente, PhD Neurology Unit, Department of Pathophysiology and Transplantation, Centro Dino Ferrari, University of Milan, Fondazione Cà Granda, IRCCS Ospedale Maggiore Policlinico, Milan, Italy

Laura Serra, PhD Neuroimaging Laboratory, Santa Lucia Foundation IRCCS, Rome, Italy

Vincenzo Silani, MD Department of Neurology and Laboratory of Neuroscience, IRCCS Istituto Auxologico Italiano, Milan, Italy

Department of Neurology–Stroke Unit and Neuroscience, IRCCS Istituto Auxologico Italiano, Milan, Italy

Sandro Sorbi, MD Department of Neuroscience, Psychology, Drug Research and Child Health (NEUROFARBA), University of Florence, Florence, Italy

Gregorio Spagnolin, MD Department of Psychiatry, University of Milan, Fondazione IRCCS Cà Granda, Ospedale Maggiore Policlinico, Milan, Italy

Neva E. Suardi, MD Department of Psychiatry, University of Milan, Fondazione IRCCS Cà Granda, Ospedale Maggiore Policlinico, Milan, Italy

Nicola Ticozzi, MD Department of Neurology and Laboratory of Neuroscience, IRCCS Istituto Auxologico Italiano, Milan, Italy

Department of Pathophysiology and Transplantation, Dino Ferrari Center, Universita degli Studi di Milano, Milan, Italy

Ana Verdelho, PhD, MD Department of Neurosciences, Hospital de Santa Maria, University of Lisbon, Lisbon, Portugal

Henrik Zetterberg, MD, PhD Clinical Neurochemistry Laboratory, Institute of Neuroscience and Physiology, The Sahlgrenska Academy at the University of Gothenburg, Mölndal, Sweden

Chapter 1
Genetics and Epigenetics: Basic Concepts

Chiara Fenoglio

Abstract Genetics is fundamental to our understanding of human variation, and by linking medical and evolutionary themes, it enables us to understand the origins and impacts of our genomic differences. The types of genetic variations used in genetic studies have changed over the last 20 years and can be classified into five major classes: RFLP (restriction fragment length polymorphism), VNTR (variable number of tandem repeat), STR (short tandem repeat or microsatellite), SNP (single-nucleotide polymorphism), and CNV (copy-number variation). Genetic linkage analysis using these tools helped to map and discover genes responsible for hundreds of hereditary diseases. Furthermore, construction of the international SNP database and recent development of high-throughput SNP typing platforms enabled us to perform genome-wide association studies, which have identified genes (or genetic variations) susceptible to common diseases. Moreover, in recent years genome-wide sequencing of individual DNAs is gaining relevant scope.

Likewise, epigenetic factors determined by gene-environment interactions, including systematic exposures or chance encounters with environmental factors in one's surroundings, add even more complexity to individual disease risk and the pattern of disease inheritance.

Epigenetics comprises the investigation of chemical modifications in the DNA and histones that regulates the gene expression or cellular phenotype.

Genetics and epigenetics, together with their newly designed technologies capable to analyze changes, have disclosed an appealing scenario that will offer for the biomedical sciences new insight for the study of neurodegenerative diseases, multifactorial complex diseases, and rare diseases. In this chapter, the main genetic and epigenetic variations will be overviewed together with the technologies adapted for their study, and the use of their modifications as possible biomarkers in several diseases will be summarized.

C. Fenoglio, PhD
Department of Pathophysiology and Transplantation, Neurology Unit, University of Milan, Fondazione Cà Granda, IRCCS Ospedale Policlinico,
F. Sforza 35, Milan 20122, Italy
e-mail: chiara.fenoglio@unimi.it

D. Galimberti, E. Scarpini (eds.), *Neurodegenerative Diseases*,
DOI 10.1007/978-1-4471-6380-0_1, © Springer-Verlag London 2014

Keywords Frontotemporal lobar degeneration • Tau • Progranulin (*GRN*) • *C9ORF72* • Genetics • Risk factor

Introduction

Most of neurological disorders could be considered multifactorial diseases. Many of these diseases "run in families," as they seem to recur in the relatives of affected individuals more frequently than in the general population. Moreover their inheritance generally does not follow a classical Mendelian pattern. Instead they are thought to result from complex interactions between a number of genetic and environmental factors, and therefore, they are said to follow a multifactorial (or complex) inheritance pattern. The familial clustering can be explained by recognizing that family members share a greater proportion of their genetic information and environmental exposures than do individuals chosen at random in the population. Thus, the relatives of an affected individual are more likely to experience the same gene-gene and gene-environment interactions that led to disease in the first place than are individuals who are unrelated to the proband. The multifactorial inheritance pattern that results represents an interaction between the collective effects of the genotype at one or, more commonly, multiple loci (polygenic or multigenic effects) either to raise or to lower susceptibility to disease, combined with a variety of environmental exposures that may trigger, accelerate, or protect against the disease process.

The gene-gene interactions in polygenic inheritance may be simply additive or much more complicated.

Gene-environment interactions, including systematic exposures or chance encounters with environmental factors in one's surroundings, add even more complexity to individual disease risk and the pattern of disease inheritance.

Herein, the main genetic variations will be described with regard to their use in unraveling the genetic basis of complex disorders. A brief overview on the evolution of techniques in genetic studies will be given. Lastly epigenetic changes will be considered with particular regard to their contribution in influencing neurological diseases.

Genetics

The human genome is constituted of around 6 billion base pairs (bp) of DNA stored in 23 chromosome pairs in the diploid organisms. Although today the exact number of genes remains unknown, it is thought that there are approximately 21,000 protein-coding genes (1–2 %) contained in the human genome. The remainder of the genome consists of RNA genes, regulatory sequences, and repetitive DNA in which the function is still not well clarified.

Human genetic variations are the differences in DNA sequences within the genome of individuals within populations. These variations can take many forms, including single-nucleotide variants or substitutions (SNPs); tandem repeats (short tandem repeats and variable number of tandem repeats); small indels (insertions and deletions of a short DNA sequence); duplications or deletions that change the copy number of a larger segment of a DNA sequence ($\geq$1 kb), i.e., copy-number variations (CNVs); and other chromosomal rearrangements such as inversions and translocations (also known as copy-neutral variations) [1–3]. The amount of genetic variation in the human genome is more abundant than previously thought, and this has been further corroborated with the findings from whole-genome resequencing studies where several millions of SNPs and several hundred thousand indels and structural variants were identified [4–6]. In addition to SNPs [7, 8], other genetic variations have also been found to be associated with various complex diseases and traits [9–11].

These genetic variations are typically referred to as either common or rare, to denote the frequency of the minor allele in the human population. Common variants are synonymous with polymorphisms, defined as genetic variants with a minor allele frequency (MAF) of at least one percent in the population, whereas rare variants have a MAF of less than 1 % [2].

The large majority of genetic variants are hypothesized to be neutral [12] (i.e., they do not contribute to phenotypic variation), achieving significant frequencies in the human population simply by chance.

Complex phenotypes of multifactorial disorders could be divided into two major categories: qualitative and quantitative traits. A genetic disease that is either present or absent is referred to as a discrete or qualitative trait. Conversely, the quantitative traits, are measurable physiological or biochemical quantities such as height, blood pressure.

Classes of Human Genetic DNA Variations

Restriction Fragment Length Polymorphisms

The importance of RFLP in medical research was first suggested by Botstein et al. [13]. RFLP is generated by differences in the size of the DNA fragment digested by a certain restriction endonuclease due to the base substitutions at the site recognized by the endonuclease. Before the discovery of polymerase chain reaction (PCR), RFLP was detected by Southern analysis and required a huge amount of non-degraded genomic DNAs. RFLP patterns show codominant Mendelian inheritance and can help distinguish between the parental alleles of the particular loci in our genome.

Variable Number of Tandem Repeats

The VNTRs were initially reported as one type of RFLP by Nakamura et al. [14], but they are highly polymorphic, with high heterozygosity in specific populations because of a great variety of the copy number of tandem repeat DNA sequences.

Isolation of VNTR markers was performed by an extension of the report of hypervariable "minisatellite" sequences by Jeffreys et al. [15]. Minisatellite probes identified very variable multiple loci simultaneously, but each VNTR marker (also called a single-locus minisatellite marker) identified a single highly polymorphic locus in the genome. Because of their highly polymorphic nature [16], both VNTRs and minisatellite markers were applied in forensic studies [17] and also involved in clinics [18].

Single-nucleotide polymorphism is the most prevalent class of genetic variation among individuals, and it has been estimated that the human genome contains at least 11 million SNPs, with 7 million of these occurring with an MAF of over 5 % [18] and the remaining having MAFs between 1 and 5 %. Depending on where a SNP occurs, it might have different consequences at the phenotypic level. SNP in the coding regions of genes that alter the function or structure of the encoded proteins is a necessary and sufficient cause of most of the known recessively or dominantly inherited monogenic disorders. In particular a variant may result in an amino acid change or may alter exon-intron splicing, thereby directly modifying the relevant protein, or it may exist in a regulatory region, altering the expression level. Alternatively an SNP may be in linkage disequilibrium (LD) with the "true" functional variant. LD, also known as allelic association, exists when alleles at two distinct loci of the genome are more highly associated than expected. To this end, the development of SNP-based LD maps could facilitate whole-genome association studies, leading to more efficient detection of candidate susceptibility genes.

Copy-Number Variation

Copy-number variation is defined as a form of genomic structural variation and refers to differences in the copy number of a particular genomic region. CNVs involve extensive genomic structural variation, ranging in size from kilobases (kb) to megabases (Mb), which are not identifiable by conventional chromosomal banding [19–21]. Deletions, duplications, duplications, triplications, insertions, and translocations can all result in CNVs. In addition, balanced genomic inversions leading to DNA structural variations that do not cause CNV can nevertheless contribute significantly to genome instability. Despite extensive studies, the total number, position, size, gene content, and population distribution of CNVs remain elusive. Recent analyses revealed 11.700 CNVs overlapping over 1,000 genes [22, 23]. It is now estimated that CNVs may account for 13 % of the human genome. CNVs can be inherited or sporadic; large de novo CNVs are thought more likely to be disease causative. However, the phenotypic effects of CNVs are sometimes unclear and depend mainly on whether dosage-sensitive genes or regulatory sequences are affected by the genomic rearrangement.

The Evolution of the Genetic Analysis of Complex Traits

Genetic Linkage Analyses

By using the polymorphic DNA markers, a genetic linkage map of all human chromosome was constructed, and also these DNA markers were made freely available to scientific communities. The interesting study from Botstein et al. [13] indicated that genetic loci responsible for genetic diseases could be mapped by linkage analysis if an appropriate number of families with certain genetic diseases were available. As polymorphic DNA markers allow us to distinguish a chromosome as of maternal or paternal origin, it can be examined whether each polymorphic "marker" allele co-segregates with the inheritance of a disease. In fact, Gusella et al. [24] determined the genetic locus of Huntington's disease to the short arm of chromosome 4. Subsequent to the increase of available DNA polymorphic markers in the late 1980s, many genes for relatively common genetic diseases, such as cystic fibrosis and neurofibromatosis type 1, were discovered and their responsible genes identified a few years later. These studies proved that linkage analysis by using DNA polymorphic markers in families with genetic diseases in any inheritance model was a powerful tool in the discovering of responsible genes even without any knowledge about the biological or biochemical mechanisms. This approach is known as the "reverse genetics" method.

It is now widely accepted that the PCR methods drastically changed thanks to the development of PCR systems. After 1990, scientists switched from RFLP analysis based on Southern technology to microsatellite analysis based on PCR technology. Microsatellite markers were first described by Weber and May [25] and are short segments of two or more base pairs repeated tandemly in tracts. Unlike VNTR loci, which number a few thousand in our genome, microsatellite loci are present at more than 100,000 regions that cover most of the genomic regions. Microsatellite markers have been successfully used for linkage analysis or population genetics because they can be easily adapted to the high-throughput system and require a very small amount of DNA. Due to their high levels of heterozygosity, they can distinguish paternal alleles with high probability and are very informative in linkage analysis.

One of the advantages of the approach using microsatellite markers is the possibility of applying linkage analysis (also known as homozygosity mapping) to recessive diseases with very low incidence for which only a very small number of patients can be collected for study [26].

Another advantage of microsatellite markers involves the application in a sib-pair analysis or transmission disequilibrium test (TDT), both of which are useful for searching genetic loci associated with common diseases such as multiple sclerosis (MS) [27].

Although these kinds of genetic approaches were used widely, they were not so successful in identifying genes or loci associated with common diseases, because a huge number of siblings or families were required to determine the genetic factors with very modest effects that increase the risk of disease.

Genome-Wide Association Study

In 2000 it was planned to develop millions of SNP markers covering an entire genome and construct a high-density SNP (also haplotype) map. On the basis of the "common variant-common disease" association hypothesis, SNPs were considered to be very useful as a tool for population genetics, particularly for identification of genes (or genetic variations) susceptible to various diseases. In 2003 multiple large-scale SNP genotyping platforms were developed, and it constructed an international consortium for making a SNP database for three major populations. The international HapMap project including six countries (Canada, China, Japan, Nigeria, the UK, and the USA) was settled in 2003. The main aims were (1) determination of the common patterns of one million or more DNA sequence variations in the human genome using DNA samples from populations with ancestry from parts of Africa, Asia, and Europe; (2) construction of the LD map for all chromosomes; and (3) making such information freely available to the scientific community. The HapMap results are expected to allow the discovery of sequence variants that affect common diseases, facilitate development of diagnostic tools, and enhance the ability to choose targets for therapeutic intervention [28]; after the very extensive efforts for the participating groups, the consortium constructed a database consisting of more than one million SNPs in 2005 [29] and subsequently reported and extended the database in 2007 [30]. Although the majority of genetic variations were commonly shared among three major populations, a small subset of variations was detected in one particular population.

However, there were many criticisms and skepticisms about the "common variation-common diseases" approach for identifying genes susceptible to common diseases at the beginning of the International HapMap project, but many published papers have shown the usefulness of the GWAS to uncover various genetic factors associated with various diseases.

Internationally, systematic GWAS was started in 2006 based on the accumulation of a large set of SNP information through the International HapMap project, as well as the development of cheap, commercially available, and accurate high-throughput SNP analysis platforms.

Next-Generation Sequencing

DNA sequencing in the laboratory has been possible since the 1970s, when the Sanger method was first developed and improved over time. However, this technique still remains too laborious and expensive for routine sequencing of whole genomes. In recent years a number of new sequencing technologies have been developed and have significantly reduced the cost and time required for sequencing. These post-Sanger technologies are collectively described as next-generation (NGS) technologies [31]. These technologies can be used for a wide range of applications, such as targeted resequencing and RNA sequencing.

NGS platforms have allowed for massive parallelization of sequencing reactions. This massive parallel sequencing has now allowed for an unprecedented interrogation of the variation in the human genome. For example, the 1000 Genomes Project, launched in 2008, is an international collaborative research project involving the Wellcome Trust Sanger Institute (England), the Beijing Genomics Institute (China), and the National Human Genome Research Institute (the USA), whose goal is to establish by far the most detailed catalog of human genetic variation [32]. The plan is to sequence the genomes of 250 anonymous participants from a number of different ethnic groups worldwide using a combination of methods. The results of a pilot study comparing different strategies for sequencing have already been published, and the sequencing of more than 1,000 genomes was completed in May 2011. This resource is publically available and can be used by researchers to identify variants in regions that are suspected of being associated with disease.

Genetics and Neurological Disorders

As we said previously, the genome of any given individual will contain millions of sequence variants of which the vast majority will have no effect (neutral variation) or will represent normal differences in phenotype (e.g., hair color). However, some may harbor pathogenic mutations that cause or predispose to disease. Determining if a single variant is associated with a disease can be a slow process, especially if the effect is subtle. Monogenic gene disorders are usually associated with rare, highly penetrant genetic mutations that have a profound effect on the function of a gene (e.g., by changing the coding sequence). However, the severity and penetrance of the phenotype can vary widely, and this could be due to the influence of other modifier genes. Such single-gene disorders tend to run in families with a clear inheritance pattern. In addition to rare, highly penetrant mutations, common variants in the population contribute to the susceptibility to common, complex neurological disease. These variants exert small effects on risk and are usually found in the noncoding portion of the genome. Assessing disease risk at the individual level based on these variants is challenging and, generally speaking, has limited clinical utility. Many other neurodegenerative disorders show an extensive family history. For example, Alzheimer's disease, frontotemporal dementia, and amyotrophic lateral sclerosis show rare but significant familial inherence, Mendelian forms of diseases, and lower-penetrance variants associated with the more common sporadic forms of disease [33].

Epigenetics

Epigenetics is the study of mechanisms able to alter the expression of genes without altering the DNA sequence. DNA methylation, histone modification, and miRNA-associated posttranscriptional gene silencing are the three most investigated epigenetic mechanisms. Although epigenetic changes are passed from parent to offspring through the germ line and are retained through successive cell divisions, they can be reversed and are highly sensitive to environmental influences [34, 35]. Hence, different epigenetic mechanisms involved in some pathogenic processes will be described.

DNA Methylation

DNA methylation involves the addition of a methyl group to the carbon-5 of a cytosine residue in DNA and is carried out by one of the several DNA methyltransferase (DNMT) enzymes. DNMT1 is the enzyme responsible for the maintenance of DNA methylation patterns during DNA replication. DNMT1 localizes to the DNA replication fork, where it methylates nascent DNA strands at the same locations as in the template strand [36]. DNMT3a and DNMT3b are involved in the de novo methylation of unmethylated and hemimethylated sites in nuclear and mitochondrial DNA, respectively [36, 37]. In mammals DNA methylation occurs predominantly at CpG sites—locations where a cytosine nucleotide is followed by a guanine nucleotide. CpG sites can occur in concentrations of up to several hundred dinucleotide repeats, called CpG islands, which are frequently found in gene promoter regions. The methylation or hypermethylation of CpG islands in promoter regions usually prevents the expression of the associated gene [38]. DNA methylation is currently the best-understood epigenetic mechanism and is known to have a crucial role in normal development, cell proliferation, and genome stability [39].

The design and development of techniques for the identification, quantification, and positioning of individual CpG methylation across the genome is a milestone that needs to be accomplished in order to provide a reliable characterization of the human epigenome.

Some research groups have shown the applicability of the methylation pattern of promoter genes analysis for clinical monitoring and its potential for diagnostics. Zhang et al. [40, 41] described the importance of screening high-risk populations for preclinical detection of occult liver neoplasms using epigenetic biomarkers such as DNA methylation of p16, p15, and RASSF1A genes. By using noninvasive techniques to collect DNA samples from serum, these authors used a methylation PCR-based assay to analyze the methylation of these genes and evaluate the methylation pattern to predict hepatocellular carcinoma susceptibility [40]. This approach, which is based on analyzing the methylation pattern of these genes, was proven to be more economical, sensitive, and reproducible than other techniques. Conversely, Wong et al. analyzed the methylation of p15/p16 promoters in plasma samples from patients with hepatocellular carcinoma and concluded that these promoters were hypermethylated in the 92 % of patients, thus suggesting the presence of circulating tumor cells [41]. The detection of circulating nucleic acids (ccfs) or isolation of the DNA from circulating cells has disclosed appealing potentialities in several diseases, with particular focus in cancer testing and prenatal diagnosis. In this last case, the discovery of epigenetic differences between the maternal and the fetal DNA methylation has opened new perspectives for clinical diagnosis using epigenetic markers in maternal plasma [42].

Histone Modification

In mammalian cells, histone proteins interact with DNA to form chromatin, the packaged form of DNA. Histones are octamer consisting of two copies of each of the four histone proteins: H2A, H2B, H3, and H4. Each histone octamer constitutes in 146 bp of the DNA strand wound around it to make up one nucleosome, which is the basic unit of chromatin. Histone proteins can be modified by posttranslational changes; among those there are acetylation, methylation, phosphorylation, ubiquitination, and citrullination. These histone modifications induce changes to the structure of chromatin and thereby affect the accessibility of the DNA strand to transcriptional enzymes, resulting in activation or repression of genes associated with the modified histone [43]. The best-understood histone modification is acetylation, which is mediated by histone acetyltransferases and deacetylases. Acetylation of histones is usually associated with upregulated transcriptional activity of the associated gene, whereas deacetylation of histones to transcriptional silencing [44].

Therapeutic strategies are designed to target epigenetic modifiers, such as histone deacetylases. Affecting the activity of this enzyme has been shown to be effective in myelodysplastic syndrome and acute myelogenous leukemia [45]. Thus, the analysis of histone acetylation levels on specific genes by chromatin immunoprecipitation (ChIP)-based technologies may be an interesting approach to monitor potential therapeutic strategies or follow the response of the patients to this therapy. Another recent interesting study has shown that histones released into the plasma enhance thrombin generation, a process that may contribute to microvascular thrombosis at sites of severe inflammation [46]. Under this point of view, the analysis of circulating histones in plasma may offer reliable information about the inflammation process. It has also been previously described that histones produce damage in endothelial cells and organ failure when injected into mice [47].

MicroRNA-Associated Gene Silencing

Single-stranded, noncoding micro (mi)RNAs are abundant in plants and animals and are conserved across species [48]. The raw transcripts undergo several nuclear and cytoplasmic posttranslational processing steps to generate mature, functional miRNAs. In the cytoplasm, mature miRNAs associate with other proteins to form the RNA-induced silencing complex (RISC), enabling the miRNA to imperfectly pair with cognate miRNA transcripts. The target mRNA is then degraded by the RISC, preventing its translation into protein [49, 50]. miRNA-mediated repression of translation is involved in many cellular processes, such as differentiation, proliferation, and apoptosis, as well as other key cellular mechanisms [51, 52].

Although the origin of some neurological diseases is still not well understood, it is increasingly acknowledged that epigenetics may contribute significantly to the diagnostics of these pathologies. miRNAs have been found in the central nervous system, and their functions have been well-established during neurogenesis and in some neurodegenerative disorders. One example is the analysis of miRNAs miR-9/miR-9* expression, which targets the transcription factor REST (repressor element-1 silencing transcription factor) involved in silencing neuronal gene expression in nonneuronal cells. Davidson et al. described that the miR-9/miR-9* expression assessed by quantitative PCR was downregulated in Huntington patients as compared with healthy controls [53], thus supporting the hypothesis that this analysis is a promising tool for monitoring Huntington's disease. Moreover, in MS there are now many evidences of an involvement of miRNAs in the disease pathogenesis [54]. Moreover, in the last 5 years, many evidences have put forward the implication of miRNAs in the development of many cancers, by oncogenes or suppressor gene-related mechanisms [55, 56]. A number of miRNAs involved in metastasis and invasion of breast cancer have been described [57, 58], and in the next years a battery of miRNAs that will serve as biomarkers for diagnosis and prognosis of several diseases may be available. Massagué et al. performed an Affymetrix HG-u133 plus 2.0 array to evaluate miRNAs expression in a breast cancer derivative cell line [57], observing that the microRNA miR-335 is involved in the inhibition of tumor reinitiation being downregulated in metastatic process. This effect was also observed by Png et al. [59]. Another promising miRNA is miR-206, which can inhibit notch3, inducing apoptosis and inhibiting tumor cell migration [60]. Thus, multiple miRNAs that regulate genes involved in different diseases are continuously emerging. Zampetaki and colleagues have observed, for example, downregulation of miR-126 in type 2 diabetes [61].

This breakthrough in the identification of miRNAs related to pathological process has opened new promising strategies for diagnostics. The relative stability of miRNAs in serum, plasma, urine, saliva, and other fluids [62] makes them suitable molecules to be analyzed in a clinical laboratory. Few evidences of the use of circulating miRNAs as possible disease biomarkers are now available. For example, we recently found a dysregulation of circulating miRNAs in serum from multiple sclerosis patients compared with controls [63].

Therefore, the study of miRNAs offers the opportunity to investigate whether specific alterations of miRNAs profile participates in metastatic pathway and will be useful for assessing the effectiveness of different therapies.

Beyond Genetic Variation: A Role for Epigenetics in Neurodegenerative Diseases?

A big outbreak in the application of epigenetic markers has occurred in cancer, autoimmune diseases, and psychiatric and neurodegenerative diseases over the past decades. The analysis of DNA methylation, histone modifications, and miRNAs has appeared as a new epigenetic field for prognosis and diagnosis. As such, increasing efforts are being focused for developing new methodologies and tests to set epigenetic biomarkers and their monitoring in clinical laboratories.

It is well-known that epigenetic mechanisms orchestrate a different range of important neurobiological and cognitive processes in the brain, for example, neurogenesis and brain development [64], neuronal activity [65], learning memory [66], and circadian rhythm [67], and disruption to these processes is likely to play a profound role in health and disease. Aberrant patterns of DNA methylation, for example, have been hypothesized to be involved in an increasing number of human neurobiological disease phenotypes including autism [68], psychosis [69], major depressive disorder [70], and recently Alzheimer's disease [71–74].

Technologies Used in Epigenetic Studies

Most of the innovative technologies used for epigenetic studies have been developed from conventional assays. For example, the classical method of DNA methylation analysis was based on the capability of two restriction enzyme pairs (*Hpa*II-*Msp*I and *Sma*I-*Xma*I) to recognize or discriminate methylated regions. However, this method has some weak points that depend on the efficiency of the enzymes, the step of Southern blot hybridization, and the expertise of user.

A major advance in DNA methylation analysis was the development of a method for sodium bisulfate modification of DNA to convert unmethylated cytosines to uracil, leaving methylated cytosines unchanged.

This method was the precursor of most of the new technologies to analyze DNA methylation. In the case of the classical method after bisulfite conversion, PCR amplification is performed followed by determination of the sequences of amplification. However, any researcher also needs to be familiarized with different techniques to obtain good results. In addition, it is too difficult processing larger amounts of samples manually and the critical step of bisulfate treatment could be not well performed, thus affecting the final results [75]. However, automated methods offer several advantages versus classical procedures. Among them automated analysis allows processing a large number of samples by a single technician. On the other hand, the automated technologies standardize the procedures, the results, and the analysis of the data. Moreover, automation and the use of these technologies deliver in high-throughput experiments, fast assays, and high reproducibility. Finally, the software of these systems offer high amount of information easy to interpret and analyze by the user.

It is essential that the epigenetic biomarkers that are applied to preclinical testing, diagnosis, disease progression, or treatment monitoring exhibit good sensitivity and reproducibility. Clearly, these technologies will allow us to discover epigenetic biomarkers for disease in the forthcoming years. They will also help identify or classify diseases and finally monitoring disease progression or the efficacy of a drug in those diseases in which genetics alone cannot give definitive answers.

References

1. Nakamura Y. DNA variations in human and medical genetics: 25 years of my experience. J Hum Genet. 2009;54:1–8.
2. Frazer KA, Murray SS, Schork NJ, Topol EJ. Human genetic variation and its contribution to complex traits. Nat Rev Genet. 2009;10:241–51.
3. Ku CS, Loy EY, Salim A, Pawitan Y, Chia KS. The discovery of human genetic variations and their use as disease markers: past, present and future. J Hum Genet. 2010;55:403–15.
4. Wheeler DA, Srinivasan M, Egholm M, Shen Y, Chen L, McGuire A, et al. The complete genome of an individual by massively parallel DNA sequencing. Nature. 2008;452:872–6.
5. Bentley DR, Balasubramanian S, Swerdlow HP, Smith GP, Milton J, Brown CG, et al. Accurate whole human genome sequencing using reversible terminator chemistry. Nature. 2008;456:53–9.
6. Kim JI, Ju YS, Park H, Kim S, Lee S, Yi JH, et al. A highly annotated whole genome sequence of a Korean individual. Nature. 2009;460:1011–5.
7. Altshuler D, Daly MJ, Lander ES. Genetic mapping in human disease. Science. 2008;322:881–8.
8. Hindorff LA, Sethupathy P, Junkins HA, Ramos EM, Mehta JP, Collins FS, Manolio TA. Potential etiologic and functional implications of genome-wide association loci for human diseases and traits. Proc Natl Acad Sci U S A. 2009;106:9362–7.
9. Haberman Y, Amariglio N, Rechavi G, Eisenberg E. Trinucleotide repeats are prevalent among cancer-related genes. Trends Genet. 2008;24:14–8.
10. Hannan AJ. Tandem repeat polymorphisms: modulators of disease susceptibility and candidates for 'missing heritability'. Trends Genet. 2010;26:59–65.
11. Stankiewicz P, Lupski JR. Structural variation in the human genome and its role in disease. Annu Rev Med. 2010;61:437–55.
12. Kimura M. Evolutionary rate at the molecular level. Nature. 1968;217:624–6.
13. Botstein D, White RL, Skolnick M, Davis RW. Construction of a genetic linkage map in man using restriction fragment length polymorphisms. Am J Hum Genet. 1980;32:314–31.
14. Nakamura Y, Leppert M, O'Connell P, Wolff R, Holm T, Culver M. Variable number of tandem repeat (VNTR) markers for human gene mapping. Science. 1987;235:1616–22.
15. Jeffreys AJ, Wilson V, Thein SL. Hypervariable 'minisatellite' regions in human DNA. Nature (London). 1985;314:67–73.
16. Odelberg SJ, Plaetke R, Eldridge JR, Ballard L, O'Connell P, Nakamura Y. Characterization of eight VNTR loci by agarose gel electrophoresis: implications for parentage testing and forensic individualization. Genomics. 1989;5:915–24.
17. Gatti R, Nakamura Y, Nussmeier M, Susi E, Shan W, Grody W. Informativeness of VNTR genetic markers for detecting chimerism after bone marrow transplantation. Dis Markers. 1989;7:105–12.

18. Vogelstein B, Fearon ER, Kern SE, Hamilton SR, Preisinger AC, Nakamura Y. Allelotype of colorectal carcinomas. Science. 1989;244:207–11.
19. Sebat J, Lakshmi B, Troge J, Alexander J, Young J, Lundin P, et al. Large-scale copy number polymorphism in the human genome. Science. 2004;305:525–8.
20. Iafrate AJ, Feuk L, Rivera MN, Listewnik ML, Donahoe PK, Qi Y, Scherer SW, Lee C. Detection of large-scale variation in the human genome. Nat Genet. 2004;36:949–51.
21. Redon R, Ishikawa S, Fitch KR. Global variation in copy number in the human genome. Nature. 2006;444:444–54.
22. Vogelstein B, Fearon ER, Hamilton SR, Kern S, Presinger AC, Leppert M. Genetic alterations during colorectal tumor development. N Engl J Med. 1988;319:525–32.
23. Baker SJ, Fearon ER, Nigro JM, Hamilton SR, Preisinger AC, Jessup JM, et al. Chromosome 17 deletions and p53 mutations in colorectal carcinomas. Science. 1989;244:217–21.
24. Gusella JF, Wexler NS, Connelly PM, Naylor SL, Anderson MA, Tanzi RE. A polymorphic DNA marker genetically linked to Huntington's disease. Nature. 1983;306:234–8.
25. Weber JL, May PE. Abundant class of human DNA polymorphisms which can be typed using the polymerase chain reaction. Am J Hum Genet. 1989;44:388–96.
26. Lander ES, Botstein D. Homozygosity mapping—a way to map human recessive traits with the DNA of inbred children. Science. 1987;236:1567–70.
27. The International HapMap Consortium. The international HapMap project. Nature. 2003;426:789–96.
28. Sawcer S. The complex genetics of multiple sclerosis: pitfalls and prospects. Brain. 2008;131:3118–31.
29. International HapMap Consortium. A haplotype map of the human genome. Nature. 2005;437:1299–320.
30. The International HapMap Consortium. A second generation human haplotype map of over 3.1 million SNPs. Nature. 2007;449:851–61.
31. Metzker ML. Sequencing technologies—the next generation. Nat Rev Genet. 2010;11(1):31–46.
32. Abecasis GR, Altshuler D, Auton A, 1000 Genomes Project Consortium. A map of human genome variation from population-scale sequencing. Nature. 2010;467(7319):1061–73.
33. Lill CM, Bertram L. Towards unveiling the genetics of neurodegenerative diseases. Semin Neurol. 2011;31(5):531–41.
34. Jaenisch R, Bird A. Epigenetic regulation of gene expression: how the genome integrates intrinsic and environmental signals. Nat Genet. 2003;33(Suppl):245–54.
35. Skinner MK, Manikkam M, Guerrero-Bosagna C. Epigenetic transgenerational actions of environmental factors in disease etiology. Trends Endocrinol Metab. 2010;21:214–22.
36. Goll MG, Bestor TH. Eukaryotic cytosine methyltransferases. Annu Rev Biochem. 2005;74:481–514.
37. Okano M, Bell DW, Haber DA, Li E. DNA methyltransferases Dnmt3a and Dnmt3b are essential for de novo methylation and mammalian development. Cell. 1999;99:247–57.
38. Klose RJ, Bird AP. Genomic DNA methylation: the mark and its mediators. Trends Biochem Sci. 2006;31:89–97.
39. Weber M, Schübeler D. Genomic patterns of DNA methylation: targets and function of an epigenetic mark. Curr Opin Cell Biol. 2007;19:273–80.
40. Zhang YJ, Wu HC, Shen J, Ahsan H, Tsai WY, Yang HI, et al. Predicting hepatocellular carcinoma by detection of aberrant promoter methylation in serum DNA. Clin Cancer Res. 2007;13:2378–84.
41. Wong IH, Lo YM, Yeo W, Lau WY, Johnson PJ. Frequent p15 promoter methylation in tumor and peripheral blood from hepatocellular carcinoma patients. Clin Cancer Res. 2000;6:3516–21.
42. Tong YK, Lo YM. Plasma epigenetic markers for cancer detection and prenatal diagnosis. Front Biosci. 2006;11:2647–56.
43. Dieker J, Muller S. Epigenetic histone code and autoimmunity. Clin Rev Allergy Immunol. 2010;39:78–84.

44. Brooks WH, Le Dantec C, Pers JO, Youinou P, Renaudineau Y. Epigenetics and autoimmunity. J Autoimmun. 2010;34:J207–19.
45. McDevitt MA. Clinical applications of epigenetic markers and epigenetic profiling in myeloid malignancies. Semin Oncol. 2012;39:109–22.
46. Ammollo CT, Semeraro F, Xu J, Esmon NL, Esmon CT. Extracellular histones increase plasma thrombin generation by impairing thrombomodulin-dependent protein C activation. J Thromb Haemost. 2011;9:1795–803.
47. Xu J, Zhang X, Pelayo R, Monestier M, Ammollo CT, Semeraro F, et al. Extracellular histones are major mediators of death in sepsis. Nat Med. 2009;15:1318–21.
48. Bernstein E, Allis CD. RNA meets chromatin. Genes Dev. 2005;19:1635–55.
49. Hwang HW, Mendell JT. MicroRNAs in cell proliferation, cell death, and tumorigenesis. Br J Cancer. 2006;94:776–80.
50. Sevignani C, Calin GA, Siracusa LD, Croce CM. Mammalian microRNAs: a small world for fine-tuning gene expression. Mamm Genome. 2006;17:189–202.
51. Chang TC, Mendell JT. MicroRNAs in vertebrate physiology and human disease. Annu Rev Genomics Hum Genet. 2007;8:215–39.
52. Fabbri M, Ivan M, Cimmino A, Negrini M, Calin GA. Regulatory mechanisms of microRNAs involvement in cancer. Expert Opin Biol Ther. 2007;7:1009–19.
53. Packer AN, Xing Y, Harper SQ, Jones L, Davidson BL. The bifunctional microRNA miR-9/miR-9* regulates REST and CoREST and is downregulated in Huntington's disease. J Neurosci. 2008;28:14341–6.
54. Fenoglio C, Ridolfi E, Galimberti D, Scarpini E. MicroRNAs as active players in the pathogenesis of multiple sclerosis. Int J Mol Sci. 2012;13(10):13227–39.
55. Esquela-Kerscher A. MicroRNAs function as tumor suppressor genes and oncogenes. In: Slack F, editor. MicroRNAs in development and cancer. London: Imperial College Press; 2010.
56. Esquela-Kerscher A, Slack FJ. Oncomirs—microRNAs with a role in cancer. Nat Rev Cancer. 2006;6:259–69.
57. Tavazoie SF, Alarcon C, Oskarsson T, Padua D, Wang Q, Bos PD, et al. Endogenous human microRNAs that suppress breast cancer metastasis. Nature. 2008;451:147–52.
58. Ma L, Teruya-Feldstein J, Weinberg RA. Tumour invasion and metastasis initiated by microRNA-10b in breast cancer. Nature. 2007;449:682–8.
59. Png KJ, Yoshida M, Zhang XH, Shu W, Lee H, Rimner A, et al. MicroRNA-335 inhibits tumor reinitiation and is silenced through genetic and epigenetic mechanisms in human breast cancer. Genes Dev. 2011;25:226–31.
60. Song G, Zhang Y, Wang L. MicroRNA-206 targets notch3, activates apoptosis, and inhibits tumor cell migration and focus formation. J Biol Chem. 2009;284:31921–7.
61. Zampetaki A, Kiechl S, Drozdov I, Willeit P, Mayr U, Prokopi M, et al. Plasma microRNA profiling reveals loss of endothelial miR-126 and other microRNAs in type 2 diabetes. Circ Res. 2010;107:810–7.
62. Zen K, Zhang CY. Circulating MicroRNAs: a novel class of biomarkers to diagnose and monitor human cancers. Med Res Rev. 2012;32:326–48.
63. Fenoglio C, Ridolfi E, Cantoni C, De Riz M, Bonsi R, Serpente M, et al. Decreased circulating miRNA levels in patients with primary progressive multiple sclerosis. Mult Scler. 2013;19(14):1938–42.
64. Ma DK, Marchetto MC, Guo JU, Ming GL, Gage FH, Song H. Epigenetic choreographers of neurogenesis in the adult mammalian brain. Nat Neurosci. 2010;13(11):1338–44.
65. Guo JU, Ma DK, Mo H, Ball MP, Jang MH, Bonaguidi MA, et al. Neuronal activity modifies the DNA methylation landscape in the adult brain. Nat Neurosci. 2011;14(10):1345–51.
66. Lubin FD, Roth TL, Sweatt JD. Epigenetic regulation of BDNF gene transcription in the consolidation of fear memory. J Neurosci. 2008;28(42):10576–86.
67. Nakahata Y, Grimaldi B, Sahar S, Hirayama J, Sassone-Corsi P. Signaling to the circadian clock: plasticity by chromatin remodeling. Curr Opin Cell Biol. 2007;19(2):230–7.

68. Wong C, Meaburn E, Ronald A, Price J, Jeffries A, Schalkwy L, et al. Methylomic analysis of monozygotic twins discordant for autism spectrum disorder (ASD) and related behavioural traits. Mol Psychiatry 2013. doi:10.1038/mp.2013.41. [Epub ahead of print].
69. Mill J, Tang T, Kaminsky Z, Khare T, Yazdanpanah S, Bouchard L, et al. Epigenomic profiling reveals DNA-methylation changes associated with major psychosis. Am J Hum Genet. 2008;82(3):696–711.
70. Mill J, Petronis A. Molecular studies of major depressive disorder: the epigenetic perspective. Mol Psychiatry. 2007;12(9):799–814.
71. Chouliaras L, Rutten BP, Kenis G, Peerbooms O, Visser PJ, Verhey F, et al. Epigenetic regulation in the pathophysiology of Alzheimer's disease. Prog Neurobiol. 2010;90(4):498–510.
72. Balazs R, Vernon J, Hardy J. Epigenetic mechanisms in Alzheimer's disease: progress but much to do. Neurobiol Aging. 2011;32(7):1181–7.
73. Mastroeni D, Grover A, Delvaux E, Whiteside C, Coleman PD, Rogers J. Epigenetic mechanisms in Alzheimer's disease. Neurobiol Aging. 2011;32(7):1161–80.
74. Mill J. Toward an integrated genetic and epigenetic approach to Alzheimer's disease. Neurobiol Aging. 2011;32(7):1188–91.
75. Grunau C, Clark SJ, Rosenthal A. Bisulfite genomic sequencing: systematic investigation of critical experimental parameters. Nucleic Acids Res. 2001;29:e65.

Chapter 2
Alzheimer's Disease: Clinical Aspects and Treatments

Laura Ghezzi

Abstract Alzheimer's disease (AD) is an age-dependent neurodegenerative disorder and the most common cause of dementia with aging. The early stages of AD are characterized by short-term memory loss. Once the disease progresses, patients experience difficulties in sense of direction, oral communication, calculation, ability to learn, and cognitive thinking. In addition, patients may develop language deficits, depression, aggressive behavior, and psychosis during the late stages, and eventually they need total care from caregivers. Currently diagnosis of AD is based either on clinical presentation or on biological biomarkers, in particular radiological and cerebrospinal fluid Amyloid, tau and phospho-tau levels. Here, the main clinical aspects and diagnostic tools for AD are revised; atypical AD presentations and possible diagnostic pitfalls are also discussed.

Keywords Alzheimer's disease • Biomarkers • Neurodegeneration

Introduction and Epidemiology

Alzheimer's disease (AD) is the most common cause of dementia in the elderly, with a prevalence of 5 % after 65 years of age, increasing to about 30 % in people aged 85 years or older. The highest prevalence is observed in North America and Western Europe (6.4 and 5.4 %, respectively), followed by Latin America (4.9 %) and China and western Pacific (4.0 %). Indeed the annual regional dementia incidence rates (per 1,000 individuals in the population) are estimated to be 10.5 for North America, 8.8 for Western Europe, 9.2 for Latin America, and 8.0 for China western Pacific [1, 2].

L. Ghezzi, MD
Neurology Unit, Department of Pathophysiology and Transplantation, University of Milan, Fondazione Cà Granda, IRCCS Ospedale Policlinico, Via F. Sforza 35, Milan 20122, Italy
e-mail: lauraghezzi@me.com

D. Galimberti, E. Scarpini (eds.), *Neurodegenerative Diseases*,
DOI 10.1007/978-1-4471-6380-0_2, © Springer-Verlag London 2014

Originally described by Alois Alzheimer in 1906, the disease is characterized clinically by progressive cognitive impairment, including impaired judgment, decision-making, and orientation, often accompanied, in later stages, by psychobehavioral disturbances as well as language impairment [3].

It is generally believed that the condition develops from multiple factors, with increasing age bringing the greatest risk. Up to 3 % of cases are linked to genetic causes; medical history and lifestyle are also contributing factors [4].

Clinical Aspects

Disease onset is usually characterized by memory loss for recent events, associated with repetitive questioning and loss of ability to learn. Past memories are usually conserved, instead recent information, such as daily agenda or objects location, are lost (Ribot's law: recent memories are more likely to be lost than the more remote memories). Patient's awareness of memory lost generates depression and anxiety, but consciousness is quickly replaced by anosognosia and the patient loses his critical abilities. With disease progression, visuospatial deficits and dyscalculia appear. Caregivers report episodes of disorientation in known places, such as the patient's neighborhood or even his own home. Dressing apraxia thwart patients' ability to dress themselves: they are neither able to choose the correct cloth nor to wear it; they need assistance even to put on a pair of trousers. In the late stages of the disease, apraxia affects every task of daily life, making impossible even the simplest action, such as taking a shower. Dyscalculia causes troubles with money, in particular, cash. Patients cannot distinguish between 50 and 500; in this phase they often lose money and are victims of cheaters. Prosopagnosia completes the clinical picture at the late stages of the disease. The patient is unable to recognize his friends or relatives' faces, making coping with their disease even more difficult. Communication also becomes a problem as vocabulary shrinks and fluency falters. Neuropsychiatric symptoms might appear too, such as wandering, irritability, disinhibition, apathy, psychosis, and affective and hyperactive behaviors. Differently from other type of dementia, such as frontotemporal lobar degeneration and primary progressive aphasia, language and/or behavioral symptoms are rarely present at the beginning of the disease. Unfortunately, with disease progression, agitation and aggressiveness are frequent. These symptoms are referred to as behavioral and psychological symptoms of dementia (BPSD) and occur in the majority of persons with dementia over the course of the disease and in 35–75 % of patients with mild cognitive impairment (MCI). Their identification and treatment is essential because they are associated with declining in cognitive and functional ability, decreased quality of life, and increased institutionalization [5]. BPSD represent a heavy burden for caregivers and the society; their treatment is difficult and often requires antipsychotic drugs, even if their use is related to an increased cerebrovascular risk in demented patients [6].

Sometimes somatic comorbidity and environmental triggers can be identified as a reversible cause, but often there is no external trigger and progressive dementia is the only cause.

Life expectancy of the population with the disease is reduced; the mean life expectancy following diagnosis is approximately 7 years. Pneumonia and dehydration are the most frequent immediate causes of death, while cancer is a less frequent cause of death than in the general population [7].

Mild Cognitive Impairment

The term "mild cognitive impairment" refers to a state in which cognitive deficits are present, but they do not hinder daily life activities [8]. New diagnostic criteria better define this "pre-dementia state" (see after) and distinguish between MCI patients those at higher risk of developing AD [9].

This distinction is particularly important because it has, or will have in the future, important therapeutic implication. Indeed all the newly developed drugs aim at this stage in order to prevent neuropathological changes and not only to halt them.

Diagnostic Criteria

In 1984, McKhann and colleagues published the first NINCDS-ADRDA criteria [10]. These widely accepted criteria supported a probabilistic diagnosis of AD within a clinical context where there is no definitive diagnostic biomarker. A definite diagnosis of AD was only made with histopathological confirmation of the clinical diagnosis [11].

Today diagnosis of AD is based on clinical symptoms and radiological and biomarker findings. Pathophysiological markers mirror the two degenerative processes characteristic of Alzheimer pathology: the deposition of amyloid-beta in neuritic plaques and the tauopathy path to neurofibrillary tangles. Moreover pathological levels of cerebrospinal fluid (CSF) biomarkers (low amyloid-beta, high tau and phospho-tau, and, even more specifically, abnormal ratio of tau to amyloid-beta) are associated with very high rates of progression from amnestic MCI to AD [9].

Following the 2007 revision of National Institute of Neurological and Communicative Diseases and Stroke/Alzheimer's Disease and Related Disorders Association (NINCDS-ADRDA) criteria (Table 2.1), the term prodromal AD is used to refer to the early pre-dementia phase of AD in which clinical symptoms are present, but not sufficiently severe to affect instrumental activities of daily living, and biomarker evidence from CSF or imaging is supportive of the presence of AD pathological changes. The state in which evidence of amyloidosis in the brain (with retention of specific positron emission tomography [PET] amyloid tracers)

Table 2.1 Revised NINCDS-ADRDA diagnostic criteria for Alzheimer's disease

Probable AD: A plus one or more supportive features B, C, D, or E

Core diagnostic criteria

1. Presence of an early and significant episodic memory impairment that includes the following features:
   (a) Gradual and progressive change in memory function reported by patients or informants over more than 6 months
   (b) Objective evidence of significantly impaired episodic memory on testing: this generally consists of recall deficit that does not improve significantly or does not normalize with cueing or recognition testing and after effective encoding of information has been previously controlled
   (c) The episodic memory impairment can be isolated or associated with other cognitive changes at the onset of AD or as AD advances

Supportive features

1. Presence of medial temporal lobe atrophy
2. Volume loss of hippocampi, entorhinal cortex, amygdala evidenced on MRI with
3. Qualitative ratings using visual scoring (referenced to well-characterized population with age norms) or quantitative volumetry of regions of interest (referenced to well-characterized population with age norms)
4. Abnormal cerebrospinal fluid biomarker
5. Low amyloid β1-42 concentrations, increased total tau concentrations, increased phospho-tau concentrations, or combinations of the three
6. Other well-validated markers to be discovered in the future
7. Specific pattern on functional neuroimaging with PET
8. Reduced glucose metabolism in bilateral temporal parietal regions
9. Other well-validated ligands, including those that foreseeably will emerge such as Pittsburg compound B or FDDNP
10. Proven AD autosomal dominant mutation within the immediate family

Exclusion criteria

1. History
2. Sudden onset
3. Early occurrence of the following symptoms: gait disturbances, seizures, behavioral changes, clinical features
4. Focal neurological features including hemiparesis, sensory loss, visual field deficits
5. Early extrapyramidal signs
6. Other medical disorders severe enough to account for memory and related symptoms
7. Non-AD dementia
8. Major depression
9. Cerebrovascular disease
10. Toxic and metabolic abnormalities, all of which may require specific investigations
11. MRI FLAIR or T2 signal abnormalities in the medial temporal lobe that are consistent with infectious or vascular insults

Criteria for definite AD

AD is considered definite if the following are present:

1. Both clinical and histopathological (brain biopsy or autopsy) evidence of the disease, as required by the NIA-Reagan criteria for the postmortem diagnosis of AD; criteria must both be present
2. Both clinical and genetic evidence (mutation on chromosome 1, 14, or 21) of AD; criteria must both be present

or in the CSF (with changes in amyloid β, tau, and phospho-tau concentrations) is not associated with any neuropsychological deficit is referred to as "preclinical AD" [12]. In the 2010 revision of the NINCDS-ADRDA criteria and definitions of AD clinical spectrum, "preclinical AD" has been split in two subgroups: the "asymptomatic at-risk state for AD" and the "presymptomatic AD." The first refers to cognitive normal people with positive AD biomarkers; it is important to stress the "at risk" since we do not know much about the value of these biological changes to predict the further development of the disease. Instead the term "presymptomatic AD" applies to individuals who will develop AD. This can be ascertained only in families that are affected by rare autosomal dominant monogenic AD mutations (monogenic AD) [8].

In 2010 the EFNS (European Federation of Neurological Society) published other revised criteria and guidelines for the diagnosis and management of AD. Following these criteria, in patients with suspected AD, other causes of dementia should be ruled out; screening for vitamin B12, folate, thyroid-stimulating hormone, calcium, glucose, complete blood cell count, and renal and liver function abnormalities should be performed. Serological tests for syphilis, Borrelia, and human immunodeficiency virus (HIV) should be considered in individual cases at high risk. Neuropsychological examination is mandatory because the diagnosis of dementia requires evidence of multiple cognitive defects, and initial stages of all principal forms of dementia have a selective anatomical localization reflected by typical patterns of neuropsychological impairment. Episodic memory is the function most commonly impaired in early AD; retrieval, which depends on frontal lobe and subcortical structures, is less affected. A predominance of executive dysfunction over episodic memory impairment is typical for frontotemporal lobar degeneration (FTLD) and vascular dementia (VaD) and is more frequent in early-onset AD. Language (speech comprehension and production, reading, and writing), praxis, and visuospatial abilities can be variably affected according to type and stage of dementia. Brain structural imaging (computed tomography [CT] scan or better magnetic resonance imaging [MRI]) shows prominent hippocampal atrophy. Fluorodeoxyglucose positron emission tomography (FDG-PET) may reveal reduced glucose metabolism in the parietal and superior/posterior temporal regions, posterior cingulate cortex, and precuneus. In advanced stages of AD, frontal lobe defects are also seen. Routine CSF cell count, protein, glucose and protein electrophoresis assessment is mandatory when vasculitis, inflammatory, hematologic, or demyelinating disease is suspected and in cases of suspected Creutzfeldt-Jakob disease (CJD). The elevation of the 14-3-3 protein reflects acute neuronal loss and supports diagnosis of CJD, while high to very high levels of total tau yield high specificity for CJD. In AD decreased levels of amyloid-beta 42 (Aβ42) and increased total tau and phospho-tau in CSF are frequently found.

In EFNS diagnostic panel CSF biomarkers aren't mandatory for the diagnosis of AD, and their sensitivity and specificity are reported to be too low to be reliable [13].

Finally, also the *Diagnostic and Statistical Manual of Mental Disorders, 4th Edition, Text Revision* (DSM-IV-TR) provides diagnostic criteria for AD. The

DSM-IV criteria for AD require a gradual onset and progressive impairment in memory function and at least one other cognitive domain that results in impairment of social and occupational function; such cognitive impairment would not be explained by other psychiatric, neurological, or systemic diseases [14].

Treatment

Pharmacological treatment of AD actually involves anticholinesterase inhibitors (AChEIs) and memantine, which provide mainly symptomatic short-term benefits without counteracting the progression of the disease.

The Cholinergic Hypothesis

In the past few decades, treatment for AD has largely involved replacement of neurotransmitters known to be lacking in AD, mostly based on the "cholinergic hypothesis" of AD. This hypothesis states that a deficit in central cholinergic transmission caused by degeneration of the basal forebrain nuclei is an important pathological and neurochemical feature of AD. A progressive loss of nicotinic receptors over the disease course of AD has also been described, and there is evidence of a role for these receptors in memory and cognition deficits [15].

To improve cholinergic neurotransmission, different strategies have been investigated including increasing acetylcholine synthesis, the augmentation of presynaptic acetylcholine release, the stimulation of cholinergic postsynaptic muscarinic and nicotinic receptors, and the reduction of acetylcholine synaptic degradation with cholinesterase inhibitors. Current data do not support the use of precursors of acetylcholinesterase, presynaptic releasing agents, or muscarinic agonists because of a lack of efficacy and unacceptable side effects.

Anticholinesterase inhibitors act by restricting the cholinesterase enzyme from breaking down acetylcholine, thus increasing the concentration and duration of acetylcholine at sites of neurotransmission. So far, three AChEIs are approved for the treatment of patients with mild-to-moderately severe AD: donepezil, rivastigmine, and galantamine.

Donepezil is a reversible, specific AChEI. It is easily absorbed by the body and can be taken once a day, initially at 5 mg and then, after 4 weeks' use, titrated up to 10 mg/day. It was approved for the treatment of mild-to-moderate AD in 1996/1997 [15]. Possible side effects include bradycardia (particularly in people with sick sinus syndrome or other supraventricular cardiac conduction conditions), seizures, nausea, vomiting, diarrhea, muscle cramp, urinary incontinence, fatigue, insomnia, and dizziness [16].

Rivastigmine tartrate is a selective inhibitor of acetylcholinesterase and also butyrylcholinesterase. Owing to its short half-life (1.5 h), it has to be taken twice a day. Doses start at 3 mg/day and increase gradually to between 6 and 12 mg/day. It can be taken orally or by a transdermal patch, with doses of either 4.6 or 9.5 mg/24 h. Care should be used with people with renal disease, mild or moderate liver disease, sick sinus syndrome, conduction abnormalities, gastric or duodenal ulcers, and a history of asthma or obstructive pulmonary disease. The main possible side effects found are nausea and vomiting, usually in the dose escalation phase [17].

Galantamine was originally made from snowdrop and narcissus bulbs but is now synthetically produced. It is a reversible AChEI, with a half-life of about 7 h, indicating that it should be taken twice a day at the recommended dose of 16–24 mg each time. An alternative version is taken once a day at doses of 8, 16, or 24 mg. The side effects from galantamine are similar to those of the other AChEIs and are mainly gastrointestinal (abdominal pain, diarrhea, nausea, and vomiting), although bradycardia and dizziness have been also reported [18].

Memantine is a voltage-dependent, moderate-affinity, uncompetitive *N-methyl-D*-aspartate (NMDA) receptor antagonist that blocks the effects of pathologically elevated tonic levels of glutamate, which may lead to neuronal dysfunction. It is approved for the treatment of moderate-to-severe AD (measured by a Mini-Mental State Exam [MMSE] score of ≤20). Memantine is taken orally twice a day. The starting dose is 10 mg/day and can be increased to a maximum daily dose of 20 mg/day. Caution should be used when prescribing memantine for people with renal failure or epilepsy; it is also contraindicated for people with severe renal impairment. Side effects may include dizziness, confusion, headache, and incontinence [17].

The Amyloid Hypothesis

On the contrary, the amyloid-beta hypothesis suggests that amyloid-beta deposition leads to tau pathology, as well as additional pathogenic mechanisms such as inflammation and oxidative damage, that results in cell death. Recent evidence suggests that the neurotoxic form of amyloid is soluble oligomers rather than monomers or the fibrillar form found in plaques.

New therapeutic strategies aim to interfere with amyloid deposition, either influencing its formation, or trying to remove it once deposited in senile plaques (SP), including mainly vaccination and passive immunization [4].

Atypical Forms of Alzheimer's Disease

The initial presentation of AD can also be atypical, with non-amnestic focal cortical cognitive symptoms. These syndromes are rare and often underestimated. The most common is posterior cortical atrophy (PCA), also known as Benson's syndrome. The prevalence and incidence of PCA are currently unknown; age of onset is 50–69 years old, much younger than typical amnestic AD. Patients often face considerable delays in diagnosis owing to the young age at onset and unusual symptoms at presentation. The neuropsychological deficits cited most frequently in individuals with PCA are visuospatial and visuoperceptual impairments, with individuals describing difficulties reading lines of text, judging distances, identifying static objects within the visual field, or having problems with stairs and escalators. Visual symptoms such as light sensitivity or visual distortions can be mistaken for migraine. Alexia and features of Balint's and/or Gerstmann's syndrome can be part of the picture, but they are rarely reported spontaneously by the patient. Although higher-order visual problems are reported more often than are basic visual impairments, a recent study by Lehmann et al. demonstrates that such deficits are due to deficits in more basic visual processing (form, motion, color, and point localization). Many patients with PCA also present positive perceptual phenomena, such as prolonged color afterimages, reverse size phenomena, and perception of movement of static stimuli. Deficits in working memory and limb apraxia have also been described. Moreover Snowden and colleagues reported extrapyramidal signs and myoclonus with a frequency of 41 and 24 % in their case histories. Indeed physical examination in most cases of PCA is unremarkable.

Voxel-based morphometry has shown the most widespread gray matter reduction in regions of the occipital and parietal lobes followed by areas in the temporal lobe. By 5 years of symptom duration, atrophy is widespread through the cortex. FDG-PET identifies areas of hypometabolism in the parieto-occipital areas and in the frontal eyes fields. Data from single-photon emission computed tomography (SPECT) usually confirm these findings.

Several studies confirm that AD is the most common pathology underlying PCA. However, some cases are attributable to other causes, such as corticobasal degeneration, dementia with Lewy bodies, and prion disease. Renner et al. reported pathological studies from 21 cases of PCA; of these 14 had Alzheimer's disease, 3 had Lewy body disease, 2 had corticobasal degeneration, and 2 had prion disease. As for the distribution of pathological changes, unfortunately there are only a small number of studies on very few patients so results are not consistent. It is reasonable to think that there are differences between PCA and typical AD as some of these studies show, but results have to be confirmed by larger studies. All studies report higher density of neurofibrillary tangle and senile plaques in the occipital lobe, but findings in other cortical regions are discrepant [19].

Diagnostic criteria for PCA have been proposed but have not been validated so far. Core features include insidious onset and gradual progression; presentation of visual deficits in the absence of ocular disease; relatively preserved episodic memory, verbal fluency, and personal insight; presence of symptoms including visual agnosia, simultanagnosia, optic ataxia, ocular apraxia, dyspraxia, and environmental disorientation; and absence of stroke or tumor. Supportive features include alexia, ideomotor apraxia, agraphia, acalculia, onset before the age of 65 years, and neuroimaging evidence of PCA or hypoperfusion [20].

As far as we know, clinical trials to assess the efficacy of AchEIs in PCA have not been performed so far. However, these drugs are often used by specialist with some benefit. Antidepressant drugs might be appropriate in patients with persistent low mood, and levodopa or carbidopa could be useful in individuals with parkinsonism [21].

Frontal variant of AD is even more rare than PCA. It is characterized by prominent behavioral symptoms from the beginning and frontal lobe atrophy at the neuroimaging. There are only few studies on the pathology of this disease, but it seems that CSF biomarker profile is always consistent with AD.

Logopenic aphasia is the most recent described variant of primary progressive aphasia (PPA). Like the other variant of PPA, the core clinical feature is difficulty with language, with impairment in daily life activities requiring speech (e.g., using a telephone, asking for information). In Mesulam et al.'s diagnostic criteria for PPA, language deficit must be the at symptom onset and for the initial phases of the disease. Word retrieval and sentence repetition deficits are the core features of the logopenic variant. Spontaneous speech is characterized by slow rate, with frequent pauses due to significant word-finding problems, but there is no frank agrammatism. Other diagnostic features include phonologic paraphasias in spontaneous speech and naming. The sound substitutions that result in phonologic paraphasias in logopenic patients are usually well articulated, without distortions. Lack of frank agrammatic errors and preservation of articulation and prosody help distinguish the logopenic from the nonfluent variants. Imaging abnormalities in the left temporoparietal junction area are necessary to make a diagnosis of imaging-supported logopenic variant. Recent evidence shows that AD might be the most common underlying pathology [22].

References

1. Ferri CP, Prince M, Brayne C, Brodaty H, Fratiglioni L, Ganguli M, et al. Global prevalence of dementia: a Delphi consensus study. Lancet. 2005;366:2112–7.
2. Reitz C, Brayne C, Mayeux R. Epidemiology of Alzheimer disease. Nat Rev Neurol. 2011;7(3):137–52.
3. Ritchie K, Lovestone S. The dementias. Lancet. 2002;360:1759–66.
4. Galimberti D, Scarpini E, Ghezzi L. Immunotherapy against amyloid pathology in Alzheimer's disease. J Neurol Sci. 2013;333(1–2):50–4.
5. Apostolova LG, Cummings JL. Neuropsychiatric manifestations in mild cognitive impairment: a systematic review of the literature. Dement Geriatr Cogn Disord. 2008;25:115–26.
6. Arnaiz E, Almkvist O. Neuropsychological features of mild cognitive impairment and preclinical Alzheimer's disease. Acta Neurol Scand. 2003;179(Suppl):34–41.
7. Fitzpatrick AL, Kuller LH, Lopez OL, Kawas CH, Jagust W. Survival following dementia onset: Alzheimer's disease and vascular dementia. J Neurol Sci. 2005;229:43–9.
8. Dubois B, Feldman HH, Jacova C, Cummings JL, Dekosky ST, Barberger-Gateau P, et al. Revising the definition of Alzheimer's disease: a new lexicon. Lancet Neurol. 2010;9(11): 1118–27.
9. Petersen RC, Roberts RO, Knopman DS, et al. Mild cognitive impairment: ten years later. Arch Neurol. 2009;66(12):1447–55.
10. McKhann G, Drachman DA, Folstein M, Katzman R, Price D, Stadlan EM. Clinical diagnosis of Alzheimer's disease report of the NINCDS–ADRDA Work Group under the auspices of Department of Health and Human Services Task Force on Alzheimer's disease. Neurology. 1984;34:939–44.

11. Dubois B, Feldman HH, Jacova C, Dekosky ST, Barberger-Gateau P, Cummings J, et al. Research criteria for the diagnosis of Alzheimer's disease: revising the NINCDS-ADRDA criteria. Lancet Neurol. 2007;6:734–46.
12. Hort J, O'Brien JT, Gainotti G, Pirttili T, Popescu BO, Rektorova I, et al. EFNS guidelines for the diagnosis and management of Alzheimer's disease. Eur J Neurol. 2010;17(10):1236–48.
13. American Psychiatric Association. Diagnostic and statistical manual of mental disorders (IV-TR). 4th ed, text revised. Washington, DC: American Psychiatric Association; 2000.
14. Bartus RT, Dean III RL, Beer B, Lippa AS. The cholinergic hypothesis of geriatric memory dysfunction. Science. 1982;217:408–14.
15. Knowles J. Donepezil in Alzheimer's disease: an evidence-based review of its impact on clinical and economic outcomes. Core Evid. 2006;1(3):195–219.
16. Courtney C, Farrell D, Gray R, Hills R, Lynch L, Sellwood E, et al. AD2000 Collaborative Group. Long-term donepezil treatment in 565 patients with Alzheimer's disease (AD2000): randomised double-blind trial. Lancet. 2004;363:2105–15.
17. Birks J, Evans JG, Iakovidou V, Tsolaki M. Rivastigmine for Alzheimer's disease. Cochrane Database Syst Rev. 2009;2, CD001191.
18. Qaseem A, Snow V, Cross Jr TJ, Forciea MA, Hopkins Jr R, Shekelle P, et al. Current pharmacologic treatment of dementia: a clinical practice guideline from the American College of Physicians and the American Academy of Family Physicians. Ann Intern Med. 2008; 148:370–8.
19. Crutch SJ, Lehmann M, Schott JM, Rabinovici GD, Rossor MN, Fox NC. Posterior cortical atrophy. Lancet Neurol. 2012;11:170–8.
20. Mendez MF, Ghajarania M, Perryman KM. Posterior cortical atrophy: clinical characteristics and differences compared to Alzheimer's disease. Dement Geriatr Cogn Disord. 2002;14:33–40.
21. Tang-Wai DF, Graff-Radford NR, Boeve BF, Dickson DW, Parisi JE, Crook R, et al. Clinical, genetic, and neuropathologic characteristics of posterior cortical atrophy. Neurology. 2004;63:1168–74.
22. Gorno-Tempini ML, Hillis AE, Weintraub S, Kertesz A, Mendez M, Cappa SF, et al. Classification of primary progressive aphasia and its variants. Neurology. 2011;76(11):1006–14.

Chapter 3
Autosomal Dominant Alzheimer's Disease: Underlying Causes

Mahdi Ghani and Ekaterina Rogaeva

Abstract Knowledge about genetics of Alzheimer's disease (AD), the most prevalent form of dementia, is important to manage challenges of aging populations. So far, genetic analyses of families with autosomal dominant AD, presenting with early-onset dementia (<65 years of age), have found three causal genes: *APP*, *PSEN1*, and *PSEN2*. The possibility to detect carriers of causal mutations could help to evaluate the efficacy of different treatments at either asymptomatic or early stages of dementia. Such individuals are currently enrolled in a longitudinal clinical trial, named the Dominantly Inherited Alzheimer Network (DIAN). We provide an overview for the molecular genetic findings available for causal AD genes, discuss how this knowledge can be applied in clinical practice, and highlight the strategies to detect novel AD genes (e.g., *TREM2* and *PLD3*).

Keywords Gene • Alzheimer's disease • APP • PSEN1 • PSEN2 • TREM2 • PLD3

Introduction

By 2050, 22 % of the global population is predicted to be above 60 years old; hence, a dramatic increase in the prevalence of aging-associated diseases is expected. While in 2010, there were 35.6 million individuals with dementia worldwide, this number will nearly double every 20 years, to an estimated 66 million by 2030 [1].

M. Ghani, MD, PhD
Tanz Centre for Research in Neurodegenerative Diseases,
University of Toronto, Toronto, ON, Canada

E. Rogaeva, PhD (✉)
Faculty of Medicine, Neurology Division,
Tanz Centre for Research in Neurodegenerative Diseases, University of Toronto,
Krembil Discovery Tower, 60 Leonard Avenue, 4th Floor – 4KD481, Toronto, ON, Canada
e-mail: ekaterina.rogaeva@utoronto.ca

D. Galimberti, E. Scarpini (eds.), *Neurodegenerative Diseases*,
DOI 10.1007/978-1-4471-6380-0_3, © Springer-Verlag London 2014

Dementia can be caused by many different reasons, such as head injury, alcohol intoxication, metabolic disorders, and also vascular dementia or frontotemporal dementia (FTD) (MIM: 600274) [2]. However, the most prevalent form of dementia is Alzheimer's disease (AD) (MIM: 104300). Almost two-thirds of dementia patients over 80 years of age are diagnosed with AD [3, 4]. The yearly cost of care for AD patients is expected to increase to about one trillion dollars in the USA alone [5].

The more we learn about the genetic factors that increase or decrease risk of AD, the better we will manage the challenges of aging populations in the future. Noticeably, the certainty of disease development in carriers of mutations in causal AD genes makes it even more critical for patient care. For instance, early genetic diagnosis of AD, while minimum comorbidities are present, could open a window of opportunity to slow down disease progression with potential therapies.

Brain pathology of familial AD is similar to sporadic form of the disease and is characterized by progressive neuronal loss; neurofibrillary tangles, consisting of hyper-phosphorylated tau protein encoded by *MAPT*; and amyloid plaques, mainly consisting of amyloid beta (Aβ(beta)) 40/42 peptides generated by the cleavage of the amyloid precursor protein (APP) [6]. The accumulation of Aβ(beta) peptides appears to be an early event that triggers a series of downstream events (e.g., mis-processing of the tau protein and brain inflammation) [7].

In this chapter we present the current state of knowledge related to the autosomal dominant form of AD, defined as an inheritance pattern where an affected individual has just one copy of a mutant allele on an autosomal chromosome. In contrast, recessive disease requires two copies of the mutant allele. Patients with an autosomal dominant disease have a 50 % chance of passing the mutation (and hence, the disorder) to their offspring. So far, genetic analyses of pedigrees with autosomal dominant AD, mainly presenting with early onset dementia (<65 years of age), have found three causal genes, *APP* (MIM:104760) [8], presenilin 1 (*PSEN1*) (MIM:607822) [9], and presenilin 2 (*PSEN2*) (MIM:600759) [10, 11].

PSEN1 mutations are the most frequent genetic defects in autosomal dominant AD (69 %) followed by *APP* duplications (7.5 %) or missense mutations (1 %) and *PSEN2* mutations (2 %) [12]. In addition to the aberrations in these causative genes, there are several genetic variants which could increase the risk of sporadic AD or modify the age at onset of the disease; among them the e4-allele of the apolipoprotein E gene (*APOE*; MIM:107741) has the largest risk effect [13]. Apart from *APOE*, large genome-wide association studies (GWAS) have identified the link between sporadic late-onset AD (>65 years of age) and common polymorphisms in nine loci (*CLU*, *PICALM*, *BIN1*, *MS4A4/MS4A6E*, *CR1*, *CD2AP*, *CD33*, *EPHA1*, *ABCA7*) [14–18], and a reproducible association between *SORL1* and sporadic AD was discovered by a candidate-gene approach [19]. Furthermore, the largest GWAS to date (meta-analysis of 74,538 individuals from published GWASs) has confirmed the AD risk associated with *SORL1* variations and identified 11 additional loci at genome-wide significance (*HLA-DRB5/HLA-DRB1*, *PTK2B*, *SLC24A4/RIN3*, *DSG2*, *INPP5D*, *MEF2C*, *NME8*, *ZCWPW1*, *CELF1*, *FERMT2*, and *CASS4*) [20] (Fig. 3.1).

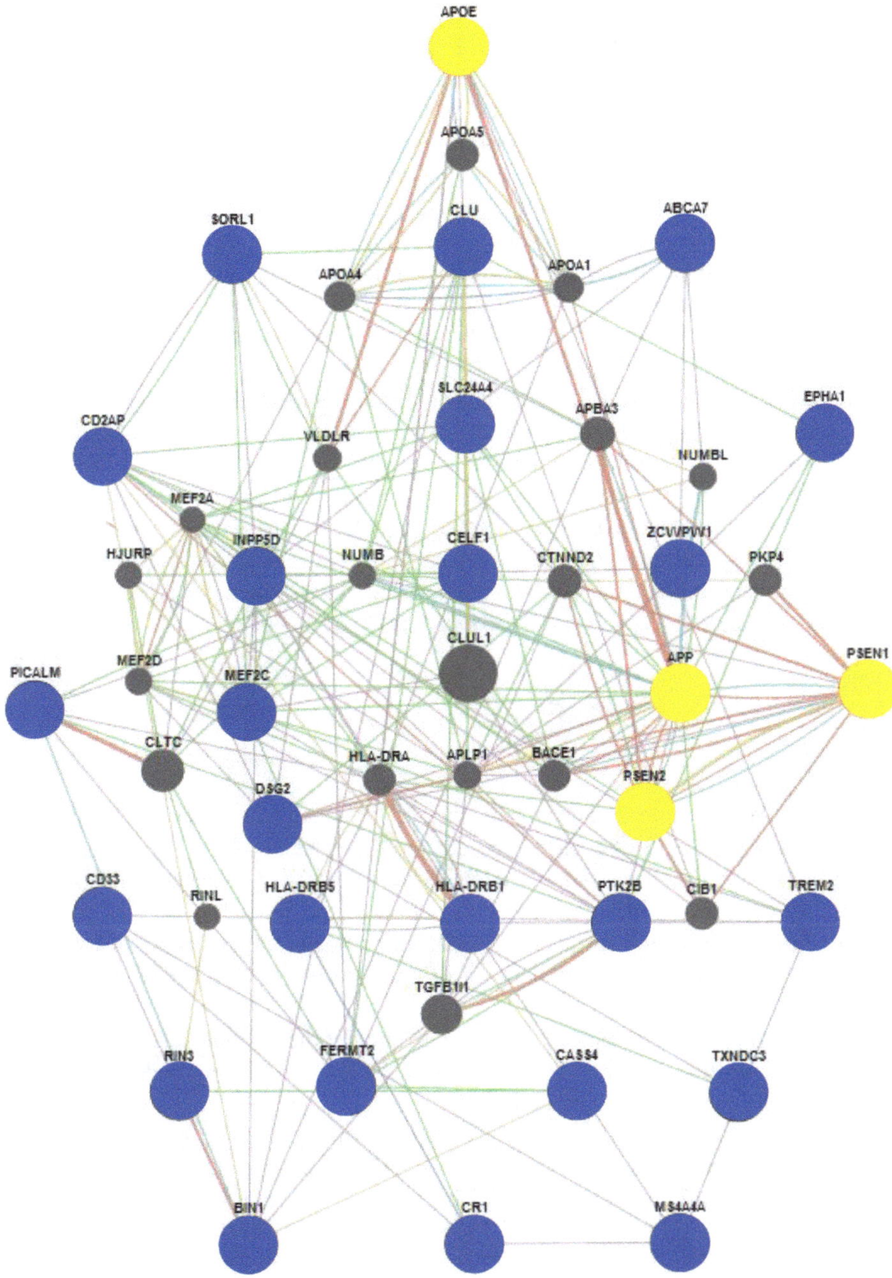

Fig. 3.1 Network of 28 well-confirmed AD genes using the GeneMANIA program that suggests connections between the genes (*lines*) based on information related to gene co-expression, physical interactions, shared protein domains, pathways, genetic interactions, and co-localization (http://www.genemania.org/). The genes with the strongest effect on AD risk are represented by *yellow circles* (*APP*, *PSEN1*, *PSEN2*, and *APOE*), while the rest of the AD genes marked by *blue circles*. In addition, the figure shows genes that might be critical functional nodes in the network of AD genes (*gray circles*)

It remains to be determined whether the AD genes identified by GWASs could modify the phenotype of patients with autosomal dominant early onset AD. Management of such a task will be possible once the functional variations explaining the GWAS signals are detected, using next-generation targeted sequencing or whole exome/genome studies. Determining which genes or gene networks contribute to AD risk could reveal basic pathogenic mechanisms important for drug development (Fig. 3.1). The functional connections between known AD genes are currently ambiguous; however, these genes could be subdivided into general functional categories (e.g., Aβ(beta) production, lipid/cholesterol metabolism, inflammation, vesicular trafficking, and synaptic function). Some of these genes could fit into several categories (e.g., *CLU* is implicated in both the cholesterol and inflammation pathways).

The goal of this chapter is to provide an overview for the molecular genetic findings currently available for the three causal AD genes (*APP*, *PSEN1*, and *PSEN2*) that are responsible for almost all known large early-onset AD families. In addition, we will indicate how this knowledge can be applied in clinical practice and discuss the strategies to detect novel AD genes. Indeed, ~50 % of early-onset AD cases without or limited family history, as well as many late-onset AD families, remain genetically unexplained and are currently being investigated by whole genome/exome sequencing that could reveal novel Mendelian AD genes.

APP

APP is a ~300-kb gene located on the minus strand of chromosome 21 at position chr21:27,252,861-27,543,138 according to the February 2009 human reference sequence (GRCh37). It is composed of 18 exons and encodes the APP protein, processing of which generates Aβ(beta) peptides, the basis of amyloid plaques. At least 11 APP isoforms are currently predicted to be produced by alternative splicing according to the UniProt database of protein sequences; the major isoform has a length of 770 amino acids (AA).

Knowledge about APP biology is critical to understanding the genetic discoveries in AD and vice versa. Currently, the normal function of APP is not well understood. In part, APP is expressed as a cell surface receptor and performs several physiological functions on the surface of neurons relevant to neurite growth, neuronal adhesion, and axonogenesis [21]. In addition, APP was implicated in cell mobility and transcriptional regulation. APP also accumulates in secretory transport vesicles leaving the late Golgi compartment and returning to the cell surface [19, 21]. During maturation, APP (N-glycosylated in the endoplasmic reticulum) moves to the Golgi complex where complete maturation occurs (O-glycosylated and sulfated) [21]. The key component of amyloid plaques (Aβ(beta)) is derived from the sequential cleavage of APP by β(beta)- and γ(gamma)-secretase. However, the major secretory pathway is nonamyloidogenic. Cleavage by either α(alpha)-, β(beta)-, or θ(theta)-secretase leads to the generation and extracellular release of soluble APP fragments,

S-APP-alpha (18–687 AA) and S-APP-beta (18–671 AA), and the retention of the corresponding membrane-anchored C-terminal fragments C83 (688–770 AA) and C99 (672–770 AA). Subsequent processing of C83 by γ(gamma)-secretase yields P3 peptides (688–713 or 688–711 AA) (http://www.uniprot.org). Several neuroprotective properties were reported for the soluble APP ectodomain fragments released from the cell surface by the action of α(alpha)-secretases [22], which are zinc metalloproteinases that include members of the adamalysin (ADAM) protein family [23].

The first amyloidogenic cleavage of APP is conducted by β(beta)-secretase, also known as beta-site APP-cleaving enzyme (BACE), that creates the C99 fragment. Next, γ(gamma)-secretase processing of C99 releases the Aβ(beta)40 (672–711 AA) or Aβ(beta)42 (672–713 AA), in addition to the cytotoxic C-terminal fragments γ(gamma)-CTF50 (721–770 AA), γ(gamma)-CTF57 (714–770 AA), and γ(gamma)-CTF59 (712–770 AA). Notably, direct sequencing, linkage, and association studies suggest that *BACE* is not genetically involved in AD, including the early-onset form of the disease [24, 25], which is in contrast to γ(gamma)-secretase, as discussed below. Yet, pathological *APP* mutations adjacent to the β(beta)-secretase cleavage site (671–672 AA) upregulate cleavage by β(beta)-secretase, increasing the generation of both Aβ(beta)40 and Aβ(beta)42 [26].

Aβ(beta)42 is not an abundant peptide, occurring in about a tenth of the amount of Aβ(beta)40, but is considered to be more pathogenic, since Aβ(beta)42 aggregates faster than Aβ(beta)40 and is apparently more toxic in cell culture assays [27, 28]. Of note, *APP* mutations around the γ(gamma)-secretase cleavage sites (711–721 AA) result in a modification of enzyme activity, enhancing the production of Aβ(beta)42 [26]. Many other minor Aβ(beta) peptides, including the Aβ(beta)X-15 peptides, are produced by α(alpha)-secretase cleavage (at 687–688 AA). In addition, cleavage at Asp-739 by either caspase-6, -8, or –9 results in the production of the neurotoxic C31 peptide and increased production of the Aβ(beta) peptides [29, 30].

The first AD mutation in *APP* (Val717Ile) affecting the transmembrane (TM) domain of APP (700–723 AA) was reported in 1991 [8]. It segregated with AD in a large autosomal dominant UK family. Currently, there are 33 different pathological *APP* mutations observed in 90 families, according to the Alzheimer Disease & Frontotemporal Dementia Mutation Database (http://www.molgen.ua.ac.be/ADMutations). These *APP* mutations include 24 missense mutations and 9 duplications overlapping *APP* with up to 5 neighboring genes. Of note, all pathologic missense mutations affect exon 16 or 17 of *APP* (around the secretase cleavage sites) (Fig. 3.2). Hence, the mutant APP molecules serve as an improved substrate for γ(gamma)- or β(beta)-secretase, leading to the overproduction of Aβ(beta) peptides [31].

In general, the Mutation Database indicated above (http://www.molgen.ua.ac.be/ADMutations) provides comprehensive and useful information on genes causing Mendelian forms of common neurodegenerative diseases (e.g., AD) [32]. It could help in assessing the pathological significance of genetic variations. For instance, the database allows the quick evaluation of known mutations for evidence of co-segregation with AD, its frequency, and the biological consequences in cell culture.

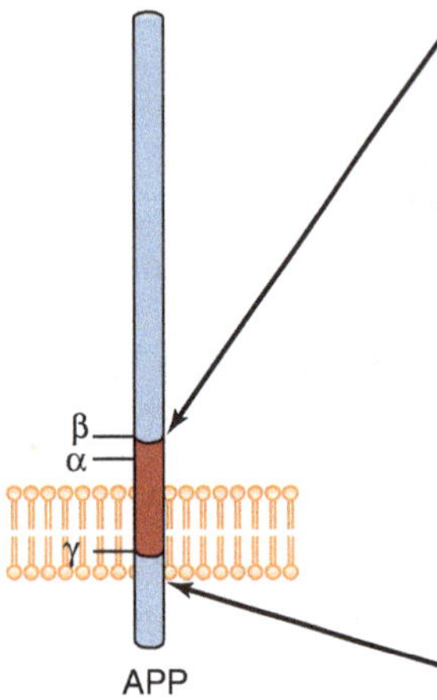

Codon	Mutation	Phenotype
670; 671	Lys→Asn; Met→Leu	AD
673	Ala→Thr	Protective
692	Ala→Gly	AD
693	Glu→Gln	Haemorrhage
694	Asp→Asn	Haemorrhage
713	Ala→Thr	AD
714	Thr→Ile	AD
715	Val→Met	AD
716	Ile→Val	AD
717	Val→Ile/Phe/Gly	AD
723	Leu→Pro	AD
Entire Locus	Duplication	AD/Storke

Fig. 3.2 Alzheimer disease mutations in *APP* are localized in or around the Aβ(beta) peptide domain

Most of the mutations (61 %) implicated in neurodegenerative disorders are reported in a single family and only ~6 % of them described in more than 10 families [32]. However, the number of families affected by the mutations is underestimated, since the literature is biased toward novel findings.

There is an intriguing connection between APP-linked disorders and cerebral amyloid angiopathy (amyloid deposits in brain blood vessels). For instance, prior to the discovery of the first AD mutation in *APP* (Val717Ile) [8], independent studies had identified the *APP* Dutch mutation (Glu693Gln) as a responsible mutation for hereditary cerebral hemorrhage with amyloidosis-Dutch type (HCHWA-D) [33–35]. This disease is often associated with stroke and characterized by the deposition of Aβ(beta) in the leptomeningeal arteries and cortical arterioles, in addition to amyloid plaques in the brain parenchyma [36]. Notably, another mutation at the same codon (Glu693Lys) has been described in Italian families and was associated with multiple strokes followed by epilepsy and cognitive decline [37]. The Arctic mutation (Glu693Gly), also affecting the same codon, has been shown to enhance Aβ(beta) protofibril formation [38, 39]. Notably, a deletion of Glu693 has recently been reported in some Japanese AD pedigrees and shown to enhance Aβ(beta) oligomerization rather than Aβ(beta) fibrillization [40]. Furthermore, missense *APP* substitutions at residues adjacent to Glu693, such as the Flemish (Ala692Gly) and Iowa (Asp694Asn) mutations, are also associated with cerebral amyloid angiopathy [41–45]. Moreover, in several families with different duplications at the *APP* locus, the AD frequently gets complicated by stroke [46, 47]. At least five early-onset AD families with such duplications were reported to be associated with cerebral amyloid angiopathy (http://www.molgen.ua.ac.be/ADMutations). However, *APP* aberrations are not considered a common cause of sporadic cerebral amyloid angiopathy [48].

The pathological consequences of different *APP* mutations often depend on their position relative to the secretase sites (Fig. 3.2). For example, 11 *APP* mutations that occur between residues 714 and 717 (http://www.molgen.ua.ac.be/ADMutations) affect the site of γ(gamma)-secretase cleavage and cause AD by increasing the level of Aβ(beta)42 [49]. Indeed, the Thr714Ile, Val715Met, Val715Ala, Ile716Val, Ile716Phe, and Val717Ile mutations are all known to decrease Aβ(beta)40 but increase Aβ(beta)42 levels and as a result raise the Aβ(beta)42/Aβ(beta)40 ratio [50–54]. In contrast, an increase in the Aβ(beta)42/Aβ(beta)40 ratio was not observed for the Swedish mutation (Lys670Asn/Met671Leu) at the β(beta)-secretase site [39], since it elevates both Aβ(beta)40 and Aβ(beta)42 levels [55, 56]. The Swedish mutation has been suggested to increase the rate of Aβ(beta) fibrillization [57]. This mutation was described in 1992 and so far has only been observed in two large early-onset AD families from Sweden [58]. Despite the extremely rare frequency of the Swedish variant, the mouse models generated based on it have been extensively studied worldwide and contributed to the discovery of several potential treatments for AD.

Of note, the *APP* Ala713Val variant located at the γ(gamma)-secretase cleavage site is not considered to be pathogenic. Hence, only specific conformational changes in the protein structure of APP lead to AD. This could explain the rarity of AD cases explained by *APP* mutations. Another important observation is that the Ala673Thr substitution, adjacent to the β(beta)-secretase site, appears to have a strong protective effect against AD (Fig. 3.2), with a ~40 % reduction in the formation of amyloidogenic peptides in vitro, which supports the hypothesis that reducing β(beta)-secretase cleavage of APP may protect against the disease [3]. The Ala673Thr variant was detected by a whole-genome sequencing of 1,795 Icelanders, and the Thr-allele was significantly more frequent in controls (0.6 %) than in AD cases (0.1 %) ($p = 5 \times 10^{-7}$). However, the investigation of independent datasets is needed to validate the protective effect of the Thr-allele.

Intriguingly, the Ala673Val substitution in *APP* (at the same codon with the protective Ala673Thr variant) has a dominant negative effect on amyloidogenesis and causes AD only when inherited in a homozygous state [59]. So far, that is the only known recessive AD mutation. However, it was recently revealed that an increased genome-wide average length of runs of homozygosity (that could harbor novel recessive mutations) is significantly associated with AD among an inbred Caribbean Hispanic population [60]. Importantly, the AD frequency in Caribbean Hispanics is up to three times greater compared to non-Hispanics in the same community. In the USA, Hispanics represent the fastest growing minority group and include ~30 % Caribbean Hispanics. The ongoing deep sequencing of significant loci affected by runs of homozygosity could detect novel recessive AD genes, which will then be tested in different ethnic groups.

As mentioned earlier, the AD in some families is explained by duplications overlapping the entire *APP* locus (ranging from 0.6 to 6.4 Mb) [46, 47]. However, *APP* duplications are rare events (in a UK study, only 5 out of 1,531 individuals with early-onset AD showed an *APP* duplication) [61]. The high level of APP could also be the result of variations in the promoter region of *APP*. For instance, a Belgian study reported that such *APP* mutations were even more frequent than coding substitutions (3 different promoter variations were identified in 7 out of 750 AD patients) [62]. While the −479C>T variant only mildly increases *APP* expression (1.2-fold), the −369C>G and −534G>A variants result in a twofold increase of APP levels (like an *APP* duplication) and were observed in probands of families with dementia. Importantly, the level of APP expression has been found to be inversely correlated with age of AD onset [62].

In several neuropsychiatric disorders, genome-wide global burden measurements of copy-number variations (CNVs), defined as genomic deletions or duplications ranging from 1 Kb to several Mb, are known to be important disease contributors. However, the role of CNVs has only recently begun to be systematically explored in AD [63]. For instance, an 18-Kb insertion in *CR1* (responsible for the CR1-S isoform) increases AD risk by almost twofolds and explains the GWAS signals at the *CR1* locus [64]. Recently, a study of early-onset AD dataset, including 261 families, revealed 5 deletions and 5 duplications that segregated with dementia [65]. Among them there were two CNVs encompassing FTD genes (deletion of *CHMP2B* and duplication of *MAPT*); however, such findings could reflect the presence of FTD cases that are clinically misdiagnosed with AD. For 6 of the 10 CNVs, the *APOE* e4-allele also co-segregated with AD, suggesting that the genes affected by the CNVs and *APOE* could act together to modify AD risk. There is also some evidence that *APOE* alleles can modify the severity of AD in cases with *APP* mutations [66]. In agreement with this, cultured cells with an e4/e4 genotype were more vulnerable to Aβ(beta) than cultures with an e3/e3 or e3/e4 genotype [67].

PSEN1

The associated locus on 14q24 was linked to autosomal dominant AD in 1992 [68, 69] and 3 years later was explained by mutations in *PSEN1* [9] that spans ~87 Kb on chr14:73,603,143-73,690,399 (GRCh37), including 10 coding and 2 noncoding exons. The PSEN1 protein (467 AA for the longest isoform) is expressed in a wide range of tissues (e.g., brain, liver, spleen, and lymph nodes) [70] and has several highly conserved transmembrane (TM) domains and a less conserved cytoplasmic hydrophilic loop domain.

PSEN1 is catalytic subunit of the γ(gamma)-secretase complex [71] that cleaves multiple integral membrane proteins, including APP and Notch receptors. It is incorporated into the γ(gamma)-secretase complex together with three other critical components: nicastrin (NCSTN), APH1, and presenilin enhancer 2 (PEN2) [72]. Several endogenous proteins have been reported to selectively modulate the function of this complex: transmembrane trafficking protein 21-KD (TMP21), CD147 antigen (basigin), the γ(gamma)-secretase-activating protein (gSAP), and the orphan G-protein-coupled receptor 3 [73]. AD-associated *PSEN1* mutations alter the conformation of the γ(gamma)-secretase complex leading to increased production of Aβ(beta)42 [74]. Postmortem studies of AD cases have shown that pathological *PSEN1* and *PSEN2* mutations are related to higher levels of insoluble Aβ(beta)42 (and to a lesser extent insoluble Aβ(beta)40) compared to sporadic AD [75–77].

PSEN1, PSEN2, Signal Peptide Peptidases (SPPs) are proteases with a highly conserved GlyXGlyAsp motif including membrane-embedded Aspartate (Asp) residues that are critical for enzymatic activity [78]. PSEN1 and PSEN2 cleave the TM region of multiple type 1 membrane proteins, while SPPs cleave type 2 membrane proteins. An AD mutation of the Gly-residue (Gly384Ala), next to the critical Asp in the GlyXGlyAsp motif of PSEN1, causes a selective loss of function by slowing Aβ(beta)40 production, while the generation of Aβ(beta)42 remains unaffected [79]. Such a disease mechanism is consistent with the consequences of several *APP* mutations that result in increased Aβ(beta)42/Aβ(beta)40 ratio [50–54].

Thus far, 185 different *PSEN1* mutations causing autosomal dominant AD have been reported in ~400 families (http://www.molgen.ua.ac.be/ADMutations/); majority of them are missense substitutions with only a few in-frame deletions or insertions [80–82]. Remarkably, sequencing of the DNA isolated from histopathological slides of the first reported AD patient (Auguste Deter), described by Dr. Alzheimer more than a century ago, has revealed a Phe176Leu mutation in *PSEN1* [83]. Such finding did explain the early onset of AD in this patient who was admitted to psychiatric service at age 51 years.

Most of the reported coding *PSEN1* variations are in fact pathogenic with just a few exceptions (e.g., Arg35Gln, Phe175Ser, and Val191Ala). However, the pathogenicity of the *PSEN1* variations that are currently recognized as innocent polymorphisms might depend on an interaction with other AD risk factor(s). For instance, the Glu318Gly variation used to be categorized as nonpathogenic due to its presence in controls (at a frequency of ~3 %) and the absence of convincing evidence for co-segregation with AD. However, a recent investigation of AD patients with extreme levels of cerebrospinal fluid biomarkers (e.g., Aβ(beta)-42, tau, or ptau) revealed that the Glu318Gly variation does indeed increase the risk for AD through a gene-gene interaction with APOE [84]. Glu318Gly carriers who are also heterozygous for the APOEe4 allele have an AD risk similar to APOEe4 homozygotes. Such individuals have higher levels of neuronal degeneration and Aβ(beta) deposition and faster cognitive decline.

On average, *PSEN1* carriers have an earlier onset and shorter duration of AD compared to carriers of *APP* or *PSEN2* mutations. AD symptoms in *PSEN1* carriers could appear as early as the third decade of life (e.g., Ser170Phe) [85]. In contrast to *APP*, the aberrations in *PSEN1* are broadly distributed throughout the gene with the exception of the first three exons, while exons 5–8 are the most frequently affected fragments. Also, genomic deletions of exon 9 and intronic mutations leading to its aberrant splicing are relatively frequent *PSEN1* mutations (http://www.molgen.ua.ac.be/ADMutations/). Another splicing aberration in *PSEN1* is caused by the mutation at the donor consensus site of intron 4 [81, 82].

Identification of the common founder mutation (Ala431Glu) among early-onset AD patients from the Jalisco state of Mexico has in fact clinically helped to provide genetic counseling advice to Mexican patients with familial AD and their relatives

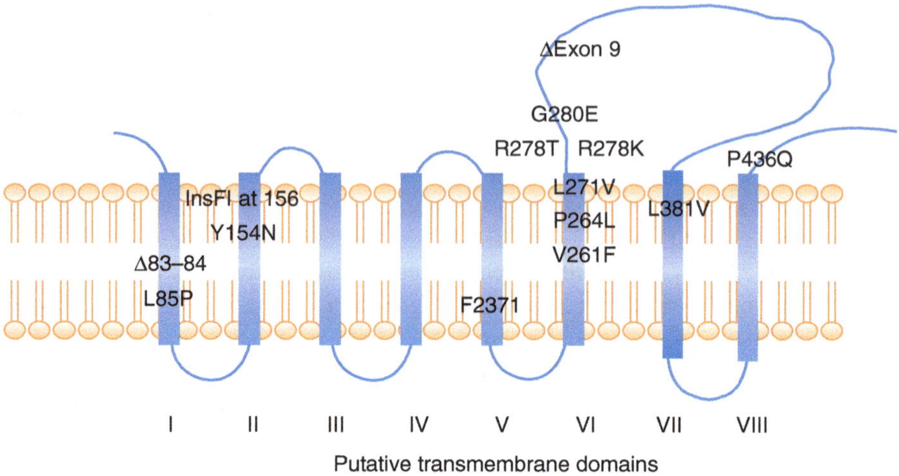

Fig. 3.3 The *PSEN1* mutations reported to be associated with variant Alzheimer disease presenting with spastic paraparesis

[86, 87]. Another frequent *PSEN1* mutation is Glu206Ala, which was observed in 42 % of early-onset AD families of Caribbean Hispanic origin [88]. Yet, the high incidence of late-onset AD in Caribbean Hispanics is not explained by *PSEN1* [88, 89] and might be related to recessive mutations hidden within genomic runs of homozygosity, as discussed previously [60].

The phenotypic heterogeneity in *PSEN1* patients includes variant AD (vAD), in which dementia is accompanied by spastic paraparesis (SP; MIM:607822), a progressive spastic weakness of the lower limbs with axonal degeneration in the corticospinal tract and dorsal columns [90]. The brain pathology of these cases differs from typical AD and characterized by large, abundant, diffuse, Aβ(beta)-positive "cotton wool plaques" without features of mature plaques (e.g., a congophilic core, neuritic pathology, and signs of inflammation) [80]. Of note, the "cotton wool plaques" can be also observed in rare cases of late-onset AD [91] or FTD syndrome [92].

The mutations associated with vAD (Fig. 3.3) are broadly distributed within *PSEN1* (codons 83–436) and reported in families with variable ages at onset (24–51 years) [90]. Importantly, identical *PSEN1* mutations have been reported in families with either classical AD or vAD (e.g., the deletion of exon 9 in families of the same Finnish origin [80, 93]). Surprisingly, even within a single family, a spectrum of disease phenotypes have been reported (e.g., AD, vAD, and pure SP) [94]. Cumulatively these observations argue in favor of a genetic modifiers responsible for vAD, the search for them is still ongoing. Currently, a modifier effect for coding variations in several SP genes has been excluded, including *ZFYVE26* which was selected as a promising gene candidate because it maps to ~5 Mb upstream of *PSEN1* and could explain a phenotypic range in some vAD families (by limited recombination events between AD and SP genes) [95].

In addition to autosomal dominant AD, *PSEN1* mutations were implicated in several other disorders, suggesting that γ(gamma)-secretase function could be critical in different tissues (e.g., brain, heart, and hair follicles). For instance, the Gly183Val mutation was reported in patients with Pick-type tauopathy, while Ile211Met was associated with posterior cortical atrophy (a dementia with symptoms of cortical visual dysfunction) [96, 97]. Furthermore, some *PSEN1* mutations have been reported in primary progressive aphasia and autosomal dominant spinocerebellar ataxia [98, 99]. Intriguingly, the Asp333Gly mutation was found in the familial form of dilated cardiomyopathy (MIM: 613694), a disorder characterized by heart ventricular dilation resulting in congestive heart failure and arrhythmia [100].

Of note, mutations in autosomal dominant FTD genes (e.g., *GRN*) can clinically manifest as AD-like dementia. For instance, a study of early-onset familial FTD showed that both the Cys139Arg mutation in *GRN* and the Val412Ile mutation in *PSEN1* manifest with almost the same phenotype [101]. However, without autopsy results the above observations reflect only a clinical similarity between AD and FTD. For example, the insArg352 in *PSEN1* was originally detected in a patient from a referral-based series of AD cases [81]. Later, the diagnosis of this patient was specified as FTD, and an in vitro study demonstrated that the insArg352 variation acts as a dominant negative mutation, inhibiting γ(gamma)-secretase cleavage of APP and Notch [102]. Therefore, it was suggested that chronic inhibition of γ(gamma)-secretase may result in neurodegeneration. Nevertheless, subsequent analysis of *GRN* identified the pathological frameshift mutation responsible for the disease in this patient, and the autopsy result confirmed the diagnosis of FTD [103]. Hence, the insArg352 is most likely a rare benign polymorphism with regard to neurodegeneration; however, it might contribute to a different disorder.

Surprisingly, loss-of-function mutations in *PSEN1* and other genes encoding the components of the γ(gamma)-secretase complex (*NCSTN* and *PEN2*) were reported in several Chinese families with an inflammatory autosomal dominant disease of the hair follicles, which develops after puberty (acne inversa; MIM:613737) [104]. Notably, patients with such mutations (15–81 years old) showed no symptoms of dementia. Acne inversa is likely the result of compromised Notch signaling due to a 50 % decrease in γ(gamma)-secretase activity in carriers of heterozygous loss-of-function mutations. The worldwide prevalence of acne inversa is 1–4 % including up to 40 % familial cases. The recent reports of γ-secretase-activating protein (gSAP) mutations in several ethnic groups confirm that familial acne inversa is an allelic disorder of early-onset familial AD.

PSEN2

PSEN2 is a ~26 Kb gene with 10 coding and 3 noncoding exons. It is located on chr1:227,058,273-227,083,804 (GRCh37) at the 1q42 locus and encodes the PSEN2 protein (448 AA) that shares substantial structural and sequence similarities with

PSEN1, apart from the cytoplasmic hydrophilic loop domain. PSEN2 also has the conserved GlyXGlyAsp motif with the Asp366 critical for γ(gamma)-secretase activity [105]. The second critical Asp is located at codon 263.

PSEN2 acts as a catalytic subunit of a γ(gamma)-secretase complex independent of PSEN1 but includes the three other critical components (PEN2, NCSTN, and APH1). Notably, PSEN2-dependant γ(gamma)-secretase activity is predominant in microglia and modulates the release of proinflammatory cytokines [106]. In addition, PSEN2 is implicated in intracellular signaling and gene expression. Furthermore, both PSEN1 and PSEN2 are localized to the nuclear membrane and are considered to be involved in cell cycle regulation and mitosis [107].

PSEN2 was cloned right after *PSEN1* in 1995 [10, 11]; however, currently only 13 different *PSEN2* mutations among 22 families have been reported worldwide, affecting exons 4–7 and exon 12 (http://www.molgen.ua.ac.be/ADMutations). Carriers of *PSEN2* mutations have a less severe AD compared to individuals with *PSEN1* aberrations [9, 108], which in part could be explained by the lower brain expression of *PSEN2* vs *PSEN1* [11]. Some *PSEN2* mutations are found in patients with late-onset AD (e.g., Val148Ile) [109]. AD families segregating the same *PSEN2* mutation could exhibit a wide range of age at onset. Even within a single family, the Thr430Met mutation was reported to be associated with a variable clinical presentation suggesting the action of unknown modifier gene(s) affecting disease severity [110]. Recently, a genome-wide search for the loci influencing the age at onset within nine families affected by the most common *PSEN2* substitution (Asn141Ile; Volga German founder mutation) has revealed several candidate modifier loci in addition to *APOE* (1q23.3, 17p13.2, 7q33, and 11p14.2) [111].

A number of reports have documented the phenotypic heterogeneity in *PSEN2* patients. For instance, the Ser130Leu substitution segregates with dilated cardiomyopathy in two families (as the Asp333Gly in *PSEN1*) [100]. In addition, the Ala85Val mutation was detected in a patient with a clinical and neuropathological phenotype indicative of dementia with Lewy bodies (DLB; MIM:127750) [112], which is closely associated with both AD and Parkinson disease. Lewy bodies are neuronal inclusions mainly consisting of α(alpha)-synuclein. Of note, it was reported that the brain pathology of 63 % of *PSEN1* and *APP* mutation carriers is associated with the presence of Lewy bodies [113]. However, the functional connection between the aberrations in autosomal dominant AD genes and α(alpha)-synuclein pathology remains to be explained.

Genetic Testing and Search for Novel AD Genes

Mutation analyses for AD-causing genes are usually performed by Sanger sequencing of either the entire coding region (for *PSEN1* and *PSEN2*) or selected exons accompanied by gene-dosage assessment (for *APP*). *PSEN1* mutations are fully penetrant and relatively frequent, which make this gene most suitable for genetic testing. Diagnostic testing for the three causal AD genes is currently established only in a few certified laboratories, according to the NCBI genetic testing registry (http://www.ncbi.nlm.nih.gov/gtr/); but health insurance organizations may cover the cost of the tests available abroad (e.g., at Athena Diagnostics, USA).

Considering the variable clinical presentation associated with mutations in the causal AD genes, there is still a need to improve understanding of the genotype-phenotype correlation in order to provide the best medical advice to mutation carriers. Furthermore, there is a concern for genetic counseling if novel mutations are detected. Without evidence of co-segregation of the mutation with AD, the analysis of Aβ(beta) levels in cell culture could provide support for the pathological impact of the mutation.

The possibility to detect carriers of causal AD mutations could help to evaluate the efficacy of different treatments at either the asymptomatic phase or early stages of dementia. Since 2013, about 160 selected carriers of *APP*, *PSEN1*, or *PSEN2* mutations have been enrolled in a longitudinal clinical trial, named DIAN (http://clinicaltrials.gov/ct2/show/NCT01760005). The goal of this trial is to assess the tolerability and biomarker efficacy of two potential modifying AD treatments, which are based on antibodies that either bind to aggregated Aβ(beta) (Gantenerumab) or soluble Aβ(beta) (Solanezumab). It includes individuals 18–80 years of age (within 10–15 years of the anticipated age of onset) who either know that they have an AD mutation or unaware of their genetic status but have a 50 % chance of inheriting such a mutation (e.g., siblings of a mutation carrier). The outcome of the early intervention in the DIAN study (e.g., prevention of the loss of cognitive function) and/or discovery of AD biomarkers could have important implications for the treatment of the common sporadic form of AD.

Nevertheless, half of AD heritability remains to be explained. The source of the missing heritability could be rare variants (with allele frequency <1 %) that are not captured by GWAS. Rare variants are broadly distributed throughout the human genome; hence, genome-wide approaches are needed to determine which of them are related to AD. Of note, rare variants are not limited to Mendelian AD but, along with common variants, may also affect risk for late-onset AD.

Recent advances in next-generation sequencing technology have provided a cost-effective approach to large-scale re-sequencing of the entire genome as a strategy for identifying protective or risk alleles [114, 115]. Notably, coding variations constitute ~85 % of known disease-causing mutations [116] and can be potentially evaluated with exome sequencing that covers ~1 % of the human genome (30 Mb; 180,000 exons) reducing the time/cost of searching for highly penetrant variants on a genome-wide scale [117, 118]. For instance, AD-associated coding variants in *TREM2* were recently identified by exome sequencing [119]. Furthermore, exome sequencing of 29 unrelated index cases from early-onset families consistent with autosomal dominant inheritance of AD revealed 7 patients with novel nonsense or missense mutations in *SORL1* [120]. However, interpretation of this finding is currently limited due to the absence of segregation data.

A recent study illustrates how extended AD kindreds may help detect rare but large-risk variants [121]. Whole-exome sequencing in 14 large families discovered a rare Val232Met mutation in *PLD3* segregating with AD in two autosomal dominant families. Analyses of case–control datasets revealed that this variant doubled the risk for AD. The frequency of Val232Met was higher in familial than sporadic cases (~3 % vs. ~1 %). *PLD3* is mapped to ~5 Mb upstream of *APOE*; however, the association of the Val232Met with AD was independent of *APOE* genotype. Gene-based burden analyses, using the sequencing results of all *PLD3* exons in multiple cases and controls, revealed that 14 *PLD3* variants increase AD risk (including 9 variants found only in cases). Most of them are missense substitutions, except the Ala442Ala variant, which affects splicing. Although *PLD3*-related AD is rare, this discovery provides novel insights into disease pathogenesis. PLD3 encodes a member of the PLD superfamily of phospholipases. PLD3 is highly expressed in brain regions susceptible to AD pathology (neurons from AD brains express less PLD3 than controls). Overexpression of PLD3 leads to a decrease in intracellular APP and extracellular Aβ(beta)40/42, while knockdown of PLD3 increases extracellular Aβ(beta)40/42 [121]. This resembles the disease mechanism of early-onset autosomal dominant AD genes (*APP*, *PSEN1*, and *PSEN2*).

Importantly, the amount of raw data produced by next-generation sequencing is enormous, and many computational steps are required to translate this output into reliable variant calls. Surprisingly, a large number of loss-of-function variants (~100 per genome) are identified in the apparently healthy individuals [122]. Hence, the identified variants must be systematically filtered using high-quality catalogues of the variants present in the genomes of healthy individuals (e.g., from the 1000 Genomes Project: http://www.1000genomes.org/). However, the age of individuals included in public databases is an important consideration for diseases with late onset (e.g., AD). Some additional tools such as SIFT and Polyphen are also used to annotate the variants and predict their functional consequences, which could help to prioritize the variants selected for follow-up association or segregation studies.

Conclusion

The huge expansion in knowledge provided by the human genome project has resulted in great progress in medicine and introduced the new field of genomic medicine. Genetics is now widely applied to AD diagnosis, monitoring, and search for a potential treatment. Mutations in *APP*, *PSEN1*, and *PSEN2* are tested for the purposes of diagnosis and presymptomatic screening of individuals with high risk of AD. In the near future, next-generation sequencing will likely uncover novel AD genes.

Acknowledgments We are grateful for the support from the Canadian Institutes of Health Research, W. Garfield Weston Foundation, Ontario Research Fund, and Alzheimer Society of Ontario.

References

1. Dementia statistics. Available at: http://www.alz.co.uk/research/statistics. Accessed 3 Jan 2014.
2. Miyoshi K. What is 'early onset dementia'? Psychogeriatrics. 2009;9(2):67–72 [Review].
3. Jonsson T, Atwal JK, Steinberg S, Snaedal J, Jonsson PV, Bjornsson S, et al. A mutation in APP protects against Alzheimer's disease and age-related cognitive decline. Nature. 2012;488(7409):96–9.
4. Mayeux R. Epidemiology of neurodegeneration. Annu Rev Neurosci. 2003;26:81–104 [Review].
5. Stefanacci RG. The costs of Alzheimer's disease and the value of effective therapies. Am J Manag Care. 2011;17 Suppl 13:S356–62 [Review].
6. Hardy JA, Higgins GA. Alzheimer's disease: the amyloid cascade hypothesis. Science. 1992;256(5054):184–5 [Review].
7. St George-Hyslop PH, Petit A. Molecular biology and genetics of Alzheimer's disease. C R Biol. 2005;328(2):119–30 [Review].
8. Goate A, Chartier-Harlin MC, Mullan M, Brown J, Crawford F, Fidani L, et al. Segregation of a missense mutation in the amyloid precursor protein gene with familial Alzheimer's disease. Nature. 1991;349(6311):704–6.
9. Sherrington R, Rogaev EI, Liang Y, Rogaeva EA, Levesque G, Ikeda M, et al. Cloning of a gene bearing missense mutations in early-onset familial Alzheimer's disease. Nature. 1995;375(6534):754–60.
10. Levy-Lahad E, Wasco W, Poorkaj P, Romano DM, Oshima J, Pettingell WH, et al. Candidate gene for the chromosome 1 familial Alzheimer's disease locus. Science. 1995;269(5226):973–7.
11. Rogaev EI, Sherrington R, Rogaeva EA, Levesque G, Ikeda M, Liang Y, et al. Familial Alzheimer's disease in kindreds with missense mutations in a gene on chromosome 1 related to the Alzheimer's disease type 3 gene. Nature. 1995;376(6543):775–8.
12. Guyant-Marechal L, Campion D, Hannequin D. Alzheimer disease: autosomal dominant forms. Rev Neurol (Paris). 2009;165(3):223–31.
13. Raber J, Huang Y, Ashford JW. ApoE genotype accounts for the vast majority of AD risk and AD pathology. Neurobiol Aging. 2004;25(5):641–50.
14. Harold D, Abraham R, Hollingworth P, Sims R, Gerrish A, Hamshere ML, et al. Genome-wide association study identifies variants at CLU and PICALM associated with Alzheimer's disease. Nat Genet. 2009;41(10):1088–93.
15. Lambert JC, Heath S, Even G, Campion D, Sleegers K, Hiltunen M, et al. Genome-wide association study identifies variants at CLU and CR1 associated with Alzheimer's disease. Nat Genet. 2009;41(10):1094–9.
16. Carrasquillo MM, Belbin O, Hunter TA, Ma L, Bisceglio GD, Zou F, et al. Replication of CLU, CR1, and PICALM associations with Alzheimer disease. Arch Neurol. 2010;67(8):961–4.
17. Naj AC, Jun G, Beecham GW, Wang LS, Vardarajan BN, Buros J, et al. Common variants at MS4A4/MS4A6E, CD2AP, CD33 and EPHA1 are associated with late-onset alzheimer's disease. Nat Genet. 2011;43(5):436–41.
18. Hollingworth P, Harold D, Sims R, Gerrish A, Lambert JC, Carrasquillo MM, et al. Common variants at ABCA7, MS4A6A/MS4A4E, EPHA1, CD33 and CD2AP are associated with Alzheimer's disease. Nat Genet. 2011;43(5):429–35.
19. Rogaeva E, Meng Y, Lee JH, Gu Y, Kawarai T, Zou F, et al. The neuronal sortilin-related receptor SORL1 is genetically associated with Alzheimer disease. Nat Genet. 2007;39(2): 168–77.
20. Lambert JC, Ibrahim-Verbaas CA, Harold D, Naj AC, Sims R, et al. Extended meta-analysis of 74,538 individuals identifies 11 new susceptibility loci for Alzheimer's disease. Nat Genet. 2013;45(12):1452–8.

21. Bignante EA, Heredia F, Morfini G, Lorenzo A. Amyloid beta precursor protein as a molecular target for amyloid beta-induced neuronal degeneration in Alzheimer's disease. Neurobiol Aging. 2013;34:2525–37.
22. Postina R. Activation of alpha-secretase cleavage. J Neurochem. 2012;120 Suppl 1:46–54 [Review].
23. Allinson TM, Parkin ET, Turner AJ, Hooper NM. ADAMs family members as amyloid precursor protein alpha-secretases. J Neurosci Res. 2003;74(3):342–52.
24. Nicolaou M, Song YQ, Sato CA, Orlacchio A, Kawarai T, Medeiros H, et al. Mutations in the open reading frame of the beta-site APP cleaving enzyme (BACE) locus are not a common cause of Alzheimer's disease. Neurogenetics. 2001;3(4):203–6.
25. Cruts M, Dermaut B, Rademakers R, Roks G, Van den Broeck M, Munteanu G, et al. Amyloid beta secretase gene (BACE) is neither mutated in nor associated with early-onset alzheimer's disease. Neurosci Lett. 2001;313(1–2):105–7.
26. Citron M, Oltersdorf T, Haass C, McConlogue L, Hung AY, Seubert P, et al. Mutation of the beta-amyloid precursor protein in familial Alzheimer's disease increases beta-protein production. Nature. 1992;360(6405):672–4.
27. Wolfe MS. The gamma-secretase complex: membrane-embedded proteolytic ensemble. Biochemistry. 2006;45(26):7931–9 [Review].
28. Selkoe DJ, Wolfe MS. Presenilin: running with scissors in the membrane. Cell. 2007;131(2):215–21 [Review].
29. Gervais FG, Xu D, Robertson GS, Vaillancourt JP, Zhu Y, Huang J, et al. Involvement of caspases in proteolytic cleavage of Alzheimer's amyloid-beta precursor protein and amyloidogenic A beta peptide formation. Cell. 1999;97(3):395–406.
30. Lu DC, Rabizadeh S, Chandra S, Shayya RF, Ellerby LM, Ye X, et al. A second cytotoxic proteolytic peptide derived from amyloid beta-protein precursor. Nat Med. 2000;6(4):397–404.
31. Sisodia SS, St George-Hyslop PH. gamma-Secretase, Notch, Abeta and Alzheimer's disease: where do the presenilins fit in? Nat Rev Neurosci. 2002;3(4):281–90.
32. Cruts M, Theuns J, Van Broeckhoven C. Locus-specific mutation databases for neurodegenerative brain diseases. Hum Mutat. 2012;33(9):1340–4.
33. Levy E, Carman MD, Fernandez-Madrid IJ, Power MD, Lieberburg I, van Duinen SG, et al. Mutation of the Alzheimer's disease amyloid gene in hereditary cerebral hemorrhage, Dutch type. Science. 1990;248(4959):1124–6.
34. Van Broeckhoven C, Haan J, Bakker E, Hardy JA, Van Hul W, Wehnert A, et al. Amyloid beta protein precursor gene and hereditary cerebral hemorrhage with amyloidosis (Dutch). Science. 1990;248(4959):1120–2.
35. Fernandez-Madrid I, Levy E, Marder K, Frangione B. Codon 618 variant of Alzheimer amyloid gene associated with inherited cerebral hemorrhage. Ann Neurol. 1991;30(5):730–3.
36. Bornebroek M, Haan J, Maat-Schieman ML, Van Duinen SG, Roos RA. Hereditary cerebral hemorrhage with amyloidosis-Dutch type (HCHWA-D): I–A review of clinical, radiologic and genetic aspects. Brain Pathol. 1996;6(2):111–4 [Review].
37. Bugiani O, Giaccone G, Rossi G, Mangieri M, Capobianco R, Morbin M, et al. Hereditary cerebral hemorrhage with amyloidosis associated with the E693K mutation of APP. Arch Neurol. 2010;67(8):987–95.
38. Kamino K, Orr HT, Payami H, Wijsman EM, Alonso ME, Pulst SM, et al. Linkage and mutational analysis of familial Alzheimer disease kindreds for the APP gene region. Am J Hum Genet. 1992;51(5):998–1014.
39. Nilsberth C, Westlind-Danielsson A, Eckman CB, Condron MM, Axelman K, Forsell C, et al. The 'Arctic' APP mutation (E693G) causes Alzheimer's disease by enhanced Abeta protofibril formation. Nat Neurosci. 2001;4(9):887–93.
40. Tomiyama T, Nagata T, Shimada H, Teraoka R, Fukushima A, Kanemitsu H, et al. A new amyloid beta variant favoring oligomerization in Alzheimer's-type dementia. Ann Neurol. 2008;63(3):377–87.
41. Grabowski TJ, Cho HS, Vonsattel JP, Rebeck GW, Greenberg SM. Novel amyloid precursor protein mutation in an Iowa family with dementia and severe cerebral amyloid angiopathy. Ann Neurol. 2001;49(6):697–705.

42. Greenberg SM, Shin Y, Grabowski TJ, Cooper GE, Rebeck GW, Iglesias S, et al. Hemorrhagic stroke associated with the Iowa amyloid precursor protein mutation. Neurology. 2003;60(6):1020–2.
43. Hendriks L, van Duijn CM, Cras P, Cruts M, Van Hul W, van Harskamp F, et al. Presenile dementia and cerebral haemorrhage linked to a mutation at codon 692 of the beta-amyloid precursor protein gene. Nat Genet. 1992;1(3):218–21.
44. Roks G, Van Harskamp F, De Koning I, Cruts M, De Jonghe C, Kumar-Singh S, et al. Presentation of amyloidosis in carriers of the codon 692 mutation in the amyloid precursor protein gene (APP692). Brain. 2000;123(Pt 10):2130–40.
45. Kumar-Singh S, Cras P, Wang R, Kros JM, van Swieten J, Lubke U, et al. Dense-core senile plaques in the Flemish variant of Alzheimer's disease are vasocentric. Am J Pathol. 2002;161(2):507–20.
46. Rovelet-Lecrux A, Hannequin D, Raux G, Le Meur N, Laquerriere A, Vital A, et al. APP locus duplication causes autosomal dominant early-onset alzheimer disease with cerebral amyloid angiopathy. Nat Genet. 2006;38(1):24–6.
47. Guyant-Marechal I, Berger E, Laquerriere A, Rovelet-Lecrux A, Viennet G, Frebourg T, et al. Intrafamilial diversity of phenotype associated with app duplication. Neurology. 2008;71(23):1925–6.
48. Domingues-Montanari S, Pares M, Hernandez-Guillamon M, Fernandez-Cadenas I, Mendioroz M, Ortega G, et al. No evidence of APP point mutation and locus duplication in individuals with cerebral amyloid angiopathy. Eur J Neurol. 2011;18(10):1279–81.
49. Haass C, Hung AY, Selkoe DJ, Teplow DB. Mutations associated with a locus for familial Alzheimer's disease result in alternative processing of amyloid beta-protein precursor. J Biol Chem. 1994;269(26):17741–8.
50. Kumar-Singh S, De Jonghe C, Cruts M, Kleinert R, Wang R, Mercken M, et al. Nonfibrillar diffuse amyloid deposition due to a gamma(42)-secretase site mutation points to an essential role for N-truncated A beta(42) in Alzheimer's disease. Hum Mol Genet. 2000;9(18): 2589–98.
51. De Jonghe C, Esselens C, Kumar-Singh S, Craessaerts K, Serneels S, Checler F, et al. Pathogenic APP mutations near the gamma-secretase cleavage site differentially affect Abeta secretion and APP C-terminal fragment stability. Hum Mol Genet. 2001;10(16):1665–71.
52. Cruts M, Dermaut B, Rademakers R, Van den Broeck M, Stogbauer F, Van Broeckhoven C. Novel APP mutation V715A associated with presenile Alzheimer's disease in a German family. J Neurol. 2003;250(11):1374–5.
53. Eckman CB, Mehta ND, Crook R, Perez-tur J, Prihar G, Pfeiffer E, et al. A new pathogenic mutation in the APP gene (I716V) increases the relative proportion of A beta 42(43). Hum Mol Genet. 1997;6(12):2087–9.
54. Herl L, Thomas AV, Lill CM, Banks M, Deng A, Jones PB, et al. Mutations in amyloid precursor protein affect its interactions with presenilin/gamma-secretase. Mol Cell Neurosci. 2009;41(2):166–74.
55. Citron M, Vigo-Pelfrey C, Teplow DB, Miller C, Schenk D, Johnston J, et al. Excessive production of amyloid beta-protein by peripheral cells of symptomatic and presymptomatic patients carrying the Swedish familial Alzheimer disease mutation. Proc Natl Acad Sci U S A. 1994;91(25):11993–7.
56. Perez RG, Squazzo SL, Koo EH. Enhanced release of amyloid beta-protein from codon 670/671 "Swedish" mutant beta-amyloid precursor protein occurs in both secretory and endocytic pathways. J Biol Chem. 1996;271(15):9100–7.
57. Kirkitadze MD, Condron MM, Teplow DB. Identification and characterization of key kinetic intermediates in amyloid beta-protein fibrillogenesis. J Mol Biol. 2001;312(5):1103–19.
58. Mullan M, Crawford F, Axelman K, Houlden H, Lilius L, Winblad B, et al. A pathogenic mutation for probable Alzheimer's disease in the APP gene at the N-terminus of beta-amyloid. Nat Genet. 1992;1(5):345–7.
59. Di Fede G, Catania M, Morbin M, Rossi G, Suardi S, Mazzoleni G, et al. A recessive mutation in the APP gene with dominant-negative effect on amyloidogenesis. Science. 2009;323(5920):1473–7.

60. Ghani M, Sato C, Lee J, Reitz C, Moreno D, Mayeux R, et al. Evidence of recessive Alzheimer's disease loci in Caribbean Hispanics: genome-wide survey of runs of homozygosity. JAMA Neurol. 2013;70(10):1261–7. doi:10.1001/jamaneurol.2013.3545.
61. McNaughton D, Knight W, Guerreiro R, Ryan N, Lowe J, Poulter M, et al. Duplication of amyloid precursor protein (APP), but not prion protein (PRNP) gene is a significant cause of early onset dementia in a large UK series. Neurobiol Aging. 2012;33(2):426.e13–21.
62. Brouwers N, Sleegers K, Engelborghs S, Bogaerts V, Serneels S, Kamali K, et al. Genetic risk and transcriptional variability of amyloid precursor protein in Alzheimer's disease. Brain. 2006;129(Pt 11):2984–91.
63. Ghani M, Pinto D, Lee JH, Grinberg Y, Sato C, Moreno D, et al. Genome-wide survey of large rare copy number variants in Alzheimer's disease among Caribbean hispanics. G3 (Bethesda). 2012;2(1):71–8.
64. Hazrati LN, Van Cauwenberghe C, Brooks PL, Brouwers N, Ghani M, Sato C, et al. Genetic association of CR1 with Alzheimer's disease: a tentative disease mechanism. Neurobiol Aging. 2012;33(12):2949.e5–12.
65. Hooli BV, Kovacs-Vajna ZM, Mullin K, Blumenthal MA, Mattheisen M, Zhang C, et al. Rare autosomal copy number variations in early-onset familial Alzheimer's disease. Mol Psychiatry. 2013. doi:10.1038/mp.2013.77. [Epub ahead of print].
66. St George-Hyslop P, McLachlan DC, Tsuda T, Rogaev E, Karlinsky H, Lippa CF, et al. Alzheimer's disease and possible gene interaction. Science. 1994;263(5146):537.
67. Wilhelmus MM, Otte-Holler I, Davis J, Van Nostrand WE, de Waal RM, Verbeek MM. Apolipoprotein E genotype regulates amyloid-beta cytotoxicity. J Neurosci. 2005;25(14):3621–7.
68. Schellenberg GD, Bird TD, Wijsman EM, Orr HT, Anderson L, Nemens E, et al. Genetic linkage evidence for a familial Alzheimer's disease locus on chromosome 14. Science. 1992;258(5082):668–71.
69. St George-Hyslop P, Haines J, Rogaev E, Mortilla M, Vaula G, Pericak-Vance M, et al. Genetic evidence for a novel familial Alzheimer's disease locus on chromosome 14. Nat Genet. 1992;2(4):330–4.
70. Hruz T, Laule O, Szabo G, Wessendorp F, Bleuler S, Oertle L, et al. Genevestigator v3: a reference expression database for the meta-analysis of transcriptomes. Adv Bioinformatics. 2008;2008:420747. doi:10.1155/2008/420747. Epub 2008 Jul 8.
71. Haass C, De Strooper B. The presenilins in Alzheimer's disease—proteolysis holds the key. Science. 1999;286(5441):916–9.
72. Yu G, Nishimura M, Arawaka S, Levitan D, Zhang L, Tandon A, et al. Nicastrin modulates presenilin-mediated notch/glp-1 signal transduction and betaAPP processing. Nature. 2000;407(6800):48–54.
73. St George-Hyslop P, Fraser PE. Assembly of the presenilin gamma-/epsilon-secretase complex. J Neurochem. 2012;120 Suppl 1:84–8 [Review].
74. Borchelt DR, Thinakaran G, Eckman CB, Lee MK, Davenport F, Ratovitsky T, et al. Familial Alzheimer's disease-linked presenilin 1 variants elevate Abeta1-42/1-40 ratio in vitro and in vivo. Neuron. 1996;17(5):1005–13.
75. Lemere CA, Lopera F, Kosik KS, Lendon CL, Ossa J, Saido TC, et al. The E280A presenilin 1 Alzheimer mutation produces increased A beta 42 deposition and severe cerebellar pathology. Nat Med. 1996;2(10):1146–50.
76. Mann DM, Iwatsubo T, Cairns NJ, Lantos PL, Nochlin D, Sumi SM, et al. Amyloid beta protein (Abeta) deposition in chromosome 14-linked Alzheimer's disease: predominance of Abeta42(43). Ann Neurol. 1996;40(2):149–56.
77. Mann DM, Iwatsubo T, Nochlin D, Sumi SM, Levy-Lahad E, Bird TD. Amyloid (Abeta) deposition in chromosome 1-linked Alzheimer's disease: the Volga German families. Ann Neurol. 1997;41(1):52–7.
78. Wolfe MS, Xia W, Ostaszewski BL, Diehl TS, Kimberly WT, Selkoe DJ. Two transmembrane aspartates in presenilin-1 required for presenilin endoproteolysis and gamma-secretase activity. Nature. 1999;398(6727):513–7.

79. Fluhrer R, Fukumori A, Martin L, Grammer G, Haug-Kroper M, Klier B, et al. Intramembrane proteolysis of GXGD-type aspartyl proteases is slowed by a familial Alzheimer disease-like mutation. J Biol Chem. 2008;283(44):30121–8.
80. Crook R, Verkkoniemi A, Perez-Tur J, Mehta N, Baker M, Houlden H, et al. A variant of Alzheimer's disease with spastic paraparesis and unusual plaques due to deletion of exon 9 of presenilin 1. Nat Med. 1998;4(4):452–5.
81. Rogaeva EA, Fafel KC, Song YQ, Medeiros H, Sato C, Liang Y, et al. Screening for PS1 mutations in a referral-based series of AD cases: 21 novel mutations. Neurology. 2001;57(4): 621–5.
82. De Jonghe C, Cruts M, Rogaeva EA, Tysoe C, Singleton A, Vanderstichele H, et al. Aberrant splicing in the presenilin-1 intron 4 mutation causes presenile Alzheimer's disease by increased Abeta42 secretion. Hum Mol Genet. 1999;8(8):1529–40.
83. Muller U, Winter P, Graeber MB. A presenilin 1 mutation in the first case of Alzheimer's disease. Lancet Neurol. 2013;12(2):129–30.
84. Benitez BA, Karch CM, Cai Y, Jin SC, Cooper B, Carrell D, et al. The PSEN1, p.E318G Variant Increases the Risk of Alzheimer's Disease in APOE-epsilon4 Carriers. PLoS Genet. 2013;9(8):e1003685.
85. Snider BJ, Norton J, Coats MA, Chakraverty S, Hou CE, Jervis R, et al. Novel presenilin 1 mutation (S170F) causing Alzheimer disease with Lewy bodies in the third decade of life. Arch Neurol. 2005;62(12):1821–30.
86. Yescas P, Huertas-Vazquez A, Villarreal-Molina MT, Rasmussen A, Tusie-Luna MT, Lopez M, et al. Founder effect for the Ala431Glu mutation of the presenilin 1 gene causing early-onset alzheimer's disease in Mexican families. Neurogenetics. 2006;7(3):195–200.
87. Murrell J, Ghetti B, Cochran E, Macias-Islas MA, Medina L, Varpetian A, et al. The A431E mutation in PSEN1 causing familial Alzheimer's disease originating in Jalisco State, Mexico: an additional fifteen families. Neurogenetics. 2006;7(4):277–9.
88. Athan ES, Williamson J, Ciappa A, Santana V, Romas SN, Lee JH, et al. A founder mutation in presenilin 1 causing early-onset alzheimer disease in unrelated Caribbean Hispanic families. JAMA. 2001;286(18):2257–63.
89. Tang MX, Cross P, Andrews H, Jacobs DM, Small S, Bell K, et al. Incidence of AD in African-Americans, Caribbean Hispanics, and Caucasians in northern Manhattan. Neurology. 2001;56(1):49–56.
90. Rogaeva E. The solved and unsolved mysteries of the genetics of early-onset alzheimer's disease. Neuromolecular Med. 2002;2(1):1–10 [Review].
91. Le TV, Crook R, Hardy J, Dickson DW. Cotton wool plaques in non-familial late-onset alzheimer disease. J Neuropathol Exp Neurol. 2001;60(11):1051–61.
92. Yokota O, Terada S, Ishizu H, Ujike H, Ishihara T, Namba M, et al. Variability and heterogeneity in Alzheimer's disease with cotton wool plaques: a clinicopathological study of four autopsy cases. Acta Neuropathol. 2003;106(4):348–56.
93. Hiltunen M, Helisalmi S, Mannermaa A, Alafuzoff I, Koivisto AM, Lehtovirta M, et al. Identification of a novel 4.6-kb genomic deletion in presenilin-1 gene which results in exclusion of exon 9 in a Finnish early onset alzheimer's disease family: an Alu core sequence-stimulated recombination? Eur J Hum Genet. 2000;8(4):259–66.
94. Smith MJ, Kwok JB, McLean CA, Kril JJ, Broe GA, Nicholson GA, et al. Variable phenotype of Alzheimer's disease with spastic paraparesis. Ann Neurol. 2001;49(1):125–9.
95. Sinha N, Grimes D, Tokuhiro S, Sato C, Rogaeva E, Woulfe J. Variant Alzheimer's disease with spastic paraparesis and supranuclear gaze palsy. Can J Neurol Sci. 2013;40(2):249–51.
96. Dermaut B, Kumar-Singh S, Engelborghs S, Theuns J, Rademakers R, Saerens J, et al. A novel presenilin 1 mutation associated with Pick's disease but not beta-amyloid plaques. Ann Neurol. 2004;55(5):617–26 [Research Support, Non-U.S. Gov't].
97. Sitek EJ, Narozanska E, Peplonska B, Filipek S, Barczak A, Styczynska M, et al. A patient with posterior cortical atrophy possesses a novel mutation in the presenilin 1 gene. PLoS One. 2013;8(4):e61074.

98. Mahoney CJ, Downey LE, Beck J, Liang Y, Mead S, Perry RJ, et al. The Presenilin 1 P264L Mutation Presenting as non-Fluent/Agrammatic Primary Progressive Aphasia. J Alzheimers Dis. 2013;36:239–43.
99. Braga-Neto P, Pedroso JL, Alessi H, de Souza PV, Bertolucci PH, Barsottini OG. Early-onset familial Alzheimer's disease related to presenilin 1 mutation resembling autosomal dominant spinocerebellar ataxia. J Neurol. 2013;260(4):1177–9.
100. Li D, Parks SB, Kushner JD, Nauman D, Burgess D, Ludwigsen S, et al. Mutations of presenilin genes in dilated cardiomyopathy and heart failure. Am J Hum Genet. 2006;79(6):1030–9.
101. Bernardi L, Tomaino C, Anfossi M, Gallo M, Geracitano S, Costanzo A, et al. Novel PSEN1 and PGRN mutations in early-onset familial frontotemporal dementia. Neurobiol Aging. 2009;30(11):1825–33.
102. Amtul Z, Lewis PA, Piper S, Crook R, Baker M, Findlay K, et al. A presenilin 1 mutation associated with familial frontotemporal dementia inhibits gamma-secretase cleavage of APP and notch. Neurobiol Dis. 2002;9(2):269–73.
103. Pickering-Brown SM, Baker M, Gass J, Boeve BF, Loy CT, Brooks WS, et al. Mutations in progranulin explain atypical phenotypes with variants in MAPT. Brain. 2006;129(Pt 11): 3124–6.
104. Wang B, Yang W, Wen W, Sun J, Su B, Liu B, et al. Gamma-secretase gene mutations in familial acne inversa. Science. 2010;330(6007):1065.
105. Kimberly WT, Xia W, Rahmati T, Wolfe MS, Selkoe DJ. The transmembrane aspartates in presenilin 1 and 2 are obligatory for gamma-secretase activity and amyloid beta-protein generation. J Biol Chem. 2000;275(5):3173–8.
106. Jayadev S, Case A, Eastman AJ, Nguyen H, Pollak J, Wiley JC, et al. Presenilin 2 is the predominant gamma-secretase in microglia and modulates cytokine release. PLoS One. 2010;5(12):e15743.
107. Li J, Xu M, Zhou H, Ma J, Potter H. Alzheimer presenilins in the nuclear membrane, interphase kinetochores, and centrosomes suggest a role in chromosome segregation. Cell. 1997;90(5):917–27.
108. Bird TD, Levy-Lahad E, Poorkaj P, Sharma V, Nemens E, Lahad A, et al. Wide range in age of onset for chromosome 1–related familial Alzheimer's disease. Ann Neurol. 1996;40(6):932–6.
109. Lao JI, Beyer K, Fernandez-Novoa L, Cacabelos R. A novel mutation in the predicted TM2 domain of the presenilin 2 gene in a Spanish patient with late-onset alzheimer's disease. Neurogenetics. 1998;1(4):293–6.
110. Ezquerra M, Lleo A, Castellvi M, Queralt R, Santacruz P, Pastor P, et al. A novel mutation in the PSEN2 gene (T430M) associated with variable expression in a family with early-onset alzheimer disease. Arch Neurol. 2003;60(8):1149–51.
111. Marchani EE, Bird TD, Steinbart EJ, Rosenthal E, Yu CE, Schellenberg GD, et al. Evidence for three loci modifying age-at-onset of Alzheimer's disease in early-onset PSEN2 families. Am J Med Genet B Neuropsychiatr Genet. 2010;153B(5):1031–41.
112. Piscopo P, Marcon G, Piras MR, Crestini A, Campeggi LM, Deiana E, et al. A novel PSEN2 mutation associated with a peculiar phenotype. Neurology. 2008;70(17):1549–54.
113. Lippa CF, Fujiwara H, Mann DM, Giasson B, Baba M, Schmidt ML, et al. Lewy bodies contain altered alpha-synuclein in brains of many familial Alzheimer's disease patients with mutations in presenilin and amyloid precursor protein genes. Am J Pathol. 1998;153(5):1365–70.
114. Bertram L, Lill CM, Tanzi RE. The genetics of Alzheimer disease: back to the future. Neuron. 2010;68(2):270–81.
115. Kuwano R, Hara N. Personal genomics for Alzheimer's disease. Brain Nerve. 2013;65(3): 235–46.
116. Choi M, Scholl UI, Ji W, Liu T, Tikhonova IR, Zumbo P, et al. Genetic diagnosis by whole exome capture and massively parallel DNA sequencing. Proc Natl Acad Sci U S A. 2009;106(45):19096–101.
117. Ng SB, Turner EH, Robertson PD, Flygare SD, Bigham AW, Lee C, et al. Targeted capture and massively parallel sequencing of 12 human exomes. Nature. 2009;461(7261):272–6.

118. Pruitt KD, Harrow J, Harte RA, Wallin C, Diekhans M, Maglott DR, et al. The consensus coding sequence (CCDS) project: Identifying a common protein-coding gene set for the human and mouse genomes. Genome Res. 2009;19(7):1316–23.
119. Guerreiro R, Wojtas A, Bras J, Carrasquillo M, Rogaeva E, Majounie E, et al. TREM2 variants in Alzheimer's disease. N Engl J Med. 2013;368(2):117–27.
120. Pottier C, Hannequin D, Coutant S, Rovelet-Lecrux A, Wallon D, Rousseau S, et al. High frequency of potentially pathogenic SORL1 mutations in autosomal dominant early-onset alzheimer disease. Mol Psychiatry. 2012;17(9):87–9.
121. Cruchaga C, Karch CM, Jin SC, Benitez BA, Cai Y, Guerreiro R, et al. Rare coding variants in the phospholipase D3 gene confer risk for Alzheimer's disease. Nature. 2014;505(7484):550–4.
122. MacArthur DG, Tyler-Smith C. Loss-of-function variants in the genomes of healthy humans. Hum Mol Genet. 2011;19(R2):R125–30.

Chapter 4
Genetic Risk Factors for Alzheimer's Disease

Onofre Combarros

Abstract The rare and familial early-onset alzheimer's disease (EOAD) is linked to fully penetrant mutational genes, whereas the commonest nonfamilial and late-onset alzheimer's disease (LOAD) is the result of multiple susceptibility genes, each one contributing a small amount to the total risk. Smaller-scale genetic association studies identified the e4 allele of the apolipoprotein E (APOE) gene as the best established LOAD risk factor, increasing risk by approximately 4 or 15 times for one or two e4 alleles, respectively. Hundreds of other candidate genes have been tested for association with LOAD, and meta-analyses of conflicting results were collected in the AlzGene database. Instead of studying only a few genetic variants in small sample sizes, larger-scale genetic association studies (genome-wide association study or GWAS) make it possible to evaluate essentially all genes and intergenic regions, in large international consortia with sufficient number of cases and controls. The four largest LOAD GWAS consortia joined forces forming a mega-consortium known as the International Genomics of Alzheimer's Project (IGAP) and conducted a mega-meta-analysis of 25,500 LOAD cases and 48,500 unaffected controls. In addition to APOE e4 allele, IGAP identified 19 susceptibility loci, but the effects of all these genes on LOAD risk are exceedingly small, increasing or decreasing the risk by approximately 1.30 times, at most. It is critical to investigate the functional basis for these LOAD-associated GWAS loci and their influence on gene expression (mRNA profiling). Examining the influences of these loci on endophenotypes (cerebrospinal fluid biomarkers and neuroimaging and cognitive measures) can help to predict age at onset and rate of preclinical and clinical progression of LOAD. The proportion of heritability of LOAD unexplained by GWAS findings could be due to rare variants that may be identified by whole exome and whole genome sequencing. In addition, a part of the still elusive genetic variability in LOAD could be due to gene-gene interaction effects.

O. Combarros, MD, PhD
Neurology Service and School of Medicine, University Hospital "Marqués de Valdecilla", University of Cantabria, Santander 39008, Spain
e-mail: combarro@unican.es

D. Galimberti, E. Scarpini (eds.), *Neurodegenerative Diseases*,
DOI 10.1007/978-1-4471-6380-0_4, © Springer-Verlag London 2014

Keywords Alzheimer's disease • Risk genes • Genome-wide association study Meta-analysis • Whole exome sequencing • Epistasis

Early-Onset alzheimer's Disease and Late-Onset alzheimer's Disease: Causative and Susceptibility Genes

Multiple genetic defects involving either predictive (mutational) or susceptibility (risk) genes have been linked to the development of Alzheimer's disease (AD) [1–3]. Rare (1–2 % of all AD cases) and fully penetrant disease-causing mutations in three different genes (*APP*, amyloid-beta protein precursor; *PSEN1*, presenilin 1; and *PSEN2*, presenilin 2) lead to early-onset (patients younger than 65 years) Mendelian (familial) forms of AD (EOAD). Of note, mutations in these three genes explain disease in only about 13 % of patients with EOAD [4]. The vast majority of AD cases, the so-called sporadic AD with no apparent familial recurrence, are defined by onset age later than 65 years or late-onset AD (LOAD), and this LOAD form does not carry Mendelian-causing mutations but is believed to be the result of multiple risk genes which do not reliably cause the disease but increase an individual's susceptibility or predisposition to developing AD. Susceptibility genes are associated with the risk of LOAD, but each one contributes only a small amount to the risk. Twin studies predicted the heritability of LOAD to be as high as 80 % [5]. Susceptibility genes are identified by genetic association studies in which allele frequency for single-nucleotide polymorphisms (SNPs) at or near a gene is compared between AD cases and controls. Susceptibility genes are revealed when case and control frequencies differ significantly. There appears to be no overlap between the genes driving Mendelian versus non-Mendelian form of the disease; that is, common SNPs in *APP*, *PSEN1*, and *PSEN2* do not seem to contribute to risk for LOAD [6].

Smaller-Scale Genetic Association Studies: The Candidate Gene Approach

Apolipoprotein E

The candidate gene approach was successful for identifying the e4 allele of the apolipoprotein E (*APOE*) gene on chromosome 19q13, as the only gene variant considered to be an "established" LOAD risk factor [7]. Unlike the EOAD mutations that are fully penetrant, *APOE* ε4 allele is a genetic risk factor that is neither necessary nor sufficient for the development of LOAD [1, 3, 4, 8]. Whereas only 24–30 % of the general Caucasian population carries at least one *APOE* ε4 allele, 40–65 % of LOAD patients have at least one copy of this allele. That is, many

LOAD patients have no *APOE* ε4 allele, and individuals carrying this allele may never develop LOAD, suggesting that there are additional factors modulating the influence of the APOE e4 allele in causing the development of LOAD. People with one *APOE* ε4 allele have a roughly four times increased risk of AD, and those with two *APOE* ε4 alleles have a roughly 15 times increased risk, compared with those with the most common genotype *APOE* ε3/3 [4]. Although there is evidence of a risk effect of *APOE* ε4 in non-Europeans, the estimated effect sizes are smaller with less consistent results in African American and Hispanic subjects, which may suggest different underlying genetic or environmental factors for these ethnic groups. The effect of *APOE* ε4 appears to be age dependent: the lifetime risk for LOAD in individuals with the *APOE* ε4/4 genotype is high, estimated as 33 % for men or 32 % for women by the age of 75 years; by age 85 years, the risk climbs to 52 % for men and 68 % for women [9]. This very high-risk estimates for *APOE* ε4 carriers seem similar to those associated with autosomal dominant Mendelian genes. Therefore, *APOE* has been proposed as a moderately penetrant gene with semi-dominant inheritance: not all *APOE* ε4 carriers develop disease (hence, the ε4 allele in this gene is not fully penetrant), and heterozygous *APOE* ε4 carriers have intermediate risk compared with homozygous carriers [9]. Considering the delayed penetrance of LOAD, lack of preventive therapies, and the potential for psychological harm, genetic testing for *APOE* is not recommended. However, when LOAD prevention becomes possible, thus, this recommendation will need to be reconsidered, and genetic testing might be indicated for either high-risk groups (e.g., family members of LOAD cases) or for population screening.

Inconsistent Replication: Meta-analyses in the AlzGene Database

Since the original report of APOE as a genetic risk factor for LOAD in 1993, hundreds of genes have been tested for association with LOAD and reported in the literature. Most candidate gene association studies in LOAD have studied a few variants in only one or two genes, and despite positive initial results, inconsistent replication of original association findings has been the rule rather than the exception (except for APOE) and even for candidate genes with convincing functional data and thorough genetic assessment [1]. Multiple testing, population stratification, genotyping errors, and initial small sample size are potential reasons for false-positive findings in the original study. In addition, underpowered studies that are too small to detect a modest effect size can lead to false-negative follow-up studies. Candidate gene association studies have revealed modest estimated effect sizes with odds ratios (ORs) of less than 2.0 for risk alleles or greater than 0.5 for protective alleles. It is estimated that thousands to tens of thousands of subjects are required to have sufficient power to detect such effect sizes, a prerequisite that has typically not been fulfilled for many association studies in LOAD [10]. To address these very large numbers of conflicting results, a database (the AlzGene database) was created which systematically collected, summarized, and meta-analyzed the results for all the genetic variants studied in association with LOAD [11]. As of 18 April 2011, the 10 genes with the strongest signals for association in the database AlzGene included *APOE* and nine other candidate genes (*BIN1*, *CLU*, *ABCA7*, *CR1*, *PICALM*, *MS4A6A*, *MS4A4E*, *CD33*, and *CD2AP*), all of which came from genome-wide association studies.

Larger-Scale Genetic Association Studies: The Genome-Wide Approach

The HapMap Project and the 1000 Genomes Project

Instead of studying one or two genetic variants, recent advances now make it possible to evaluate essentially all genes and all regions between genes in a single experiment, a method called genome-wide association study (GWAS). The GWAS method represented an important advance compared to "candidate gene" studies in which sample sizes were generally smaller and the variants assayed were limited to a selected few associations that were difficult to replicate. The International HapMap Project [12, 13] launched in October 2002 led to the generation of a database of the common variants (defined as minor allele frequency of greater than 5 %) and the underlying linkage disequilibrium (LD) structure, or correlation between neighboring SNPs, providing the foundation for the GWAS. GWAS uses tagging SNPs, for example, polymorphisms in LD with each other, and this means that if one knows the genotype in one locus, one can predict with a high accuracy (dependent on the strength of the LD and the allele frequencies) the genotype occurring at linked loci [14]. Understanding LD not only allows the construction of haplotypes but also provides the ability to impute the genotypes of nearby unobserved (not genotyped) SNPs using directly observed genotypes. Imputing facilitates merging data from different genotyping platforms with incomplete overlap [10]. Until 2010, GWAS studies had almost exclusively employed the HapMap data set as the reference panel for imputation of their genetic data, which contained up to two to three million SNPs. Using genome-wide sequencing with high-throughput platforms, the 1000 Genomes Project Consortium [15] described the location, allele frequency, and local haplotype structure of approximately 15 million SNPs, 1 million short insertions and deletions, and 20,000 structural variants. Over 95 % of the currently accessible variants found in any individual are present in this data set. From 2010 onward, the 1000 Genomes Project has increased power of GWAS to detect genetic influences due to less common variants. Rigorous quality control and statistical methods coupled with sufficient sample size can lead to high reproducibility of GWAS. Disadvantages of GWAS are that signals can be in intergenic regions making assessment of the functional relevance difficult, genetic methods often cannot identify which single-nucleotide variant is pathogenic, and most signals are from small effect loci [14].

International Consortia: Meta-analyses of GWAS

A consensus has emerged that a *P* value less than 5×10^{-8} corresponds to genome-wide significance in a non-African population-based GWAS. This is a conservative Bonferroni correction based on roughly one million "effectively independent" common SNPs throughout the genome. This involves the risk of rejecting biologically valid hypotheses on purely statistical grounds, that is, false negatives. Therefore, statistical power is the main threat to GWAS, necessitating the formation of large international consortia that can provide sufficient number of cases and controls. The four largest LOAD GWAS consortia are the European Alzheimer's Disease Initiative (EADI) based in France, the US-based Alzheimer's Disease Genetics Consortium (ADGC), the Genetic and Environmental Risk in Alzheimer's Disease (GERARD) group from the UK, and the neurology subgroup of the multinational Cohorts for Heart and Aging in Genomic Epidemiology (CHARGE) consortium. The first two GWAS were published in 2009 by the GERARD [16] and EADI [17] consortia. In approximately 6,000 LOAD and 10,000 control subjects, in addition to APOE-related SNPs that revealed genome-wide significance ($P=4.9\times10^{-37}$ to 1.8×10^{-157}), the GERARD consortium found that rs11136000 in clusterin (*CLU*, $P=8.5\times10^{-10}$, OR=0.86) and rs3851179 in the phosphatidylinositol-binding clathrin assembly protein (*PICALM*, $P=1.3\times10^{-9}$, OR=0.86) were significantly associated with LOAD. Analyzing 6,000 LOAD and more than 8,000 control subjects from EADI consortium, rs11136000 in *CLU* and rs6656401 in complement component receptor 1 (*CR1*, $P=3.7\times10^{-9}$, OR=1.21) were significantly associated with LOAD. In 2010, in more than 35,000 persons, the CHARGE consortium reported strong evidence that rs744373 near bridging integrator 1 gene (*BIN1*, $P=1.59\times10^{-11}$, OR=1.13) was significantly associated with LOAD [18]. In 2011, two simultaneously published manuscripts reported meta-analyses of the findings of the ADGC, CHARGE, GERARD, and EADI consortia and described strong evidence for five new LOAD risk loci. In nearly 20,000 cases and 40,000 controls, Hollingworth et al. [19] described association with LOAD of rs3764650 in *ABCA7* ($P=5.0\times10^{-21}$, OR=1.23), rs610932 in *MS4A6A* ($P=1.2\times10^{-16}$, OR=0.91), rs9349407 in *CD2AP* ($P=8.6\times10^{-9}$, OR=1.11), rs11767557 in *EPHA1* ($P=6.0\times10^{-10}$, OR=0.90), and rs3865444 in *CD33* ($P=1.6\times10^{-9}$, OR=0.91). In approximately 19,000 cases and 29,000 controls, Naj et al. [20] confirmed that common variants at *MS4A* gene cluster, *CD2AP*, *CD33*, and *EPHA1* were associated with LOAD.

International Genomics of Alzheimer's Project: Mega-meta-analysis of GWAS

The four LOAD GWAS consortia have joined forces, forming a mega-consortium known as the International Genomics of Alzheimer's Project (IGAP). The project drew on data from a total of 74,000 people of European ancestry (25,500 LOAD and 48,500 unaffected controls) and conducted a mega-meta-analysis, working with more than 11 million SNPs with a very dense coverage of the genomic map [21]. Table 4.1 depicts the list of genes and variants associated with LOAD in this mega-meta-analysis: in addition to the already eight known GWAS-defined genes (*ABCA7*, *BIN1*, *CLU*, *CR1*, *CD2AP*, *EPHA1*, *MS4A4*, and *PICALM*) that have been confirmed (*CD33* gene did not reach here genome-wide significance), 11 new susceptibility loci have been identified in or near plausible candidate genes (*CASS4*, *CELF1*, *FERMT2*, *HLA-DRB5/DRB1*, *INPP5D*, *MEF2C*, *NME8*, *PTK2B*, *SLC24A4/RIN3*, *SORL1*, and *ZCWPW1*). The effects of all these 19 genes on risk for LOAD are exceedingly small (Table 4.1), with allelic ORs between 0.77 (*SORL1*) and 1.22 (*BIN1*); in contrast, the ORs for APOE ε4 are approximately 4 or 15 for one or two ε4 alleles, respectively. That is, one or two copies of the APOE ε4 allele increases the risk for APOE by more than 400 % or 1500 %, whereas one copy of all these non-APOE alleles merely increases or decreases the risk by approximately 30 %, at most. However, the findings from this mega-meta-analysis are, for the most part, not based on the true susceptibility variants but are reflective of their tagging markers, which may harbor greater heterogeneity than the former with respect to alleles and extent of LD. Thus, it remains a possibility that the actual functional susceptibility variants may have bigger effect sizes.

Population Attributable Fraction: Understudied Populations

The cumulative population attributable fraction (e.g., the proportion of LOAD cases in a population that would be prevented if an exposure were eliminated) at each of the 19 non-*APOE* loci identified by the IGAP (Table 4.1) was between 1.1 % (*CASS4* and *SORL1*) and 8.1 % (*BIN1*) and that of APOE was 27.3 % [21]. The remaining genetic risk for LOAD could be due to new common loci, rare variants, structural variants, and gene-gene and gene-environment interactions. Most of large GWAS have identified several variants that affect LOAD susceptibility in non-Hispanic whites of European ancestry. African Americans and other minorities are understudied, and it is unclear whether any of the recently identified loci modify risk of LOAD in racial or ethnic groups other than whites. The ADGC consortium [22] performed a GWAS among the largest sample of African Americans ever assembled for genetic study of LOAD (nearly 2,000 cases and 4,000 cognitively normal elderly controls). The *APOE* ε4 allele, previously shown to be associated with LOAD in whites, was also implicated in African Americans ($P=5.5\times10^{-47}$, OR = 2.3), and more striking was that the effect size for *ABCA7* was comparable with that observed for *APOE*. In fact, variants at the *ABCA7* gene increased the risk for LOAD approximately 1.8-fold ($P=2.21\times10^{-9}$) in individuals of African ancestry as opposed to the modest increased risk of 1.15-fold in individuals of European ancestry (Table 4.1). A number of other variants in other genes (*CR1, BIN1, PICALM, CLU, EPHA1, MS4A* cluster, *CD2AP,* and *CD33*) did not reach the *P* value cutoff for genome-wide significance in this African American population.

Table 4.1 LOAD-associated GWAS loci in the International Genomics of Alzheimer's Project mega-meta-analysis

SNP	Gene	Mutation type	MAF in controls	PAF (%)	PAF type	OR (95 % CI)	*P* value	Implicated pathway
rs4147929	*ABCA7*	Intronic	0.162	2.8	Risk	1.15 (1.11–1.19)	1.1×10^{-15}	Aβ, cholesterol, immunity
rs6733839	*BIN1*	Intergenic	0.366	8.1	Risk	1.22 (1.18–1.25)	6.9×10^{-44}	Aβ, cellular signaling, tau
rs7274581	*CASS4*	Intronic	0.088	1.1	Preventive	0.88 (0.84–0.92)	2.5×10^{-8}	Aβ, tau
rs10948363	*CD2AP*	Intronic	0.255	2.3	Risk	1.10 (1.07–1.13)	5.2×10^{-11}	Aβ, cellular signaling
rs10838725	*CELF1*	Intronic	0.312	2.4	Risk	1.08 (1.05–1.11)	1.1×10^{-8}	Cellular signaling
rs9331896	*CLU*	Intronic	0.398	5.3	Preventive	0.86 (0.84–0.89)	2.8×10^{-25}	Aβ, cholesterol, immunity
rs6656401	*CR1*	Intronic	0.191	3.7	Risk	1.18 (1.14–1.22)	5.7×10^{-24}	Immunity
rs11771145	*EPHA1*	Intergenic	0.350	3.1	Preventive	0.90 (0.88–0.93)	1.1×10^{-13}	Immunity, cellular signaling
rs17125944	*FERMT2*	Intronic	0.079	1.5	Risk	1.14 (1.09–1.19)	7.9×10^{-9}	Tau
rs9271192	*HLA-DRB5*	Intergenic	0.277	3.2	Risk	1.11 (1.08–1.15)	2.9×10^{-12}	Immunity
rs35349669	*INPP5D*	Intronic	0.462	4.6	Risk	1.08 (1.05–1.11)	3.2×10^{-8}	Immunity
rs190982	*MEF2C*	Intergenic	0.389	2.7	Preventive	0.93 (0.90–0.95)	3.2×10^{-8}	Immunity, cellular signaling
rs983392	*MS4A6A*	Intergenic	0.406	4.2	Preventive	0.90 (0.87–0.92)	6.1×10^{-16}	Immunity
rs2718058	*NME8*	Intergenic	0.368	2.9	Preventive	0.93 (0.90–0.95)	4.8×10^{-9}	Cellular signaling
rs10792832	*PICALM*	Intergenic	0.365	5.3	Preventive	0.87 (0.85–0.89)	9.3×10^{-26}	Aβ, cellular signaling, tau
rs28834970	*PTK2B*	Intronic	0.358	3.1	Risk	1.10 (1.08–1.13)	7.4×10^{-14}	Cellular signaling
rs10498633	*SLC24A4/ RIN3*	Intronic	0.212	1.5	Preventive	0.91 (0.88–0.94)	5.5×10^{-9}	Tau
rs11218343	*SORL1*	Intronic	0.044	1.1	Preventive	0.77 (0.72–0.82)	9.7×10^{-15}	Aβ, cellular signaling
rs1476679	*ZCWPW1*	Intronic	0.293	3.2	Preventive	0.91 (0.89–0.94)	5.6×10^{-10}	?

Modified from [21]

LOAD late-onset alzheimer's disease, *GWAS* genome-wide association study, *MAF* minor allele frequency, *PAF* population attributable fraction, *OR* (95 % CI) odds ratio (95 % confidence interval), *Aβ* amyloid-beta

Functional Basis for the LOAD-Associated GWAS Loci

True Functional Variants: Expression Quantitative Trait Loci

The associated SNPs identified through GWAS are unlikely to be functional variants themselves. For any disease-associated SNP, the true variant underlying the phenotype studied may be the GWAS hit itself, a known common SNP in LD with the identified GWAS hit, an unknown common or rare SNP tagged by a haplotype on which the hit occurs, or a linked copy number variant [23]. For all traits studied by GWAS, only 12 % of the associated SNPs are located in, or occur in high LD with, protein-coding regions of genes; the vast majority (80 %) of trait-associated SNPs are located in intergenic regions or noncoding introns [24]. LOAD is not different: taking into account the 19 SNPs reported in the 11 new loci and the 8 previously reported loci associated to LOAD in the IGAP mega-meta-analysis [21], 12 SNPs are located in intronic regions and 7 in intergenic regions (Table 4.1). These findings clearly indicate that follow-up studies should not only examine coding variability but should also play close attention to the potential roles of these intronic and intergenic regions in the regulation of gene expression. Therefore, GWAS follow-up studies should rely on fine mapping of the associated loci and deep re-sequencing of the associated regions in samples of interest in order to identify all possible functional variants. In addition, it is critical to characterize the novel LOAD candidate variants and genes that are being identified in LOAD-associated GWAS with respect to their influence on gene expression, also known as expression quantitative trait loci (eQTL) studies [25]. The underlying premise of these studies is that the level of the expressed gene transcript (mRNA profiling) from LOAD patients will have changes in comparison to controls, by using data generated from tissue affected by the disease (such as the temporal cortex) or peripheral immune cells [26]. SNPs that influence brain gene expression (eSNPs) constitute an important class of functional variants. In this respect, SNPs in the CLU (rs11136000) and *MS4A4A* (rs2304933/rs2304935) genes influenced their expression levels in temporal cortex [27]: the LOAD-protective *CLU* and the risky *MS4A4A* alleles both occurred in conjunction with elevated levels of brain expression, implicating regulatory genetic variation for these genes in LOAD risk. In a systematical examination of *CLU*, *CR1*, *ABCA7*, *BIN1*, *PICALM*, and *MS4A6A/MS4A6E* loci for LOAD, coding variability might explain only the *ABCA7* association with LOAD, but common coding variability did not explain any of the other loci; in addition, none of these loci had eQTL effects and the regional expression of each of the loci did not match the pattern of brain regional distribution in Alzheimer pathology [23].

Pathogenic Pathways Implicated in LOAD from GWAS Loci

The LOAD candidate genes make biological sense and have identified different pathways involved in LOAD pathogenesis [4, 21, 28]. As suggested by Table 4.1, the implicated pathways are:

- A/Amyloid-beta metabolism (production, degradation, and clearance): *APOE*, *CLU*, *ABCA7*, *PICALM*, *BIN1*, *CD2AP*, *SORL1*, *CASS4*, and *CD33* [29]
- B/Immune system function (both innate and adaptive): *CLU*, *CR1*, *ABCA7*, *MS4A* cluster, *CD33*, *EPHA1*, *HLA-DRB5/DRB1*, *INPP5D*, and *MEF2C*
- C/Cholesterol metabolism: *APOE*, *CLU*, and *ABCA7*
- D/Synaptic cell functioning mechanisms and cell membrane processes (endocytosis): *PICALM*, *BIN1*, *CD33*, *CD2AP*, *EPHA1*, *SORL1*, *CELF1*, *NME8*, *MEF2C*, and *PTK2B*
- E/Tau pathology (microtubule stability, tau phosphorylation/aggregation, and neurofibrillary tangle formation): *CASS4*, *FERMT2*, *SLC24A4/RIN3*, *BIN1* [30], and *PICALM* [31]

Exactly how *APOE* might cause LOAD is a matter of debate, and as well as being the main transporter of cholesterol and other lipids into the brain, it is also thought to remove amyloid-beta from the brain. Ultimately, the validation of the pathogenic mechanisms of all these LOAD GWAS loci will require comprehensive functional studies in in vitro systems, in vivo animal models, and clinical samples.

Examining Genetic Influences on Endophenotypes

Endophenotypes are biologically relevant, quantitative, and heritable phenotypes. There are many endophenotypes that are currently utilized or are excellent candidates for genetic studies of LOAD, including cerebrospinal fluid measures of amyloid-beta, tau and phosphorylated tau, neuroimaging measures in magnetic resonance imaging (MRI) and positron emission tomography (PET) scans (such as hippocampal volume), and cognitive measures [25]. Genetic studies of LOAD endophenotypes are an effective approach for identifying disease risk loci that are complementary to case–control association studies, and these genetic variants might be implicated not only with risk for LOAD but also with age at onset or rate of progression. Cognitive endophenotypes (e.g., level of cognitive function and rate of decline in cognition) can help to detect genetic risk factors attributable to the preclinical and subclinical change in cognition in LOAD. For example, the simultaneous consideration of the joint effects of eight non-APOE LOAD-associated GWAS loci (*ABCA7*, *BIN1*, *CD2AP*, *CLU*, *CR1*, *MS4A4E*, *MS4A6A*, and *PICALM*) aggregated as a cumulative genetic risk score predicts accelerated progression from mild cognitive impairment (MCI) to LOAD in those subjects with higher scores [32]. Moreover, MCI patients with the *APOE* ε4 allele are more likely to convert to LOAD as compared to those without the *APOE* ε4 allele [33]. No clear profile has emerged from studies of the relation between genotype and amyloid or tau phenotype in cerebrospinal fluid: whereas no evidence for association between variants in *BIN1*, *CLU*, *CR1*, and *PICALM* genes and amyloid-beta and phosphorylated tau levels in cerebrospinal fluid has been found in a study [34], APOE ε4 allele, *CLU*, and *MS4A4A* genetic variants were associated with significantly reduced amyloid-beta levels in cerebrospinal fluid in other study [35]. Investigating whether LOAD-associated GWAS loci influence MRI measures (hippocampal and amygdala volumes and entorhinal cortex and temporal pole cortex thicknesses), the *APOE* ε4 allele and *PICALM* and *CR1* genotypes have been significantly associated with these neuroimaging measures [36].

The Whole Exome and Whole Genome Sequencing Approach

Common Versus Rare Variants

A proportion of heritability (the portion of phenotypic variance in a population attributable to additive genetic factors) is apparently unexplained by GWAS findings. Explanations for this "missing heritability" include rarer variants (possibly with larger effects) that are poorly detected by available genotyping arrays that focus on variants present in 5 % or more of the population; structured variants poorly captured by existing arrays, including copy number variants such as insertions and deletions and copy neutral variation such as inversions and translocations; low power to detect gene-gene interactions; and inadequate accounting for shared environment among relatives [37]. It is likely that a substantial portion of the genetic risk underlying LOAD is actually conferred by rare sequence variants, those occurring with a frequency <1 % in the general population, and possibly of relatively large genetic effect (e.g., with odds ratios >2). Rare variants are much more likely to have functional consequence than the more common variants; in fact, regulatory regions show preferential exclusion of common variants relative to rare ones just like protein-coding sequence [38]. GWAS are by design powered to detect association with causal variants that are relatively common in the population, and current microarray technology is not designed for de novo identification of rare sequence variants. Thus, the identification of presumed disease-associated rare variants requires deep re-sequencing in suitable data sets, either small scale (e.g., previously associated GWAS regions) or large-scale (whole exome or whole genome). Whole exome sequencing is most often chosen for monogenic Mendelian diseases, largely because of its low cost compared with whole genome sequencing (the exome is 1–2 % of the whole genome) and the notion that most sequence variations leading to a severe phenotypic effect are located in the coding part of the genome [4]. Whole exome sequencing is capable of identifying not only very rare Mendelian causes of disease but also low-frequency variability with medium-effect sizes modulating disease development. A significantly proportion of EOAD is caused by autosomal dominant, fully penetrant mutations. LOAD recurs within families more often than expected by chance alone, and this observed familial recurrence could be attributed to genetic loci with large phenotypic effects and reduced penetrance (possibly recessive loci) [10]. With monogenic recessive contributions to LOAD, one would not necessarily expect to see recurrence of the disease in multiple generations, nor a high recurrence rate among siblings, and the disease would be sporadic in the population. So far, the role of recessive mutations in LOAD has been considerably overlooked.

Rare Monogenic Forms of LOAD

TREM2

Homozygous loss-of-function mutations in *TREM2* gene, encoding the triggering receptor expressed on myeloid cells 2 protein, have previously been associated with an autosomal recessive form of early-onset dementia presenting with bone cysts and consequent fractures called polycystic lipomembranous osteodysplasia with sclerosing leukoencephalopathy or Nasu-Hakola disease. Homozygous *TREM2* mutations have been recently identified in three Turkish patients presenting with a clinical phenotype associated with frontotemporal dementia and with leukodystrophy but without any bone-associated symptoms [39]. Whereas severe and early-onset disease is caused by homozygous loss-of-function mutations in *TREM2*, heterozygous loss-of-function variants are associated with LOAD. For example, in 2,000 LOAD patients and 4,000 controls, a rare missense mutation (rs75932628-T) in *TREM2*, which was predicted to result in a R47H substitution, showed a strong, highly significant association with LOAD ($P=9.0\times10^{-9}$, OR = 5.05), with a minor allele frequency among healthy controls of 0.12 % in the United States [40]. Similarly, in 2,261 Icelandic participants, the *T* allele of rs75932628 in *TREM2* was found to confer a significant risk of LOAD ($P=3.42\times10^{-10}$, OR = 2.9), with a minor allele frequency of 0.63 % in healthy controls [41]. Consequently, this R47H variant of *TREM2* is a low-prevalence variant that increases LOAD risk with a moderate-to-high effect size, similar to that of the *APOE* ε4 allele. Neurodegeneration in *TREM2-associated* LOAD is probably driven by a chronic inflammatory process with dysfunction in the microglial phagocytosis [42].

APP

About 25 coding mutations in the *APP* gene have resulted in EOAD, but until now, mutations in APP had not been implicated in LOAD. In a set of whole genome sequence data from 1,795 Icelanders, the *A* allele of rs63750847 results in an alanine to threonine substitution at position 673 in *APP* (*A673T*) and was found to be significantly more common in the elderly control group aged 85 or greater than in the LOAD group (0.62 % versus 0.13 %, $P=4.78\times10^{-7}$, OR = 0.189) [43]. In addition, the cognitive function of elderly noncarriers remained poorer than for carriers of *A673T* after removing LOAD cases. *A673T* represents the first example of a rare sequence variant conferring strong protection against LOAD and also protecting against cognitive decline in the elderly without LOAD. The *A673T* substitution is critical for reducing the production of amyloid-beta. The complete absence of the *A673T* variant in a large cohort of Asian subjects [44] suggests that this is possibly an ethnicity-specific variant.

MAPT

In a combined analysis of 15,369 subjects, re-sequencing at the gene encoding for the microtubule associated protein tau (*MAPT*) discovered that the rare substitution *A152T* within exon 7 of MAPT increases the risk for LOAD (0.69 % in patients versus 0.30 % in controls, $P=4.0\times10^{-3}$, OR=2.3) and also for frontotemporal dementia (0.89 % in patients, $P=5.0\times10^{-4}$, OR=3.0) [45]. This study emphasizes the point that statistical evaluation of the role of rare sequence variants poses a challenge, and no thresholds for rare variant significance have been established. The functional studies show that the *A152T* in *MAPT* causes a pronounced decrease in microtubule stability and might enhance the level of tau oligomers. This is another example that rare variants can increase the risk for complex diseases with heterogeneous phenotypes.

FRMD4A

In a meta-analysis of EADI and GERARD consortia and a combined analysis of five additional case–control studies (10,000 LOAD and 16,000 controls), the AAC haplotype in the *FRMD4A* locus was associated with increased LOAD risk ($P=1.1\times10^{-10}$, OR=1.68) when compared with most frequent *GGT* haplotype [46]. As the *AAC* haplotype is rare (with a mean frequency of 2 % in Caucasian populations), this might explain why the locus was not detected in previous GWAS based on single-SNPs analyses, as SNPs with low frequency are poorly imputed even when using the 1000 Genomes data set. Therefore, other complementary approach to GWAS is this example of genome-wide haplotype association study. The protein encoded by *FRMD4A* is involved in cell structure, transport, and signaling functions.

Gene-Gene Interactions (Epistasis)

Evidence is accumulating that a pronounced part of the still elusive genetic variability in complex diseases could be due to ignored epistatic effects [47]. The term epistasis is conventionally used when an increased risk is only seen in the presence of two genetic factors and not seen when they act apart. In such cases, studies that examine simple loci individually, such as most GWAS, will fail to detect an effect. To understand the causes of LOAD, one needs to study not simple factors one at a time but interactions between genetic risk factors. In the case of LOAD, epistasis is likely to play a major part, in view of the high heritability of the disease. Epistasis had previously proved hard to demonstrate, mainly because sample sets had been too small and poorly characterized and inappropriate statistical methods had been used. The Epistasis Project [48] was designed to avoid these problems, with a multinational collaboration of 7 LOAD research groups from the UK, Spain, the

Netherlands, and Germany, contributing DNA samples from 1,757 LOAD cases and 6,295 controls. A typical GWAS may examine perhaps 500,000 loci, but the number of potential two-way interactions between these 500,000 loci is >100 billion (10^{11}). In order therefore to reduce the number of potential interactions to a manageable figure, a hypothesis-driven approach is required, and consequently, a selection of gene-gene interactions should be chosen according to prior evidence of a statistical interaction and a plausible biological hypothesis [49]. The chosen interactions in the Epistasis Project were involved in various pathogenic networks that contribute to the development of LOAD (lipid metabolism, amyloid-beta metabolism, inflammation, oxidative stress, and insulin metabolism), and the "synergy factor" [50] (equivalent to the interaction term defined by two binary factors in a logistic regression model) was used to measure the gene-gene interaction. In the inflammation pathway, the Epistasis Project has demonstrated that the interaction between the interleukin-6 proinflammatory cytokine and the interleukin-10 anti-inflammatory cytokine genes [48] and the interaction between the aromatase (a rate-limiting enzyme in the synthesis of estrogens) and the interleukin-10 genes [51] are both associated with increased LOAD risk. In the oxidative stress pathway, the Epistasis Project has revealed an increased LOAD risk due to the interaction between the hemochromatosis and transferrin genes [52] and the interaction between the glutathione S-transferase and the gene cluster of the hematopoietically expressed homeobox, the insulin-degrading enzyme, and the kinesin family member 11 [53]. In the future, to achieve higher power for such gene-gene interaction studies, larger sample sizes are needed, such as that of the IGAP mega-meta-analysis of GWAS [21].

References

1. Ertekin-Taner N. Genetics of Alzheimer disease in the pre- and post-GWAS era. Alzheimers Res Ther. 2010;2:3.
2. Bertram L. Alzheimer's genetics in the GWAS era: a continuing story of 'replications and refutations'. Curr Neurol Neurosci Rep. 2011;11:246–53.
3. Tanzi RE. The genetics of Alzheimer disease. Cold Spring Harb Perspect Med. 2012;2:a006296.
4. Bettens K, Sleegers K, Van Broeckhoven C. Genetic insights in Alzheimer's disease. Lancet Neurol. 2013;12:92–104.
5. Gatz M, Reynolds CA, Fratiglioni L, Johansson B, Mortimer JA, Berg S, et al. Role of genes and environments for explaining Alzheimer disease. Arch Gen Psychiatry. 2006;63:168–74.
6. Gerrish A, Russo G, Richards A, Moskvina V, Ivanov D, Harold D, et al. The role of variation at AβPP, PSEN1, PSEN2, and MAPT in late onset alzheimer's disease. J Alzheimers Dis. 2012;28:377–87.
7. Strittmatter WJ, Saunders AM, Schmechel D, Pericak-Vance M, Enghild J, Salvesen GS, Roses AD. Apolipoprotein E: high-avidity binding to beta-amyloid and increased frequency of type 4 allele in late-onset familial Alzheimer disease. Proc Natl Acad Sci U S A. 1993;90:1977–81.
8. Bertram L, Tanzi RE. The genetics of Alzheimer's disease. In: Teplow D, editor. Progress in molecular biology and translational science, vol. 107. Burlington: Academic; 2012. p. 79–100.
9. Genin E, Hannequin D, Wallon D, Sleegers K, Hiltunen M, Combarros O, et al. APOE and Alzheimer disease: a major gene with semi-dominant inheritance. Mol Psychiatry. 2011; 16:903–7.

10. Guerreiro RJ, Gustafson DR, Hardy J. The genetic architecture of Alzheimer's disease: beyond APP, PSENs and APOE. Neurobiol Aging. 2012;33:437–56.
11. Bertram L, McQueen MB, Mullin K, Blacker D, Tanzi RE. Systematic meta-analyses of Alzheimer disease genetic association studies: the AlzGene database. Nat Genet. 2007;39:17–23.
12. International HapMap Consortium. A genome haplotype map of the human genome. Nature. 2005;437:1299–320.
13. International HapMap Consortium, Frazer KA, Ballinger DG, Cox DR, Hinds DA, Stuve LL, et al. A second generation human haplotype map of over 3.1 million SNPs. Nature. 2007;449:851–61.
14. Schellenberg GD, Montine TJ. The genetics and neuropathology of Alzheimer's disease. Acta Neuropathol. 2012;124:305–23.
15. 1000 Genomes Project Consortium, Abecasis GR, Altshuler D, Auton A, Brooks LD, Durbin RM, et al. A map of human genome variation from population-scale sequencing. Nature. 2010;467:1061–73.
16. Harold D, Abraham R, Hollingworth P, Sims R, Gerrish A, Hamshere ML, et al. Genome-wide association study identifies variants at CLU and PICALM associated with Alzheimer's disease. Nat Genet. 2009;41:1088–93.
17. Lambert JC, Heath S, Even G, Campion D, Sleegers K, Hiltunen M, et al. Genome-wide association study identifies variants at CLU and CR1 associated with Alzheimer's disease. Nat Genet. 2009;41:1094–9.
18. Seshadri S, Fitzpatrick AL, Ikram MA, DeStefano AL, Gudnason V, Boada M, et al. Genome-wide analysis of genetic loci associated with Alzheimer disease. JAMA. 2010; 303:1832–40.
19. Hollingworth P, Harold D, Sims R, Gerrish A, Lambert JC, Carrasquillo MM, et al. Common variants at ABCA7, MS4A6A/MS4A4E, EPHA1, CD33 and CD2AP are associated with Alzheimer's disease. Nat Genet. 2011;43:429–35.
20. Naj AC, Jun G, Beecham GW, Wang LS, Vardarajan BN, Buros J, et al. Common variants at MS4A4/MS4A6E, CD2AP, CD33 and EPHA1 are associated with late-onset Alzheimer's disease. Nat Genet. 2011;43:436–41. doi:10.1038/ng.2802.
21. Lambert JC, Ibrahim-Verbaas CA, Harold D, Naj AC, Sims R, Bellenguez C, et al. Meta-analysis of 74,046 individuals identifies 11 new susceptibility loci for Alzheimer´s disease. Nat Genet. 2013;45:1452–8.
22. Reitz C, Jun G, Naj A, Rajbhandary R, Vardarajan BN, Wang LS, et al. Variants in the ATP-binding cassette transporter (ABCA7), apolipoprotein E ϵ4, and the risk of late-onset alzheimer disease in African Americans. JAMA. 2013;309:1483–92.
23. Holton P, Ryten M, Nalls M, Trabzuni D, Weale ME, Hernandez D, et al. Initial assessment of the pathogenic mechanisms of the recently identified Alzheimer risk loci. Ann Hum Genet. 2013;77:85–105.
24. Manolio TA. Genome wide association studies and assessment of the risk of disease. N Engl J Med. 2010;363:166–76.
25. Ertekin-Taner N, De Jager PL, Yu L, Bennett DA. Alternative approaches in gene discovery and characterization in Alzheimer's disease. Curr Genet Med Rep. 2013;1:39–51.
26. Stranger BE, Stahl EA, Raj T. Progress and promise of genome-wide association studies for human complex trait genetics. Genetics. 2011;187:367–83.
27. Allen M, Zou F, Chai HS, Younkin CS, Crook J, Pankratz VS, et al. Novel late-onset alzheimer disease loci variants associate with brain gene expression. Neurology. 2012;79:221–8.
28. Morgan K. The three new pathways leading to Alzheimer's disease. Neuropathol Appl Neurobiol. 2011;37:353–7.
29. Griciuc A, Serrano-Pozo A, Parrado AR, Lesinski AN, Asselin CN, Mullin K, et al. Alzheimer's disease risk gene CD33 inhibits microglial uptake of amyloid beta. Neuron. 2013;78:631–43.
30. Chapuis J, Hansmannel F, Gistelinck M, Mounier A, Van Cauwenberghe C, Kolen KV, et al. Increased expression of BIN1 mediates Alzheimer genetic risk by modulating tau pathology. Mol Psychiatry. 2013;18:1225–34. doi:10.1038/mp.2013.1.

31. Ando K, Brion JP, Stygelbout V, Suain V, Authelet M, Dedecker R, et al. Clathrin adaptor CALM/PICALM is associated with neurofibrillary tangles and is cleaved in Alzheimer's brains. Acta Neuropathol. 2013;125:861–78.
32. Rodríguez-Rodríguez E, Sánchez-Juan P, Vázquez-Higuera JL, Mateo I, Pozueta A, Berciano J, et al. Genetic risk score predicting accelerated progression from mild cognitive impairment to Alzheimer's disease. J Neural Transm. 2013;120:807–12.
33. Elias-Sonnenschein LS, Viechtbauer W, Ramakers IH, Verhey FR, Visser PJ. Predictive value of APOE-ε4 allele for progression from MCI to AD-type dementia: a meta-analysis. J Neurol Neurosurg Psychiatry. 2011;82:1149–56.
34. Kauwe JS, Cruchaga C, Karch CM, Sadler B, Lee M, Mayo K, et al. Fine mapping of genetic variants in BIN1, CLU, CR1 and PICALM for association with cerebrospinal fluid biomarkers for Alzheimer's disease. PLoS One. 2011;6:e15918.
35. Elias-Sonnenschein LS, Helisalmi S, Natunen T, Hall A, Paajanen T, Herukka SK, et al. Genetic loci associated with Alzheimer's disease and cerebrospinal fluid biomarkers in a Finnish case–control cohort. PLoS One. 2013;8:e59676.
36. Biffi A, Anderson CD, Desikan RS, Sabuncu M, Cortellini L, Schmansky N, et al. Alzheimer's Disease Neuroimaging Initiative (ADNI). Genetic variation and neuroimaging measures in Alzheimer disease. Arch Neurol. 2010;67:677–85.
37. Manolio TA, Collins FS, Cox NJ, Goldstein DB, Hindorff LA, Hunter DJ, et al. Finding the missing heritability of complex diseases. Nature. 2009;461:747–53.
38. Zhu Q, Ge D, Maia JM, Zhu M, Petrovski S, Dickson SP, et al. A genome-wide comparison of the functional properties of rare and common genetic variants in humans. Am J Hum Genet. 2011;88:458–68.
39. Guerreiro RJ, Lohmann E, Brás JM, Gibbs JR, Rohrer JD, Gurunlian N, et al. Using exome sequencing to reveal mutations in TREM2 presenting as a frontotemporal dementia-like syndrome without bone involvement. JAMA Neurol. 2013;70:78–84.
40. Guerreiro R, Wojtas A, Bras J, Carrasquillo M, Rogaeva E, Majounie E, Alzheimer Genetic Analysis Group, et al. TREM2 variants in Alzheimer's disease. N Engl J Med. 2013; 368:117–27.
41. Jonsson T, Stefansson H, Steinberg S, Jonsdottir I, Jonsson PV, Snaedal J, et al. Variant of TREM2 associated with the risk of Alzheimer's disease. N Engl J Med. 2013;368: 107–16.
42. Neumann H, Daly MJ. Variant TREM2 as risk factor for Alzheimer's disease. N Engl J Med. 2013;368:182–4.
43. Jonsson T, Atwal JK, Steinberg S, Snaedal J, Jonsson PV, Bjornsson S, et al. A mutation in APP protects against Alzheimer's disease and age-related cognitive decline. Nature. 2012;488:96–9.
44. Seng Ting SK, Chong MS, Kandiah N, Hameed S, Tan L, Au WL, et al. Absence of A673T amyloid-β precursor protein variant in Alzheimer's disease and other neurological diseases. Neurobiol Aging. 2013;34:2441e7–8.
45. Coppola G, Chinnathambi S, Lee JJ, Dombroski BA, Baker MC, Soto-Ortolaza AI, et al. Evidence for a role of the rare p.A152T variant in MAPT in increasing the risk for FTD-spectrum and Alzheimer's diseases. Hum Mol Genet. 2012;21:3500–12.
46. Lambert JC, Grenier-Boley B, Harold D, Zelenika D, Chouraki V, Kamatani Y, et al. Genome-wide haplotype association study identifies the FRMD4A gene as a risk locus for Alzheimer's disease. Mol Psychiatry. 2013;18:461–70.
47. Zuk O, Hechter E, Sunyaev SR, Lander ES. The mystery of missing heritability: genetic interactions create phantom heritability. Proc Natl Acad Sci U S A. 2012;109:1193–8.
48. Combarros O, van Duijn CM, Hammond N, Belbin O, Arias-Vásquez A, Cortina-Borja M, et al. Replication by the Epistasis Project of the interaction between the genes for IL-6 and IL-10 in the risk of Alzheimer's disease. J Neuroinflammation. 2009;6:22.
49. Combarros O, Cortina-Borja M, Smith AD, Lehmann DJ. Epistasis in sporadic Alzheimer's disease. Neurobiol Aging. 2009;30:1333–49.

50. Cortina-Borja M, Smith AD, Combarros O, Lehmann DJ. The synergy factor: a statistic to measure interactions in complex diseases. BMC Res Notes. 2009;2:105.
51. Medway C, Combarros O, Cortina-Borja M, Butler HT, Ibrahim-Verbaas CA, de Bruijn RF, et al. The sex-specific associations of the aromatase gene with Alzheimer's disease and its interaction with IL10 in the Epistasis Project. Eur J Hum Genet. 2014;22(2):216–20. doi:10.1038/ejhg.2013.116.
52. Lehmann DJ, Schuur M, Warden DR, Hammond N, Belbin O, Kölsch H, et al. Transferrin and HFE genes interact in Alzheimer's disease risk: the Epistasis Project. Neurobiol Aging. 2012;33:202.e1–13.
53. Bullock JM, Medway C, Cortina-Borja M, Turton JC, Prince JA, Ibrahim-Verbaas CA, et al. Discovery by the Epistasis Project of an epistatic interaction between the GSTM3 gene and the HHEX/IDE/KIF11 locus in the risk of Alzheimer's disease. Neurobiol Aging. 2013;34:1309.e1–7.

Chapter 5
The Role of Cerebrovascular Disease in Cognitive Decline

Ana Verdelho

Abstract Vascular risk factors and cerebrovascular disease are recognized factors implicated in the evolution toward dementia, not only of vascular origin but also degenerative dementia as Alzheimer's disease. Even among nondemented subjects, hypertension, diabetes, and stroke are associated with worse performance in attention, executive functions, and speed and motor control. Influence of vascular risk factors in cognition starts early in life. Treatment and control of vascular risk factors since early ages has a key role in order to prevent cognitive impairment associated with aging. Cerebral white matter changes have gained attention in the last decades and can represent a potential outcome in experimental studies aiming to reduce cerebrovascular burden.

Keywords Vascular risk factors • Hypertension • Diabetes • Stroke • White matter changes • Lacunes • Microbleeds

Vascular risk factors and cerebrovascular disease of the brain are recognized factors that influence cognition and are implicated in the evolution toward dementia, not only of vascular origin but also degenerative dementia as Alzheimer's disease.

This chapter has two different sections. The first section covers the impact of main vascular factors in cognition and in the risk of dementia. As small vessel disease is closely linked to vascular risk factors and represents one of the consequences of several vascular risk factors measured in the brain; we approach, in the second section, the impact of cerebral small vessel disease in cognition and in dementia.

A. Verdelho, PhD, MD
Department of Neurosciences, Hospital de Santa Maria, University of Lisbon, Av prof, Egas Moniz, Lisbon, Portugal
e-mail: averdelho@fm.ul.pt

D. Galimberti, E. Scarpini (eds.), *Neurodegenerative Diseases*,
DOI 10.1007/978-1-4471-6380-0_5, © Springer-Verlag London 2014

Role of Vascular Risk Factors in Cognition

Vascular risk factors have been implicated in cognitive decline and dementia (including degenerative dementia). Among the whole spectrum of vascular risk factors, hypertension, stroke, and diabetes seem to play the most important role [1–12]. Before exploring evidence that supports the relationship between some of the major risk factors and cognitive impairment, we present two concepts that have evolved in past years. The first is that cognitive decline is insidious and slowly developing starting early in life, around the fourth decade [13]. This is probably one of the explanations for many of the controversial data concerning some of the vascular risk factors, namely, cholesterol blood levels and body mass index [14–18]. It is likely that these pathologies contribute to cognitive decline mainly when present in midlife.

The second concept is that the interaction between several cardiovascular risk factors contributes more strongly for cognitive decline than isolated risk factors [4, 16]. A systematic review stressed that the risk of dementia in diabetes is increased when associated with other vascular risk factors, phenomena that were also identified for other risk factors [4, 16, 19], mainly if they are concomitantly present in midlife [4, 20].

Role of Diabetes in Cognition

Diabetes has increasingly been identified as a risk factor for cognitive impairment and dementia [12, 21, 22], including Alzheimer's disease [23]. Among nondemented subjects, diabetics have worse cognitive performance when compared to nondiabetics [7, 22, 24] in global tests of cognition [25], attention, executive functions, processing speed and motor control, and also memory, praxis, and language [25, 26], independently of other confounders. Diabetic subjects have a twofold increase in risk for mild cognitive impairment and dementia compared with diabetics [7, 12, 27, 28].

Diabetes has several pathways to be implicated in the progression for dementia, not only due to the higher risk of vascular disease but also mediated through metabolic changes due to the insulin and glycemia pathways that are implicated in the metabolic production of beta-amyloid protein and tau protein [21], promoting neuronal degeneration [29] and thus implicated in pathogenesis of Alzheimer's disease [7, 30, 31]. Moreover, recent data suggest a genetic link between diabetes and the pathogenesis of Alzheimer's disease [32, 33].

Role of Stroke in Cognition

Stroke is a well-recognized risk factor for cognitive impairment in prospective community studies [1, 8, 28, 34, 35] and is associated with a twofold risk of dementia [35], not only for vascular dementia and vascular cognitive impairment but also for degenerative dementias such as Alzheimer's disease [35].

The higher risk of dementia in stroke survivors can be partially explained by concomitant vascular factors [36] and by prestroke dementia, but this is not the only explanation [35–37]. Nondemented stroke survivors have worse performance in tasks of attention and executive functions [25] compared with subjects without stroke. On the other hand, small vessel disease predicts vascular dementia [38], even without clinical stroke.

The clear impact of stroke on the development of degenerative types of dementia is not well established. Although a higher risk of Alzheimer's disease is associated with stroke, the pathological association between the two diseases is not clear.

Neuropathological data suggested that vascular disease could affect cognition, not only through the effects on subcortical connections and white matter disease but also exacerbating cortical atrophy [34, 39]. One of the likely explanations could be that vascular acute events anticipate incipient cognitive impairment due to concomitant amyloid pathology or otherwise have a synergistic or additive effect to develop degenerative dementia.

Role of Hypertension in Cognition

There is a considerable controversy between studies approaching some of the vascular risk factors and cognitive decline. One of the examples is the effect of hypertension. One of the most important variables that explain differences between studies considering hypertension is age of included subjects in those studies. Hypertension in midlife has been consistently associated with later development of cognitive decline and dementia. Although the strongest association is with vascular dementia, there is also an increased risk of degenerative dementia as Alzheimer's disease [1, 4, 11, 40–43]. Recently, it was indeed suggested that hypertension was associated with greater amyloid burden not only in middle-aged but also among older adults [44]. Treatment with antihypertensive treatment was associated with reduced hippocampus atrophy in hypertensive subjects [45] and with less Alzheimer's disease neuropathology [46].

However, the relationship between late-onset hypertension and cognitive decline and dementia is less clear: some studies were negative for this association [5, 6, 47] or sustain that a very low systolic and/or diastolic value was associated with higher risk of cognitive decline [41, 42].

In cross-sectional studies among nondemented subjects, hypertension in late life was associated with worse performance in several cognitive tests mainly related with executive functions and attention, digit symbol test, and word fluency [48] but also difficulties in some global cognitive functioning tests [27, 49, 50]. The most likely explanation for these discrepancies is that the deleterious effect of hypertension is due to chronic vascular damage starting in midlife that later originates cognitive impairment [43]. Results from trials focusing on the prevention of dementia using antihypertensive medication have failed to show a consistent protective effect, sustaining this explanation [51, 52]. From the six main randomized placebo-controlled studies, four were negative for a protective effect [53–56], one found a small effect on the prevention of dementia [57], and the other [58] found a protective effect only for poststroke dementia. In fact those studies were probably performed in older ages than what was desirable to prevent dementia, and, additionally, the follow-up was short.

Role of Alcohol Intake and Smoking in Cognition

The influence of alcohol intake on brain structure and cognition has been a focus of interest in late years. In the Leukoaraiosis And DISability (LADIS) study [25], among subjects with white matter changes free of dementia and living independently, mild and moderate alcohol consumption was associated with better performance on global measures of cognition compared to nondrinkers (included never drinkers), but this relation was lost overtime [25, 38]. Low or moderate alcohol intake was associated with reduced risk of Alzheimer's disease in a systematic review with meta-analysis, compared to the risk of dementia in nondrinkers [59]. In this review, nondrinkers had a small higher risk compared also with excessive drinkers. However, nondrinkers could include former excessive drinkers that stopped consuming due to health problems [59]. Recently, a study conducted among older subjects could not find evidence that moderate alcohol intake could prevent cognitive decline [60]. Considering imaging data, brain atrophy was associated with alcohol intake even for low drinkers [61], and controversial effects on white matter changes (WMC) and infarcts were associated with alcohol consumption in the same study [61].

Risk of dementia associated with smoking has also been studied. Smoking habits could have a theoretical beneficial effect in cognition, mediated through the stimulating effect of nicotine. In fact, the acute administration of nicotine in nonsmoking young adults with attentional deficit was associated with improvement in attention, executive functions, and working memory, probably mediated through the activation of the cholinergic system [62]. Indeed, in a study with elderly people from Taiwan, a better cognitive profile was observed in smokers [63]. Very recently, an improvement in measures of attention, memory, and mental processing was found after 6 months of transdermal nicotine in nonsmoking subjects with amnestic mild cognitive impairment, in a double-blind randomized trial [64]. However, the deleterious effect of smoking, mediated through oxidative stress, triggering atherogenesis and inflammation could, even indirectly, mediate increased risk for cognitive decline. In a meta-analysis of 19 observational prospective studies, smoking increased the risk for dementia, not only vascular dementia but also for degenerative dementias, an effect found mainly comparing active smokers against never smokers [65]. This risk could potentially be more pronounced among persons without the apolipoprotein E type 4 allele (*APOE-* ε 4) than among *APOE-* ε *4* carriers [66].

Role of Small Vessel Disease in Cognition

Small vessel disease is a broad concept used in several contexts and involves the cognitive, clinical, and imagiological consequences of the pathological changes of the small vessels of the brain [67]. As small vessels are not visualized in vivo, visible imagiological consequences of small vessel disease are usually considered as the marker of the disease. Clinical expression of small vessel disease is not uniform, as it includes lacunar infarcts, white matter changes, or hemorrhagic events as microbleeds (Fig. 5.1). Moreover, definition of small vessel disease definition varies between the different studies. In this section we will focus on the cognitive implications of small vessel disease.

White matter changes designate the changes of the radiological appearance of the white matter of the brain, detected through computed tomography (CT) or magnetic resonance imaging (MRI), of probable vascular etiology, that are frequently described in older subjects with or without cognitive deficit [68–79]. White matter changes do not follow specific vascular territories and are usually described as periventricular and subcortical but can also appear infratentorial in the pons. Age is the most frequent risk factor, but white matter changes are increased in subjects with hypertension and stroke [80]. Clinical manifestations of white matter changes include cognitive decline, gait disturbances, urinary dysfunction, and personality and mood changes [67]. The knowledge of an implication of white matter changes in cognition has more than a century, but it was only after the advent of brain imaging that this concept gained interest, and the term leukoaraiosis was introduced [81]. Periventricular white matter changes are frequent in demented subjects, independently of the type of dementia [71]. White matter changes are associated with worse cognitive performance among nondemented older subjects, mainly in executive functions, attention, and processing speed and motor control [25, 72, 73, 82] but also in global measures of cognition [12–14], independently of other confounders. WMC severity is implicated in higher risk of cognitive impairment and dementia [38, 75–78], and the relation is stronger with vascular dementia [38, 79–84].

Lacunes are frequently described in CT and MRI of elderly subjects and have been implicated in higher risk of dementia [85]. Similarly to white matter changes, lacunes have been implicated in worse executive functioning [86], processing speed and motor control [87] among demented and nondemented subjects, with or without previous clinical stroke. The higher frequency of lacunes in nondemented subjects [88] and the coexistence of other small vessel disease types [89] make it difficult to determine the exact influence of lacunes in cognition.

Specific locations, such as thalamic and basal ganglia lacunes, can have a specific impact in cognition [80], but further studies are needed to understand the individual effect of lacunes, even considering other concomitant confounders.

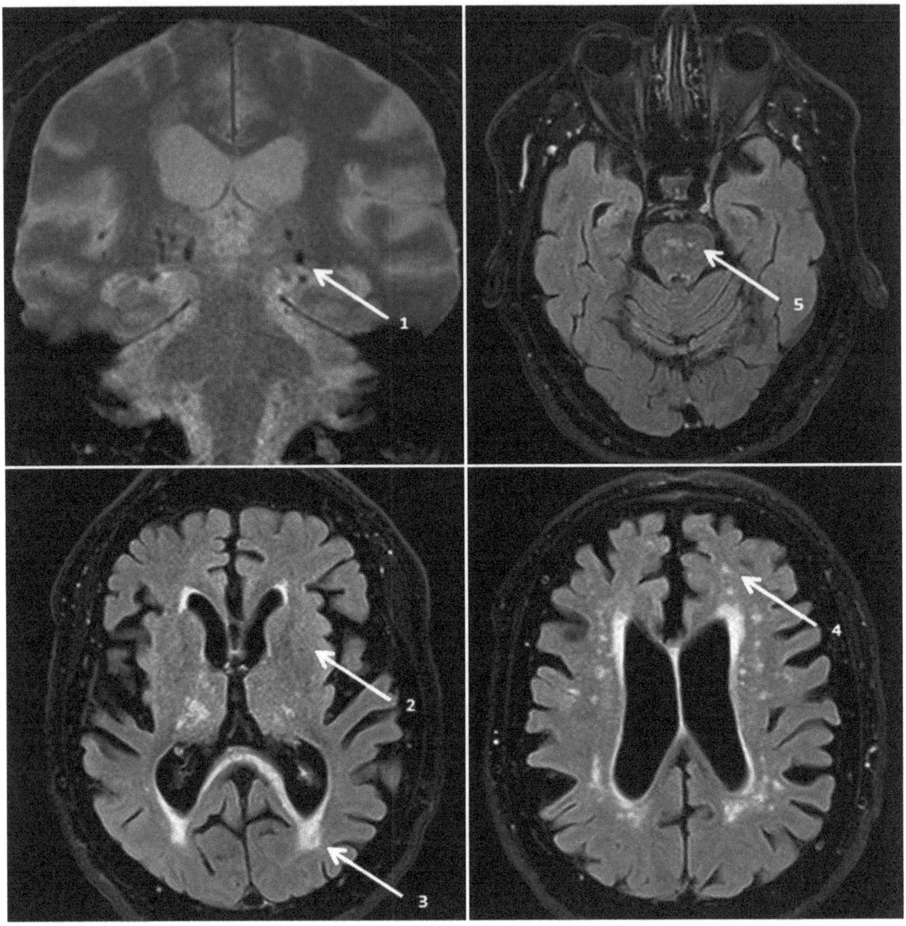

Fig. 5.1 Different expressions of small vessel disease in the same patient. *1* Microbleeds, *2* lacunes, *3* periventricular white matter changes, *4* subcortical white matter changes, *5* white matter changes in the pons

Cerebral microbleeds have been progressively described using specific susceptible MRI sequences. Prevalence data are highly variable, lower in community studies (7–36 %), higher among demented subjects, and mainly in subcortical vascular dementia (up to 85 %) [90–92].

Cerebral microbleeds have been associated with worse performance mainly in executive functions [93–95], processing and motor speed [95, 96], and attention [97], but the individual impact in cognition is not settled yet. It is not clear if different localizations are associated with specific profiles of cognitive deterioration, but increasing number of microbleeds seem to be associated with an increasing cognitive decline [95, 98].

Conclusion

Vascular risk factors are associated with an increased risk of cognitive decline and dementia, including degenerative dementia, and even among nondemented subjects are associated with worse cognitive performance. Treatment and control of vascular risk factors since at an early age has a key role in order to prevent cognitive impairment associated with aging. Nowadays, enough evidence sustains treatment of diabetes, prevention of stroke and stroke recurrence, and also treatment of hypertension in midlife, in order to prevent progression toward dementia. Further studies are needed to determine the type of intervention in each subject, considering other vascular risk factors. Small vessel disease is increased in subjects with vascular risk factors, can be monitored with brain imaging, is associated with cognitive decline, and can be used as a hallmark of cerebral vascular disease. In future studies white matter changes (and other expressions of small vessel disease) could be used as a potential end point of experimental studies.

References

1. Hénon H, Pasquier F, Leys D. Poststroke dementia. Cerebrovasc Dis. 2006;22:61–70.
2. Troncoso JC, Zonderman AB, Resnick SM, Crain B, Pletnikova O, O'Brien RJ. Effect of infarcts on dementia in the Baltimore longitudinal study of aging. Ann Neurol. 2008;64:168–76.
3. Xu WL, Qiu CX, Wahlin A, Winblad B, Fratiglioni L. Diabetes mellitus and risk of dementia in the Kungsholmen project: a 6-year follow-up study. Neurology. 2004;63:1181–6.
4. Kivipelto M, Helkala EL, Laakso MP, Hänninen T, Hallikainen M, Alhainen K, Soininen H, Tuomilehto J, Nissinen A. Midlife vascular risk factors and Alzheimer's disease in later life: longitudinal, population based study. BMJ. 2001;322:1447–51.
5. Hebert LE, Scherr PA, Bennett DA, Bienias JL, Wilson RS, Morris MC, Evans DA. Blood pressure and late-life cognitive function change. A biracial longitudinal population study. Neurology. 2004;62:2021–4.
6. Shah RC, Wilson RS, Bienias JL, Arvanitakis Z, Evans DA, Bennett DA. Relation of blood pressure to risk of incident Alzheimer's disease and change in global cognitive function in older persons. Neuroepidemiology. 2006;26:30–6.
7. Arvanitakis Z, Wilson RS, Bienias JL, Evans DA, Bennett DA. Diabetes Mellitus and risk of Alzheimer disease and decline in cognitive function. Arch Neurol. 2004;61:661–6.
8. Rastas S, Pirttilä T, Mattila K, Verkkoniemi A, Juva K, Niinistö L, Länsimies E, Sulkava R. Vascular risk factors and dementia in the general population aged >85 years. Prospective population-based study. Neurobiol Aging. 2010;31:1–7.
9. Ruitenberg A, Skoog I, Ott A, Aevarsson O, Witteman JC, Lernfelt B, van Harskamp F, Hofman A, Breteler MM. Blood pressure and risk of dementia: results from the Rotterdam study and the Gothenburg H-70 Study. Dement Geriatr Cogn Disord. 2001;12:33–9.
10. Harrington F, Saxby BK, McKeith IG, Wesnes K, Ford GA. Cognitive performance in hypertensive and normotensive older subjects. Hypertension. 2000;36:1079–82.
11. Launer LJ, Ross GW, Petrovitch H, Masaki K, Foley D, White LR, Havlik RJ. Midlife blood pressure and dementia: the Honolulu-Asia aging study. Neurobiol Aging. 2000;21:49–55.

12. Biessels GJ, Staekenborg S, Brunner E, Brayne C, Scheltens P. Risk of dementia in diabetes mellitus: a systematic review. Lancet Neurol. 2006;5:64–74.
13. Singh-Manoux A, Kivimaki M, Glymour MM, Elbaz A, Berr C, Ebmeier KP, Ferrie JE, Dugravot A. Timing of onset of cognitive decline: results from Whitehall II prospective cohort study. BMJ. 2011;344:d7622.
14. Strand BH, Langballe EM, Hjellvik V, Handal M, Næss O, Knudsen GP, Refsum H, Tambs K, Nafstad P, Schirmer H, Bergem AL, Selmer R, Engedal K, Magnus P, Bjertness E, GENIDEM-Group. Midlife vascular risk factors and their association with dementia deaths: results from a Norwegian prospective study followed up for 35 years. J Neurol Sci. 2013;324(1–2):124–30.
15. Alonso A, Jacobs Jr DR, Menotti A, Nissinen A, Dontas A, Kafatos A, Kromhout D. Cardiovascular risk factors and dementia mortality: 40 years of follow-up in the Seven Countries Study. J Neurol Sci. 2009;280(1–2):79–83.
16. Whitmer RA, Sidney S, Selby J, Johnston SC, Yaffe K. Midlife cardiovascular risk factors and risk of dementia in late life. Neurology. 2005;64(2):277–81.
17. Whitmer RA, Gustafson DR, Barrett-Connor E, Haan MN, Gunderson EP, Yaffe K. Central obesity and increased risk of dementia more than three decades later. Neurology. 2008;71:1057–64.
18. Anstey KJ, Cherbuin N, Budge M, Young J. Body mass index in midlife and late-life as a risk factor for dementia: a meta-analysis of prospective studies. Obes Rev. 2011;12(5):e426–37.
19. Purnell C, Gao S, Callahan CM, Hendrie HC. Cardiovascular risk factors and incident Alzheimer disease: a systematic review of the literature. Alzheimer Dis Assoc Disord. 2009;23:1–10.
20. Virta JJ, Heikkilä K, Perola M, Koskenvuo M, Räihä I, Rinne JO, Kaprio J. Midlife cardiovascular risk factors and late cognitive impairment. Eur J Epidemiol. 2013;28:405–16.
21. Luchsinger JA. Adiposity, hyperinsulinemia, diabetes and Alzheimer's disease: an epidemiological perspective. Eur J Pharmacol. 2008;585:119–29.
22. Euser SM, Sattar N, Witteman JC, Bollen EL, Sijbrands EJ, Hofman A, Perry IJ, Breteler MM, Westendorp RG, PROSPER and Rotterdam Study. A prospective analysis of elevated fasting glucose levels and cognitive function in older people: results from PROSPER and the Rotterdam Study. Diabetes. 2010;59:1601–7.
23. Vagelatos NT, Eslick GD. Type 2 diabetes as a risk factor for Alzheimer's disease: the confounders, interactions, and neuropathology associated with this relationship. Epidemiol Rev. 2013;35:152–60.
24. Cukierman T, Gerstein HC, Williamson JD. Cognitive decline and dementia in diabetes-systematic overview of prospective observational studies. Diabetologia. 2005;48:2460–9.
25. Verdelho A, Madureira S, Ferro JM, Basile AM, Chabriat H, Erkinjuntti T, Fazekas F, Hennerici M, O'Brien J, Pantoni L, Salvadori E, Scheltens P, Visser MC, Wahlund LO, Waldemar G, Wallin A, Inzitari D. Differential impact of cerebral white matter changes, diabetes, hypertension and stroke on cognitive performance among non-disabled elderly. The LADIS study. J Neurol Neurosurg Psychiatry. 2007;78:1325–30.
26. Manschot SM, Brands AM, van der Grond J, Kessels RP, Algra A, Kappelle LJ, Biessels GJ, Utrecht Diabetic Encephalopathy Study Group. Brain magnetic resonance imaging correlates of impaired cognition in patients with type 2 diabetes. Diabetes. 2006;55:1106–13.
27. Kilander L, Nyman H, Boberg M, Hansson L, Lithell H. Hypertension is related to cognitive impairment: a 20-year follow-up of 999 men. Hypertension. 1998;31:780–6.
28. Matthews FE, MRC Cognitive Function and Ageing Study. Risk factors for incident dementia in England and Wales: the Medical Research Council Cognitive Function and Ageing Study. A population-based nested case–control study. Age Ageing. 2006;35:154–60.
29. Folch J, Pedrós I, Patraca I, Martínez N, Sureda F, Camins A. Metabolic basis of sporadic Alzheimer's disease. Role of hormones related to energy metabolism. Curr Pharm Des. 2013;19(38):6739–48.
30. Liu F, Shi J, Tanimukai H, Gu J, Gu J, Grundke-Iqbal I, Iqbal K, Gong CX. Reduced O-GlcNAcylation links lower brain glucose metabolism and tau pathology in Alzheimer's disease. Brain. 2009;132:1820–32.

31. de la Monte SM, Wands JR. Alzheimer's disease is type 3 diabetes-evidence reviewed. J Diabetes Sci Technol. 2008;2:1101–13.
32. Mirza Z, Kamal MA, Abuzenadah AM, Al-Qahtani MH, Karim S. Establishing genomic/transcriptomic links between Alzheimer's disease and type II diabetes mellitus by meta-analysis approach. CNS Neurol Disord Drug Targets. 2013;1:1–17.
33. Abdul-Rahman O, Sasvari-Szekely M, Ver A, Rosta K, Szasz BK, Kereszturi E, Keszler G. Altered gene expression profiles in the hippocampus and prefrontal cortex of type 2 diabetic rats. BMC Genomics. 2012;13:81.
34. Reitz C, Bos MJ, Hofman A, Koudstaal PJ, Breteler MM. Prestroke cognitive performance, incident stroke, and risk of dementia: the Rotterdam Study. Stroke. 2008;39:36–41.
35. Savva GM, Stephan BC, Alzheimer's Society Vascular Dementia Systematic Review Group. Epidemiological studies of the effect of stroke on incident dementia: a systematic review. Stroke. 2010;41(1):e41–6.
36. Allan LM, Rowan EN, Firbank MJ, Thomas AJ, Parry SW, Polvikoski TM, O'Brien JT, Kalaria RN. Long term incidence of dementia, predictors of mortality and pathological diagnosis in older stroke survivors. Brain. 2011;134(Pt 12):3716–27.
37. Pendlebury ST, Rothwell PM. Prevalence, incidence, and factors associated with pre-stroke and post-stroke dementia: a systematic review and meta-analysis. Lancet Neurol. 2009;8(11):1006–18.
38. Verdelho A, Madureira S, Moleiro C, Ferro JM, Santos CO, Erkinjuntti T, Pantoni L, Fazekas F, Visser M, Waldemar G, Wallin A, Hennerici M, Inzitari D. White matter changes and diabetes predict cognitive decline in the elderly: the LADIS study. Neurology. 2010;75:160–7.
39. Jagust WJ, Zheng L, Harvey DJ, Mack WJ, Vinters HV, Weiner MW, Ellis WG, Zarow C, Mungas D, Reed BR, Kramer JH, Schuff N, DeCarli C, Chui HC. Neuropathological basis of magnetic resonance images in aging and dementia. Ann Neurol. 2008;63:72–80.
40. Capizzano AA, Acién L, Bekinschtein T, Furman M, Gomila H, Martínez A, Mizrahi R, Starkstein SE. White matter hyperintensities are significantly associated with cortical atrophy in Alzheimer's disease. J Neurol Neurosurg Psychiatry. 2004;75:822–7.
41. Qiu C, Winblad B, Fratiglioni L. Low diastolic pressure and risk of dementia in very old people: a longitudinal study. Dement Geriatr Cogn Disord. 2009;28:213–9.
42. Razay G, Williams J, King E, Smith AD, Wilcock G. Blood pressure, dementia and Alzheimer's disease: the OPTIMA longitudinal study. Dement Geriatr Cogn Disord. 2009;28:70–4.
43. Stewart R, Xue QL, Masaki K, Petrovitch H, Ross GW, White LR, Launer LJ. Change in blood pressure and incident dementia: a 32-year prospective study. Hypertension. 2009;54:233–40.
44. Rodrigue KM, Rieck JR, Kennedy KM, Devous Sr MD, Diaz-Arrastia R, Park DC. Risk factors for β-amyloid deposition in healthy aging: vascular and genetic effects. JAMA Neurol. 2013;70:600–6.
45. Korf ES, White LR, Scheltens P, Launer LJ. Midlife blood pressure and the risk of hippocampal atrophy: the Honolulu Asia Aging Study. Hypertension. 2004;44:29–34.
46. Hoffman LB, Schmeidler J, Lesser GT, Beeri MS, Purohit DP, Grossman HT, Haroutunian V. Less Alzheimer disease neuropathology in medicated hypertensive than nonhypertensive persons. Neurology. 2009;72:1720–6.
47. Di Carlo A, Baldereschi M, Amaducci L, Maggi S, Grigoletto F, Scarlato G, Inzitari D. Cognitive impairment without dementia in older people: prevalence, vascular risk factors, impact on disability. The Italian Longitudinal Study on aging. J Am Geriatr Soc. 2000;48:775–82.
48. Cerhan JR, Folsom AR, Mortimer JA, Shahar E, Knopman DS, McGovern PG, Hays MA, Crum LD, Heiss G. Correlates of cognitive function in middle-aged adults. Atherosclerosis Risk in Communities (ARIC) Study Investigators. Gerontology. 1998;44:95–105.
49. Cacciatore F, Abete P, Ferrara N, Paolisso G, Amato L, Canonico S, Maggi S, Varricchio M, Rengo F. The role of blood pressure in cognitive impairment in an elderly population. Osservatorio Geriatrico Campano Group. J Hypertens. 1997;15:135–42.
50. Budge MM, de Jager C, Hogervorst E, Smith AD. Total plasma homocysteine, age, systolic blood pressure, and cognitive performance in older people. J Am Geriatr Soc. 2002;50: 2014–8.

51. Gorelick PB, Scuteri A, Black SE, Decarli C, Greenberg SM, Iadecola C, Launer LJ, Laurent S, Lopez OL, Nyenhuis D, Petersen RC, Schneider JA, Tzourio C, Arnett DK, Bennett DA, Chui HC, Higashida RT, Lindquist R, Nilsson PM, Roman GC, Sellke FW, Seshadri S, American Heart Association Stroke Council, Council on Epidemiology and Prevention, Council on Cardiovascular Nursing, Council on Cardiovascular Radiology and Intervention, and Council on Cardiovascular Surgery and Anesthesia. Vascular contributions to cognitive impairment and dementia: a statement for healthcare professionals from the American Heart Association/American Stroke Association. Stroke. 2011;42:2672–713.
52. McGuinness B, Todd S, Passmore P, Bullock R. Blood pressure lowering in patients without prior cerebrovascular disease for prevention of cognitive impairment and dementia. Cochrane Database Syst Rev. 2009;4, CD004034.
53. SHEP Cooperative Research Group. Prevention of stroke by antihypertensive drug treatment in older persons with isolated systolic hypertension. Final results of the Systolic Hypertension in the Elderly Program (SHEP). JAMA. 1991;265:3255–64.
54. Lithell H, Hansson L, Skoog I, Elmfeldt D, Hofman A, Olofsson B, Trenkwalder P, Zanchetti A, SCOPE Study Group. The Study on Cognition and Prognosis in the Elderly (SCOPE): principal results of a randomized double-blind intervention trial. J Hypertens. 2003;21:875–86.
55. Peters R, Beckett N, Forette F, Tuomilehto J, Clarke R, Ritchie C, Waldman A, Walton I, Poulter R, Ma S, Comsa M, Burch L, Fletcher A, Bulpitt C, HYVET Investigators. Incident dementia and blood pressure lowering in the Hypertension in the Very Elderly Trial cognitive function assessment (HYVET-COG): a double-blind, placebo controlled trial. Lancet Neurol. 2008;7:683–9.
56. Yusuf S, Diener HC, Sacco RL, Cotton D, Ounpuu S, Lawton WA, Palesch Y, Martin RH, Albers GW, Bath P, Bornstein N, Chan BP, Chen ST, Cunha L, Dahlöf B, De Keyser J, Donnan GA, Estol C, Gorelick P, Gu V, Hermansson K, Hilbrich L, Kaste M, Lu C, Machnig T, Pais P, Roberts R, Skvortsova V, Teal P, Toni D, VanderMaelen C, Voigt T, Weber M, Yoon BW, PRoFESS Study Group. Telmisartan to prevent recurrent stroke and cardiovascular events. N Engl J Med. 2008;359:1225–37.
57. Forette F, Seux ML, Staessen JA, Thijs L, Birkenhager WH, Babarskiene MR, Babeanu S, Bossini A, Gil-Extremera B, Girerd X, Laks T, Lilov E, Moisseyev V, Tuomilehto J, Vanhanen H, Webster J, Yodfat Y, Fagard R. Prevention of dementia in randomised double-blind placebo-controlled Systolic Hypertension in Europe (Syst-Eur) trial. Lancet. 1998;352:1347–51.
58. Tzourio C, Anderson C, Chapman N, Woodward M, Neal B, MacMahon S, Chalmers J, PROGRESS Collaborative Group. Effects of blood pressure lowering with perindopril and indapamide therapy on dementia and cognitive decline in patients with cerebrovascular disease. Arch Intern Med. 2003;163:1069–75.
59. Anstey KJ, Mack HA, Cherbuin N. Alcohol consumption as a risk factor for dementia and cognitive decline: meta-analysis of prospective studies. Am J Geriatr Psychiatry. 2009;17:542–55.
60. Hogenkamp PS, Benedict C, Sjögren P, Kilander L, Lind L, Schiöth HB. Late-life alcohol consumption and cognitive function in elderly men. Age (Dordr). 2014;36(1):243–9.
61. Ding J, Eigenbrodt ML, Mosley Jr TH, Hutchinson RG, Folsom AR, Harris TB, Nieto FJ. Alcohol intake and cerebral abnormalities on magnetic resonance imaging in a community-based population of middle-aged adults: the Atherosclerosis Risk in Communities (ARIC) study. Stroke. 2004;35:16–21.
62. Potter AS, Newhouse PA. Acute nicotine improves cognitive deficits in young adults with attention-deficit/hyperactivity disorder. Pharmacol Biochem Behav. 2008;88:407–17.
63. Wang CC, Lu TH, Liao WC, Yuan SC, Kuo PC, Chuang HL, Lee MC, Yen CH. Cigarette smoking and cognitive impairment: a 10-year cohort study in Taiwan. Arch Gerontol Geriatr. 2010;51:143–8.
64. Newhouse P, Kellar K, Aisen P, White H, Wesnes K, Coderre E, Pfaff A, Wilkins H, Howard D, Levin ED. Nicotine treatment of mild cognitive impairment: a 6-month double-blind pilot clinical trial. Neurology. 2012;78:91–101.
65. Anstey KJ, von Sanden C, Salim A, O'Kearney R. Smoking as a risk factor for dementia and cognitive decline: a meta-analysis of prospective studies. Am J Epidemiol. 2007;166:367–78.

66. Reitz C, den Heijer T, van Duijn C, Hofman A, Breteler MM. Relation between smoking and risk of dementia and Alzheimer disease: the Rotterdam Study. Neurology. 2007;69:998–1005.
67. Pantoni L. Cerebral small vessel disease: from pathogenesis and clinical characteristics to therapeutic challenges. Lancet Neurol. 2010;9:689–701.
68. de Leeuw FE, de Groot JC, Achten E, Oudkerk M, Ramos LM, Heijboer R, Hofman A, Jolles J, van Gijn J, Breteler MM. Prevalence of cerebral white matter lesions in elderly people: a population based magnetic resonance imaging study. The Rotterdam Scan Study. J Neurol Neurosurg Psychiatry. 2001;70:9–14.
69. Longstreth WT, Manolio TA, Arnold A. Clinical correlates of white matter findings on cranial magnetic resonance imaging of 3301 elderly people: the Cardiovascular Health Study. Stroke. 1996;27:1274–82.
70. Ylikoski A, Erkinjuntti T, Raininko R, Sarna S, Sulkava R, Tilvis R. White matter hyperintensities on MRI in the neurologically nondiseased elderly. Analysis of cohorts of consecutive subjects aged 55 to 85 years living at home. Stroke. 1995;26:1171–7.
71. Schmidt R, Schmidt H, Haybaeck J, Loitfelder M, Weis S, Cavalieri M, Seiler S, Enzinger C, Ropele S, Erkinjuntti T, Pantoni L, Scheltens P, Fazekas F, Jellinger K. Heterogeneity in age-related white matter changes. Acta Neuropathol. 2011;122:171–85.
72. Skoog IBS, Johansson B, Palmertz B, Andreasson LA. The influence of white matter lesions on neuropsychological functioning in demented and non-demented 85-yeras-olds. Acta Neurol Scand. 1996;93:142–8.
73. de Leeuw FE, de Groot JC, Oudkerk M, Witteman JC, Hofman A, van Gijn J, Breteler MM. Hypertension and cerebral white matter lesions in a prospective cohort study. Brain. 2002;125:765–72.
74. Ylikoski R, Ylikoski A, Raininko R, Keskivaara P, Sulkava R, Tilvis R, Erkinjuntti T. Cardiovascular diseases, health status, brain imaging findings and neuropsychological functioning in neurologically healthy elderly individuals. Arch Gerontol Geriatr. 2000;30:115–30.
75. Inaba M, White L, Bell C, Chen R, Petrovitch H, Launer L, Abbott RD, Ross GW, Masaki K. White matter lesions on brain magnetic resonance imaging scan and 5-year cognitive decline: the Honolulu-Asia Aging Study. J Am Geriatr Soc. 2011;59:1484–9.
76. Silbert LC, Howieson DB, Dodge H, Kaye JA. Cognitive impairment risk: white matter hyperintensity progression matters. Neurology. 2009;73:120–5.
77. Jokinen H, Kalska H, Ylikoski R, Madureira S, Verdelho A, van der Flier WM, Scheltens P, Barkhof F, Visser MC, Fazekas F, Schmidt R, O'Brien J, Waldemar G, Wallin A, Chabriat H, Pantoni L, Inzitari D, Erkinjuntti T. Longitudinal cognitive decline in subcortical ischemic vascular disease – the LADIS study. Cerebrovasc Dis. 2009;27:384–91.
78. Steffens DC, Potter GG, McQuoid DR, MacFall JR, Payne ME, Burke JR, Plassman BL, Welsh-Bohmer KA. Longitudinal magnetic resonance imaging vascular changes, apolipoprotein e genotype, and development of dementia in the neurocognitive outcomes of depression in the elderly study. Am J Geriatr Psychiatry. 2007;15:839–49.
79. Kuller LH, Lopez OL, Newman A, Beauchamp NJ, Burke G, Dulberg C, Fitzpatrick A, Fried L, Haan MN. Risk factors for dementia in the cardiovascular health cognition study. Neuroepidemiology. 2003;22:13–22.
80. The LADIS Study Group. 2001–2011: a decade of the LADIS (Leukoaraiosis and DISability) Study: what have we learned about white matter changes and small-vessel disease? Cerebrovasc Dis. 2011;32:577–88.
81. Hachinski VC, Potter P, Merskey H. Leuko-araiosis: an ancient term for a new problem. Can J Neurol Sci. 1986;13:533–4.
82. Madureira S, Verdelho A, Ferro J, Basile AM, Chabriat H, Erkinjuntti T, Fazekas F, Hennerici M, O'brien J, Pantoni L, Salvadori E, Scheltens P, Visser MC, Wahlund LO, Waldemar G, Wallin A, Inzitari D, LADIS Study Group. Development of a neuropsychological battery for a multinational study: the LADIS. Neuroepidemiology. 2006;27:101–16.
83. Bombois S, Debette S, Bruandet A, Delbeuck X, Delmaire C, Leys D, Pasquier F. Vascular subcortical hyperintensities predict conversion to vascular and mixed dementia in MCI patients. Stroke. 2008;39:2046–51.

84. Meguro K, Ishii H, Kasuya M, Akanuma K, Meguro M, Kasai M, Lee E, Hashimoto R, Yamaguchi S, Asada T. Incidence of dementia and associated risk factors in Japan: the Osaki-Tajiri project. J Neurol Sci. 2007;260:175–82.
85. Loeb C, Gandolfo C, Crose R, Conti M. Dementia associated with lacunar infarction. Stroke. 1992;23:1225–9.
86. Carey CL, Kramer JH, Josephson SA, Mungas D, Reed BR, Schuff N, Weiner MW, Chui HC. Subcortical lacunes are associated with executive dysfunction in cognitively normal elderly. Stroke. 2008;39:397–402.
87. Benisty S, Gouw AA, Porcher R, Madureira S, Hernandez K, Poggesi A, van der Flier WM, Van Straaten EC, Verdelho A, Ferro J, Pantoni L, Inzitari D, Barkhof F, Fazekas F, Chabriat H. Location of lacunar infarcts correlates with cognition in a sample of non-disabled subjects with age-related white-matter changes: the LADIS Study. J Neurol Neurosurg Psychiatry. 2009;80:478–83.
88. Jellinger KA, Attems J. Incidence of cerebrovascular lesions in Alzheimer's disease: a post-mortem study. Acta Neuropathol. 2003;105:14–7.
89. Miyao S, Takano A, Teramoto J, Takahashi A. Leukoaraiosis in relation to prognosis for patients with lacunar infarction. Stroke. 1992;23:1434–8.
90. Hanyu H, Tanaka Y, Shimizu S, Takasaki M, Fujita H, Kaneko N, Yamamoto Y, Harada M. Cerebral microbleeds in Binswanger's disease: a gradient-echo t2*-weighted magnetic resonance imaging study. Neurosci Lett. 2003;340:213–6.
91. Poels MM, Vernooij MW, Ikram MA, Hofman A, Krestin GP, van der Lugt A, Breteler MM. Prevalence and risk factors of cerebral microbleeds: an update of the Rotterdam Scan Study. Stroke. 2010;41:S103–6.
92. Seo SW, Hwa Lee B, Kim EJ, Chin J, Sun Cho Y, Yoon U, Na DL. Clinical significance of microbleeds in subcortical vascular dementia. Stroke. 2007;38:1949–51.
93. Gregoire SM, Smith K, Jager HR, Benjamin M, Kallis C, Brown MM, Cipolotti L, Werring DJ. Cerebral microbleeds and long-term cognitive outcome: longitudinal cohort study of stroke clinic patients. Cerebrovasc Dis. 2012;33:430–5.
94. Werring DJ, Frazer DW, Coward LJ, Losseff NA, Watt H, Cipolotti L, Brown MM, Jager HR. Cognitive dysfunction in patients with cerebral microbleeds on t2*-weighted gradient-echo MRI. Brain. 2004;127:2265–75.
95. Qiu C, Cotch MF, Sigurdsson S, Jonsson PV, Jonsdottir MK, Sveinbjrnsdottir S, Eiriksdottir G, Klein R, Harris TB, van Buchem MA, Gudnason V, Launer LJ. Cerebral microbleeds, retinopathy, and dementia: the AGES-Reykjavik Study. Neurology. 2010;75:2221–8.
96. Poels MM, Ikram MA, van der Lugt A, Hofman A, Niessen WJ, Krestin GP, Breteler MM, Vernooij MW. Cerebral microbleeds are associated with worse cognitive function: the Rotterdam Scan Study. Neurology. 2012;78:326–33.
97. van Norden AG, van den Berg HA, de Laat KF, Gons RA, van Dijk EJ, de Leeuw FE. Frontal and temporal microbleeds are related to cognitive function: the Radboud University Nijmegen Diffusion Tensor and Magnetic Resonance Cohort (RUN DMC) Study. Stroke. 2011;42:3382–6.
98. Ayaz M, Boikov AS, Haacke EM, Kido DK, Kirsch WM. Imaging cerebral microbleeds using susceptibility weighted imaging: one step toward detecting vascular dementia. J Magn Reson Imaging. 2010;31:142–8.

Chapter 6
Nongenetic Risk Factors for Alzheimer's Disease

Francesca Clerici

Abstract This chapter summarizes the major findings concerning prevention of Alzheimer's disease (AD). The review addresses nongenetic risk and protective factors for AD, including nutritional, social, and medical factors. Although many aspects of the disease are still unclear, some interesting hypotheses have been suggested to explain the role of exposures in its pathogenesis. At the moment it is also possible to delineate some preventive strategies for AD.

Keywords Alzheimer's disease • Prevention • Risk factor • Protective factor • Nutrition • Leisure activities • Vascular risk factors

Introduction

The risk of Alzheimer's disease (AD) in late life is considered to be a result of complex interactions of genetic susceptibility, biological factors, and environmental exposures experienced over the whole life span.

Age is the strongest known risk factor for AD, with the prevalence doubling every 5 years after the age of 65 [1]. The strong association of dementia with increasing age can be, at least partially, explained by a lifetime cumulative risk of exposures.

Several exposures have been identified over the last two decades, but the overall scientific quality of the evidence is considered poor, and risk modification was generally small to moderate (i.e., odds ratios and relative risk were often substantially <2.0).

F. Clerici, MD, PhD
Department of Biomedical and Clinical Sciences,
Centre for Research and Treatment on Cognitive Dysfunctions,
Institute of Clinical Neurology, "Luigi Sacco" Hospital, University of Milan,
Via G.B. Grassi 74, 20157 Milan, Italy
e-mail: francesca.clerici1@unimi.it

D. Galimberti, E. Scarpini (eds.), *Neurodegenerative Diseases*,
DOI 10.1007/978-1-4471-6380-0_6, © Springer-Verlag London 2014

Researchers face several challenges in identifying factors that alter the risk of AD.

The primary limitation with most of these studies is the distinction between causality and association. The majority of studies on risk factors are observational and as such allow us to define associations but not to establish causal links. Moreover, in all observational studies it is impossible to control for unknown cofounders.

Unfortunately, clinical trials (RCTs) that help in defining causal relationship are quite rare in this area for many reasons including the fact that they can investigate only relatively short-duration exposure at a particular stage of life.

A key problem with associations is that they often involve factors that are themselves correlated. For instance, smoking may be a marker for an unhealthy lifestyle, including undertaking less physical activity. When more correlated factors show an association with the disease, it is difficult to distinguish whether any (or all) of them contributes to the disease.

Among the challenge of interpreting the results of the studies is distinguishing the factors associated with AD from those associated with other diseases common in elderly people. This is the case, for example, of vascular risk factors that are found to be associated both to AD and to vascular brain damage, which can cause dementia too.

Moreover, some factors may be misinterpreted as early features of the disease, as in the case of mood disorders.

Finally as the pathophysiological process of AD begins years prior to the clinical symptoms [2], the exposures may cover much of the lifespan. For some factors, there may be a limited window of time during which exposures influence the risk of AD. Similarly, interventions may also have different effects at different points throughout the life span. This seems to be the case of blood pressure: hypertension in midlife has been associated with increased risk of AD, while low blood pressure in the oldest old has been associated with reduced risk of AD.

Despite these limitations, we will try in the next pages to synthesize the evidence concerning the nongenetic risk factors of AD. The exposures evaluated in this chapter are summarized in Table 6.1.

Table 6.1 Exposures evaluated in this chapter

1. *Nutritional and dietary factors*	3. *Medical factors*
(a) Vitamins intake	3.1. *Vascular factors*
(b) Fats intake	(a) Diabetes mellitus
(c) Alcohol intake	(b) Metabolic syndrome
(d) Mediterranean diet	(c) Blood pressure (high/low)
(e) Calories intake	(d) Hypercholesterolemia
	(e) Hyperhomocysteinemia
	(f) Vascular burden
2. *Social, economical, and behavioral factors*	3.2. *Other medical factors*
(a) Childhood socioeconomic status	(a) Body mass index (BMI)
(b) Education, intelligence, and occupation	(b) Traumatic brain injury
(c) Leisure activities	(c) Depression
(d) Tobacco use	

Nutritional and Dietary Factors

The food we eat is composed of multiple components, and the meals consist of complex combination of nutrients; this makes it difficult to interpret the evidence concerning the impact of single dietary factors on AD risk. As an alternative approach, some dietary patterns have recently received much attention for their potential role in the prevention of AD.

Vitamins Intake

There is preliminary evidence from two studies [3, 4] that low folate levels are associated with increased risk of AD, while no association was found between B12 vitamin and risk of AD.

Few studies report an association between AD and higher intake of vitamin C [5, 6] and E [6, 7]. However, other studies [8, 9] did not confirm these associations. This raises some questions about the robustness of the findings and leads to the conclusion that there is little evidence supporting a preventive role of vitamins in the prevention of AD.

Theoretically, oxidative stress could be a biological mechanism linking antioxidant vitamins intake and reduced risk of AD. AD brains exhibit constant evidence of reactive oxygen species-mediated injury [10]. The highly unstable reactive oxygen species derive from normal metabolic processes. In some circumstances their production can exceed the antioxidant ability to destroy them, and oxidative stress occurs. Increasing the efficiency of antioxidant systems by vitamin intake could help to keep low the levels of reactive oxygen species. Nevertheless, further studies are needed to clarify the role of vitamins in the prevention of AD.

Fats Intake

A recent systematic review [11] of seven prospective studies concluded that the existing data do not favor a role for long-chain omega 3 fatty acids, usually estimate by dietary histories of fish consumption, in preventing dementia, including AD.

Alcohol Intake

Heavy drinking is a known risk factor for dementia [12], but light to moderate alcohol intake was found to be associated with a decreased risk of AD in a recent systematic review [13]. The pooled estimated relative risk for light to moderate drinkers versus nondrinkers was 0.72 (95 % confidence interval [CI] 0.61–0.86).

The mechanism by which low alcohol intake could be protective against AD is at present unknown. It might exert its protective effect via a reduction in vascular risk factors [14]. Alternatively, wine consumption may be protective, through the antioxidant effects of polyphenols richly represented in red wine [15].

Mediterranean Diet

The Mediterranean diet is characterized by a high intake of legumes, cereals, fruits, and vegetables and a moderately high intake of fish. It also contains high amounts of unsaturated fatty acids (i.e., olive oil), against low saturated fatty acid content. The intake of dairy products (such as cheese and yogurt) is low to moderate in this dietary pattern, but the intake of meat and poultry is poor. Finally, the Mediterranean diet is characterized by a regular but moderate amount of alcohol, primarily in the form of wine and generally with meals.

Higher adherence to the Mediterranean diet by healthy subjects was found to be associated with lower AD risk [16–19]. The studies reported significant trend effects, suggesting a dose–response pattern. Moreover, higher adherence to this diet was associated with lower risk of progression toward AD also in people at risk of dementia such those with mild cognitive impairment (MCI) [20].

The Mediterranean diet is composed of many dietary factors reported to be beneficial in reducing AD risk [21]. Olive oil [22]; wine; fruits; vegetables; vitamins C, E, and B12; and folate [23, 24] contain antioxidants and may fight oxidative stress. Increasing the intake of antioxidants should theoretically counteract the reactive oxygen species-mediated brain injury, and this could, at least partially, provide an explanation for the observed association with a lower AD risk.

An alternative explanation could be via the attenuation of the inflammation pathway. Higher adherence to the Mediterranean diet has been associated with lower levels of C-reactive protein [25, 26], an inflammatory marker detected in AD brains.

The protective role of the Mediterranean diet against AD may also be mediated by the effects on the vascular system. This dietary pattern was found to be associated with lower incidence of vascular disease-associated conditions such as the metabolic syndrome [25], hypertension [27], dyslipidemia [26, 28], and cardiovascular disease [28, 29], which are also known risk factors for AD (see below).

Calories Intake

Luchsinger et al. [30] reported that higher caloric intake was associated with higher risk of incident AD, but the hazard ratio was less than 2, which may suggest that residual confounding variables could explain the association. Moreover, this result is inconsistent with other studies (see below) showing that weight loss may precede AD onset.

In summary, it is evident that diet and nutrition have effects on a living organism, and these effects are likely to include susceptibility to disease. But it is unlikely that a single nutrient or food group is causal. It is also unlikely that only a dietary pattern is beneficial. From current studies we can reasonably conclude that a varied diet, rich in fruits, vegetables, fish, and saturated fatty acid and poor in meat and dairy, is good for health. Additional researches are needed to understand whether it is specifically good for lowering AD risk.

Social, Economical, and Behavioral Factors

Social neuroscience connects social, psychological, and neurobiological processes that are relevant to understanding how environmental factors contribute to late-onset dementias.

Childhood Socioeconomic Status

A single ecological study [31] suggests a slight association between a disadvantaged childhood and AD, but the largest cohort study [32] did not support this relation.

Education, Intelligence, and Occupation

Strong evidence supports the hypothesis that illiteracy and low education increase the risk of AD. A systematic review [33] including nine prospective cohort studies concluded that more years of education may provide protection from AD. The pooled estimated relative risk for the lowest education level versus the highest was 1.59 (95 % CI, 1.35–1.86).

It is not clear whether more years of education actually prevents AD, delays onset of the disease, or just delays the detection of cognitive decline.

The prevailing model to explain this association [34] hypothesizes a positive contribution of education to the cognitive reserve that is available to withstand the burden of neurodegenerative pathology. In persons with higher cognitive reserve, more cerebral lesions are needed to clinically express dementia. Another mechanism that has been proposed to explain the association between education and dementia risk is that education is a surrogate for intelligence. For example, in a cohort from Scotland [35], it has been shown that intelligence quotient (IQ) scores obtained at age 11 predicted risk of dementia in old age.

Occupational attainment is closely linked to both education and childhood intelligent quotient, suggesting it may be difficult to establish whether occupation in midlife influences the incidence of AD in later life, independent of either education or childhood mental ability [36]. Occupation itself might also influence the incidence of AD, the complexity of work being theoretically protective [37]. In addition, the exposure to neurotoxic agents in the workplace, such as organic solvent, may be associated with AD [38]. However, to date the preponderance of studies does not support an association between occupational level and risk of AD that is independent of the influence of educational level.

Leisure Activities

Leisure activities can be defined as the voluntary use of free time for activities outside the daily routine and are part of the huge group of theoretically modifiable protective factors for dementia. There are three main components of leisure activities: cognitive, social, and physical component. A recent systematic review [39] of population-based studies reported the protective role of these lifestyle components on the risk of developing cognitive impairment and dementia.

Particularly, physical activity is the component of healthy lifestyle that collects the strongest evidence [40] as a protective factor against dementia. An observational study [41] reports that regular physical activity may reduce the risk or delay the onset of dementia and AD, especially among genetically susceptible individuals. A meta-analysis [40] of prospective studies suggested a significant and consistent protection for all levels of physical activity against the occurrence of cognitive decline among nondemented subjects. Finally two 6-month randomized clinical trials (RCTs) [42, 43] involving subjects with MCI have demonstrated that physical exercise has positive effects on cognition, without providing results on the risk of progression to dementia.

One point to consider when interpreting these results is that physical engagement may be a marker for a generally healthier lifestyle and that these other healthy lifestyle factors may contribute to preserving cognitive function. One study [17] addressed this point by examining the combination of physical exercise and a Mediterranean diet on risk of AD. Compared with individuals neither adhering to the dietary pattern nor doing physical exercise, those both adhering to the diet and participating in physical activity had a lower risk of AD (hazard ratio [HR] 0.65; 95 % CI, 0.44–0.96). This multifactorial approach should be encouraged for future work.

Observational studies [44–46] show that also greater cognitive engagement is associated with a decreased risk of AD. The one study [47] that assessed past and current participation in cognitive activities found that current activities explained the protective association. Moreover, as cognitive, physical, and social activity levels may be correlated, this study conducted analysis using physical and social activity levels as covariate showing that findings concerning the protective role of cognitive engagement were independent from the levels of social and physical engagement.

The primary challenge in identifying the role of the social component of lifestyle on the incidence of AD is that exposure, social support, and social network were defined too heterogeneously both within and between the studies, including objective measures such as marital status, living situation, number of people in social network, as well as subjective measures such as feeling of loneliness. There is preliminary evidence that a higher degree of loneliness [48] and being single and not cohabiting with a partner in later life [49, 50] are risk factors for AD. However, further studies are needed to clarify the direction of the relationship between social engagement and AD.

Tobacco Use

A meta-analysis of 19 prospective studies [51] shows that when compared to people who have never smoked, current smokers have an increased risk of AD (relative risk [RR] 1.79; 95 % CI,1.43–2.23). However, former smokers do not appear to be at increased risk of AD. The authors of the review noted that there were insufficient data to evaluate the duration of smoking among the current and former smokers or the duration of abstinence from smoking among former smokers. Thus, questions about the amount of time it takes former smokers to return to the level of risk of a never smoker are still unanswered.

Smoking may be a marker for a poorer lifestyle, including less physical activity, drinking harmful levels of alcohol, or have a poor nutrition. Although many studies adjusted for health factors that may influence the observed association, the authors of the meta-analysis noted that there were inconsistencies among studies in the choice of covariates.

Smoking may affect AD risk via its effect on other medical conditions, and it may interact with other vascular risk factors in a synergistic or additive way [52].

Medical Factors

Vascular Factors

Although AD and vascular dementia have traditionally been considered distinct disorders, it is now generally agreed that the two rarely occur in isolation. Moreover the presence and severity of cerebrovascular pathology appear to increase the risk and stage of AD for any given level of AD neuropathology [53]. Thus, it seems likely that the modification of vascular risk might influence the risk of AD.

Diabetes Mellitus

There is convincing evidence coming from two systematic reviews [54, 55] and a meta-analysis [55] that shows an association between diabetes mellitus and incident AD. Overall the incidence of AD was increased by 50–100 % relative to people without diabetes [54].

Both neurodegenerative and neurovascular mechanisms may underlie this association. Alterations in insulin and glucose homoeostasis could affect amyloid metabolism and *tau* protein phosphorylation [56]. Insulin resistance is present in most diabetic patients and is associated with compensatory hyperinsulinemia. Insulin appears to stimulate amyloid-β secretions and inhibits the extracellular degradation of amyloid-β by competing for insulin-degrading enzyme. Another mechanism is an increase of oxidative stress secondary to hyperglycemia. Additionally, chronic exposure to hyperglycemia in diabetes might lead to microvascular changes causing an insidious ischemia of the brain [57]. Taken together, these mechanisms suggest that drugs used to ameliorate hyperglycemia may also have beneficial effects in diabetic patients with AD. A few studies have already been performed (for a review see Moreira et al. 2013 [58]), but results are still preliminary and inconclusive. Larger RCTs are needed to elucidate if antidiabetic drugs have a role in primary and secondary prevention of AD in diabetic patients.

Metabolic Syndrome

The most commonly accepted definition of the metabolic syndrome [59] requires three of the following conditions to be present: (1) elevated fasting glucose (≥110 mg/dL) or currently using antidiabetic medications, (2) elevated waist circumference (men, ≥102 cm; women, ≥88 cm), (3) elevated triglycerides (≥150 mg/dL), (4) reduced HDL ("good") cholesterol (men, <40 mg/d; women, <50 mg/dL), and (5) elevated blood pressure (≥130/85 mmHg). The metabolic syndrome was not associated with an increased risk of AD in the Honolulu-Asia Aging Study [59]. Muller and coworkers [60], using a different definition of the metabolic syndrome, came to the same conclusion.

Blood Pressure (High/Low)

Many community-based studies [52, 61–63] have focused on the putative role of hypertension as a risk factor for AD, but only two of them [62, 63] found an association between high blood pressure and AD. In the FINMONICA study [62] midlife high systolic blood pressure nearly doubled the risk of late life AD. In the Honolulu-Asia Aging Study (HAAS) [63], untreated high diastolic blood pressure increased by four times the risk of AD. Both the HAAS cohort and the FINMONICA cohort are distinguished by having the longest follow-up, 27 and 21 years, respectively. It is possible that the cohorts formed later in life [52, 61] had a selection bias in that if hypertension predisposes to AD and to death, those subjects with hypertension would have selectively died prior to cohort formation.

Interestingly, in a recent study by Li and coworkers [64], hypertension was associated with accelerated progression to AD in a cohort of MCI subjects, while antihypertensive treatment reduced the risk of progression to AD.

High blood pressure has been linked to deep white matter lesions (WMLs) of the brain [65, 66]. WMLs may lower the threshold at which AD pathology produces clinically relevant symptoms. Alternatively, WMLs may interact with pathological changes related to AD and thereby accelerate its clinical expression [67].

Several RCTs have addressed the effects of antihypertensive medication in the prevention of dementia (for a review see Valenzuela M et al. 2012 [68]), but only few data are available on AD.

The Syst-EUR trial [69] is the only RCT to provide evidence for effective AD prevention by calcium channel blockers. Nevertheless, dementia wasn't the primary end-point of the RCT, and this is a limitation of the findings.

Finally, according to the hypoperfusion hypothesis, very low blood pressure, rather than hypertension, is associated with AD risk among very old people [70]. This inverse association is not an exception in the dementia literature since the relationship between different vascular risk factors and dementia may be age dependent.

Hypercholesterolemia

The brain is the most cholesterol-rich organ. Brain cholesterol, that is almost entirely produced in situ, plays a role in the production of beta-amyloid and thus in the AD process [71]. Indeed, the allele ε4 of the APOE gene (encoding a cholesterol transporter protein) is a major genetic risk factor for AD (for details see Chap. 4).

Based on a systematic review [72], midlife hypercholesterolemia is associated with increased incidence of late-life AD. By contrast late-life cholesterol levels are not associated with incident AD. The studies included in the systematic review were considered too heterogeneous to combine in a single analysis.

Interestingly, the rate of conversion to AD was found to be higher in MCI subjects with hypercholesterolemia than in those without hypercholesterolemia [64].

Although some observational studies suggest that lipid-lowering drugs (particularly statins) can reduce the risk of AD, a meta-analysis of these studies concluded that statins did not protect against dementia [73]. Moreover, two RCTs [74, 75] failed in demonstrating a protective effect of statins on incident dementia in elderly populations with high cardiovascular risk.

Hyperhomocysteinemia

Homocysteine levels depend on folate and vitamin B levels and rise with age, renal insufficiency, coffee and heavy alcohol intake, and tobacco use. High serum homocysteine is associated with an increased risk of AD, as reported in some cohort [3, 76, 77], but not in another study [78]. The relative risk varied substantially across positive studies, from modest (1.3) to large (4.2).

Vascular Burden

Given the individual relationship of vascular risk factors with AD and the frequency of their coexistence, an additive or synergistic effect of multiple vascular risk factors on the risk of AD has been postulated. The concept of vascular burden refers to the cumulative effects of multiple vascular risk factors, vascular diseases, and vascular lesions on the aging brain [79]. We have demonstrated [80] that vascular burden accelerates the progression of mild cognitive impairment toward AD. Particularly our findings support the importance of WMLs and hypertension as predictors of the progression to AD.

Vascular risk scores have been developed to quantify the risk of dementia associated with the clustering of multiple vascular factors, but the use of such scores in clinical practice is still limited due to low predictive value [81, 82].

The vascular hypothesis is based on the concept that cerebrovascular lesions may interact with neurodegenerative lesions to produce the dementia syndrome in individuals who do not have sufficient neurodegenerative damage to clinically express dementia [79]. Other hypotheses suggest a direct role in neurodegeneration of vascular factors [83].

Further evidence is needed to determine whether progression toward AD may be slowed down by treatment addressing vascular risk factors.

Other Medical Factors

Body Mass Index

The body mass index (BMI), defined as weight in kilograms divided by height in meters squared (kg/m^2), is one of the most widely used measure of body mass and adiposity. It has several advantages (i.e., it is easy to use and no cost) together with some limitations, including the fact that it cannot differentiate between fat and lean mass. Therefore, other adiposity indices are usually combined with BMI to provide a more realistic overview of elderly body composition.

One systematic review examined the association between various measures of adiposity and the development of AD [84]. Midlife obesity was found to be associated with higher risk of incident AD (RR 1.80; 95 % CI 1.00–3.29). This association may be at least partially explained by the fact that obesity is associated with diabetes and hypertension, two recognized risk factors for AD. Therefore, midlife obesity may be a marker for one of those conditions.

Interestingly, other studies [85–87] that have focused on late-life BMI assessment found a reversal in the direction of risk as higher BMI was associated with a lower risk of developing AD. These paradoxical findings are more easily understandable in the context of a life-course approach to the study of exposures and show that the role of BMI in dementia might change during the life course. This time-dependent association suggests the hypothesis that weight loss in late life may be a marker of incipient AD.

Traumatic Brain Injury

A systematic review [88] examined the association between traumatic brain injury and the development of AD in case-control studies. The authors concluded that traumatic brain injury even in early adulthood might confer an increased risk of AD years later (OR 1.58; 95 % CI 1.21–2.06). The association was demonstrated only for males.

Depression

Depression is a critically important issue for those working with the elderly and especially those working in the field of dementias. It is out of question that depression and cognition are linked to each other in the elderly, but the direction of the association is still unclear. Depression has been associated with poor cognitive functions [89], but it is a behavioral symptom of AD too [90]. Therefore, understanding the relation of depression and AD is complicated by the possibility that depression may be a prodromal symptom of AD rather than a risk factor for the disease. A better understanding of the relation of AD and depression thus might have important clinical and research implications.

One systematic review [91] examined the association between depression and incident AD in 11 cohort studies and 9 case–control studies. The authors stated that there was a reasonably consistent association between the two conditions even if they found a high variability across studies in depression assessment, ranging from a self-reported history to hospitalization. In the four studies using the most rigorous criteria for depression and AD diagnosis, the pooled OR was 2.23 (95 % CI 1.71–3.09).

The association between AD and depression might have different interpretation. Both conditions may share risk factors for vascular disease [92]. Furthermore, the inflammatory processes may underlie depression and AD [93]. Finally possible genetic links between the two disorders have also been explored, but they have led to inconclusive results [94].

Conclusions

Few of the putative risk or protective factors covered in this chapter had sufficient evidence from which to draw firm conclusion about their effect on AD. Many of the exposures reviewed probably do not work in isolation in their effect but are more likely to work in combination with other factors. Additionally, the interplay of environmental exposures with genetic factors seems very likely. Thus, the ideal interventions should be multidimensional, combining interventions for multiple risk factors.

Using mathematical models, it has been shown [95] that small delays in the onset or the progression of AD would result in significant reduction of the global burden of the disease. A 1-year delay in both onset and progression of AD would decrease the prevalent AD cases in 2050 by 9.19 million. This reduction in the number of AD cases will be almost entirely due to fewer individuals with late-stage dementia, when the most care is needed. Thus, both from a scientific and a public health perspective, further efforts are needed to better understand which factors could be modified in order to change the trajectory of the disease.

References

1. Brookmeyer R, Gray S, Kawas C. Projections of Alzheimer's disease in the United States and the public health impact of delaying disease onset. Am J Public Health. 1998;88(9):1337–42.
2. Sperling RA, Aisen PS, Beckett LA, Bennett DA, Craft S, Fagan AM, et al. Toward defining the preclinical stages of Alzheimer's disease: recommendations from the National Institute on Aging-Alzheimer's Association workgroups on diagnostic guidelines for Alzheimer's disease. Alzheimers Dement. 2011;7(3):280–92.
3. Ravaglia G, Forti P, Maioli F, Martelli M, Servadei L, Brunetti N, et al. Homocysteine and folate as risk factors for dementia and Alzheimer disease. Am J Clin Nutr. 2005;82(3):636–43.
4. Wang HX, Wahlin A, Basun H, Fastbom J, Winblad B, Fratiglioni L. Vitamin B(12) and folate in relation to the development of Alzheimer's disease. Neurology. 2001;56(9):1188–94.
5. Engelhart MJ, Geerlings MI, Ruitenberg A, van Swieten JC, Hofman A, Witteman JC, et al. Dietary intake of antioxidants and risk of Alzheimer disease. JAMA. 2002;287(24):3223–9.
6. Morris MC, Beckett LA, Scherr PA, Hebert LE, Bennett DA, Field TS, et al. Vitamin E and vitamin C supplement use and risk of incident Alzheimer disease. Alzheimer Dis Assoc Disord. 1998;12(3):121–6.
7. Laurin D, Masaki KH, Foley DJ, White LR, Launer LJ. Midlife dietary intake of antioxidants and risk of late-life incident dementia: the Honolulu-Asia Aging Study. Am J Epidemiol. 2004;159(10):959–67.
8. Fillenbaum GG, Kuchibhatla MN, Hanlon JT, Artz MB, Pieper CF, Schmader KE, et al. Dementia and Alzheimer's disease in community-dwelling elders taking vitamin C and/or vitamin E. Ann Pharmacother. 2005;39(12):2009–14.
9. Gray SL, Anderson ML, Crane PK, Breitner JC, McCormick W, Bowen JD, et al. Antioxidant vitamin supplement use and risk of dementia or Alzheimer's disease in older adults. J Am Geriatr Soc. 2008;56(2):291–5.
10. Martins RN, Harper CG, Stokes GB, Masters CL. Increased cerebral glucose-6-phosphate dehydrogenase activity in Alzheimer's disease may reflect oxidative stress. J Neurochem. 1986;46(4):1042–5.
11. Fotuhi M, Mohassel P, Yaffe K. Fish consumption, long-chain omega-3 fatty acids and risk of cognitive decline or Alzheimer disease: a complex association. Nat Clin Pract Neurol. 2009;5(3):140–52.
12. Saunders PA, Copeland JR, Dewey ME, Davidson IA, McWilliam C, Sharma V, et al. Heavy drinking as a risk factor for depression and dementia in elderly men. Findings from the Liverpool longitudinal community study. Br J Psychiatry. 1991;159:213–6.

13. Anstey KJ, Mack HA, Cherbuin N. Alcohol consumption as a risk factor for dementia and cognitive decline: meta-analysis of prospective studies. Am J Geriatr Psychiatry. 2009;17(7):542–55.
14. Ruitenberg A, van Swieten JC, Witteman JC, Mehta KM, van Duijn CM, Hofman A, et al. Alcohol consumption and risk of dementia: the Rotterdam Study. Lancet. 2002;359(9303):281–6.
15. Howard A, Chopra M, Thurnham D, Strain J, Fuhrman B, Aviram M. Red wine consumption and inhibition of LDL oxidation: what are the important components? Med Hypotheses. 2002;59(1):101–4.
16. Scarmeas N, Stern Y, Tang MX, Mayeux R, Luchsinger JA. Mediterranean diet and risk for Alzheimer's disease. Ann Neurol. 2006;59(6):912–21.
17. Scarmeas N, Luchsinger JA, Schupf N, Brickman AM, Cosentino S, Tang MX, et al. Physical activity, diet, and risk of Alzheimer disease. JAMA. 2009;302(6):627–37.
18. Gardener S, Gu Y, Rainey-Smith SR, Keogh JB, Clifton PM, Mathieson SL, et al. Adherence to a Mediterranean diet and Alzheimer's disease risk in an Australian population. Transl Psychiatry. 2012;2:e164.
19. Feart C, Samieri C, Rondeau V, Amieva H, Portet F, Dartigues JF, et al. Adherence to a Mediterranean diet, cognitive decline, and risk of dementia. JAMA. 2009;302(6):638–48.
20. Scarmeas N, Stern Y, Mayeux R, Manly JJ, Schupf N, Luchsinger JA. Mediterranean diet and mild cognitive impairment. Arch Neurol. 2009;66(2):216–25.
21. Scarmeas N, Luchsinger JA, Mayeux R, Stern Y. Mediterranean diet and Alzheimer disease mortality. Neurology. 2007;69(11):1084–93.
22. Fito M, Cladellas M, de la Torre R, Marti J, Alcantara M, Pujadas-Bastardes M, et al. Antioxidant effect of virgin olive oil in patients with stable coronary heart disease: a randomized, crossover, controlled, clinical trial. Atherosclerosis. 2005;181(1):149–58.
23. Szeto YT, Tomlinson B, Benzie IF. Total antioxidant and ascorbic acid content of fresh fruits and vegetables: implications for dietary planning and food preservation. Br J Nutr. 2002; 87(1):55–9.
24. Stupans I, Kirlich A, Tuck KL, Hayball PJ. Comparison of radical scavenging effect, inhibition of microsomal oxygen free radical generation, and serum lipoprotein oxidation of several natural antioxidants. J Agric Food Chem. 2002;50(8):2464–9.
25. Esposito K, Marfella R, Ciotola M, Di Palo C, Giugliano F, Giugliano G, et al. Effect of a mediterranean-style diet on endothelial dysfunction and markers of vascular inflammation in the metabolic syndrome: a randomized trial. JAMA. 2004;292(12):1440–6.
26. Chrysohoou C, Panagiotakos DB, Pitsavos C, Das UN, Stefanadis C. Adherence to the Mediterranean diet attenuates inflammation and coagulation process in healthy adults: The ATTICA Study. J Am Coll Cardiol. 2004;44(1):152–8.
27. Psaltopoulou T, Naska A, Orfanos P, Trichopoulos D, Mountokalakis T, Trichopoulou A. Olive oil, the Mediterranean diet, and arterial blood pressure: the Greek European Prospective Investigation into Cancer and Nutrition (EPIC) study. Am J Clin Nutr. 2004;80(4):1012–8.
28. Singh RB, Dubnov G, Niaz MA, Ghosh S, Singh R, Rastogi SS, et al. Effect of an Indo-Mediterranean diet on progression of coronary artery disease in high risk patients (Indo-Mediterranean Diet Heart Study): a randomised single-blind trial. Lancet. 2002; 360(9344):1455–61.
29. Trichopoulou A, Costacou T, Bamia C, Trichopoulos D. Adherence to a Mediterranean diet and survival in a Greek population. N Engl J Med. 2003;348(26):2599–608.
30. Luchsinger JA, Tang MX, Shea S, Mayeux R. Caloric intake and the risk of Alzheimer disease. Arch Neurol. 2002;59(8):1258–63.
31. Moceri VM, Kukull WA, Emanual I, van Belle G, Starr JR, Schellenberg GD, et al. Using census data and birth certificates to reconstruct the early-life socioeconomic environment and the relation to the development of Alzheimer's disease. Epidemiology. 2001;12(4): 383–9.
32. Wilson RS, Scherr PA, Hoganson G, Bienias JL, Evans DA, Bennett DA. Early life socioeconomic status and late life risk of Alzheimer's disease. Neuroepidemiology. 2005;25(1):8–14.

33. Caamano-Isorna F, Corral M, Montes-Martinez A, Takkouche B. Education and dementia: a meta-analytic study. Neuroepidemiology. 2006;26(4):226–32.
34. Bennett DA, Wilson RS, Schneider JA, Evans DA, Mendes de Leon CF, Arnold SE, et al. Education modifies the relation of AD pathology to level of cognitive function in older persons. Neurology. 2003;60(12):1909–15.
35. Whalley LJ, Starr JM, Athawes R, Hunter D, Pattie A, Deary IJ. Childhood mental ability and dementia. Neurology. 2000;55(10):1455–9.
36. Karp A, Kareholt I, Qiu C, Bellander T, Winblad B, Fratiglioni L. Relation of education and occupation-based socioeconomic status to incident Alzheimer's disease. Am J Epidemiol. 2004;159(2):175–83.
37. Smyth KA, Fritsch T, Cook TB, McClendon MJ, Santillan CE, Friedland RP. Worker functions and traits associated with occupations and the development of AD. Neurology. 2004;63(3):498–503.
38. Li CY, Wu SC, Sung FC. Lifetime principal occupation and risk of cognitive impairment among the elderly. Ind Health. 2002;40(1):7–13.
39. Wang HX, Xu W, Pei JJ. Leisure activities, cognition and dementia. Biochim Biophys Acta. 2012;1822(3):482–91.
40. Sofi F, Valecchi D, Bacci D, Abbate R, Gensini GF, Casini A, et al. Physical activity and risk of cognitive decline: a meta-analysis of prospective studies. J Intern Med. 2010;269(1):107–17.
41. Rovio S, Kareholt I, Helkala EL, Viitanen M, Winblad B, Tuomilehto J, et al. Leisure-time physical activity at midlife and the risk of dementia and Alzheimer's disease. Lancet Neurol. 2005;4(11):705–11.
42. Baker LD, Frank LL, Foster-Schubert K, Green PS, Wilkinson CW, McTiernan A, et al. Effects of aerobic exercise on mild cognitive impairment: a controlled trial. Arch Neurol. 2010;67(1):71–9.
43. Nagamatsu LS, Handy TC, Hsu CL, Voss M, Liu-Ambrose T. Resistance training promotes cognitive and functional brain plasticity in seniors with probable mild cognitive impairment. Arch Intern Med. 2012;172(8):666–8.
44. Wilson RS, Mendes De Leon CF, Barnes LL, Schneider JA, Bienias JL, Evans DA, et al. Participation in cognitively stimulating activities and risk of incident Alzheimer disease. JAMA. 2002;287(6):742–8.
45. Verghese J, Lipton RB, Katz MJ, Hall CB, Derby CA, Kuslansky G, et al. Leisure activities and the risk of dementia in the elderly. N Engl J Med. 2003;348(25):2508–16.
46. Akbaraly TN, Portet F, Fustinoni S, Dartigues JF, Artero S, Rouaud O, et al. Leisure activities and the risk of dementia in the elderly: results from the Three-City Study. Neurology. 2009;73(11):854–61.
47. Wilson RS, Scherr PA, Schneider JA, Tang Y, Bennett DA. Relation of cognitive activity to risk of developing Alzheimer disease. Neurology. 2007;69(20):1911–20.
48. Wilson RS, Krueger KR, Arnold SE, Schneider JA, Kelly JF, Barnes LL, et al. Loneliness and risk of Alzheimer disease. Arch Gen Psychiatry. 2007;64(2):234–40.
49. Fratiglioni L, Wang HX, Ericsson K, Maytan M, Winblad B. Influence of social network on occurrence of dementia: a community-based longitudinal study. Lancet. 2000;355(9212):1315–9.
50. Hakansson K, Rovio S, Helkala EL, Vilska AR, Winblad B, Soininen H, et al. Association between mid-life marital status and cognitive function in later life: population based cohort study. BMJ. 2009;339:b2462.
51. Anstey KJ, von Sanden C, Salim A, O'Kearney R. Smoking as a risk factor for dementia and cognitive decline: a meta-analysis of prospective studies. Am J Epidemiol. 2007;166(4):367–78.
52. Luchsinger JA, Reitz C, Honig LS, Tang MX, Shea S, Mayeux R. Aggregation of vascular risk factors and risk of incident Alzheimer disease. Neurology. 2005;65(4):545–51.
53. Launer LJ. Demonstrating the case that AD is a vascular disease: epidemiologic evidence. Ageing Res Rev. 2002;1(1):61–77.
54. Biessels GJ, Staekenborg S, Brunner E, Brayne C, Scheltens P. Risk of dementia in diabetes mellitus: a systematic review. Lancet Neurol. 2006;5(1):64–74.
55. Lu FP, Lin KP, Kuo HK. Diabetes and the risk of multi-system aging phenotypes: a systematic review and meta-analysis. PLoS One. 2009;4(1):e4144.

56. Craft S, Watson GS. Insulin and neurodegenerative disease: shared and specific mechanisms. Lancet Neurol. 2004;3(3):169–78.
57. Biessels GJ, van der Heide LP, Kamal A, Bleys RL, Gispen WH. Ageing and diabetes: implications for brain function. Eur J Pharmacol. 2002;441(1–2):1–14.
58. Moreira RO, Campos SC, Soldera AL. Type 2 Diabetes Mellitus and Alzheimer's disease: from physiopathology to treatment implications. Diabetes Metab Res Rev. 2013. doi:10.1002/dmrr.2442. [Epub ahead of print].
59. Kalmijn S, Foley D, White L, Burchfiel CM, Curb JD, Petrovitch H, et al. Metabolic cardiovascular syndrome and risk of dementia in Japanese-American elderly men. The Honolulu-Asia aging study. Arterioscler Thromb Vasc Biol. 2000;20(10):2255–60.
60. Muller M, Tang MX, Schupf N, Manly JJ, Mayeux R, Luchsinger JA. Metabolic syndrome and dementia risk in a multiethnic elderly cohort. Dement Geriatr Cogn Disord. 2007;24(3):185–92.
61. Posner HB, Tang MX, Luchsinger J, Lantigua R, Stern Y, Mayeux R. The relationship of hypertension in the elderly to AD, vascular dementia, and cognitive function. Neurology. 2002;58(8):1175–81.
62. Kivipelto M, Ngandu T, Fratiglioni L, Viitanen M, Kareholt I, Winblad B, et al. Obesity and vascular risk factors at midlife and the risk of dementia and Alzheimer disease. Arch Neurol. 2005;62(10):1556–60.
63. Launer LJ, Ross GW, Petrovitch H, Masaki K, Foley D, White LR, et al. Midlife blood pressure and dementia: the Honolulu-Asia aging study. Neurobiol Aging. 2000;21(1):49–55.
64. Li J, Wang YJ, Zhang M, Xu ZQ, Gao CY, Fang CQ, et al. Vascular risk factors promote conversion from mild cognitive impairment to Alzheimer disease. Neurology. 2011;76(17):1485–91.
65. Kuller LH, Margolis KL, Gaussoin SA, Bryan NR, Kerwin D, Limacher M, et al. Relationship of hypertension, blood pressure, and blood pressure control with white matter abnormalities in the Women's Health Initiative Memory Study (WHIMS)-MRI trial. J Clin Hypertens (Greenwich). 2010;12(3):203–12.
66. Skoog I. The relationship between blood pressure and dementia: a review. Biomed Pharmacother. 1997;51(9):367–75.
67. Zlokovic BV. Neurovascular mechanisms of Alzheimer's neurodegeneration. Trends Neurosci. 2005;28(4):202–8.
68. Valenzuela M, Esler M, Ritchie K, Brodaty H. Antihypertensives for combating dementia? A perspective on candidate molecular mechanisms and population-based prevention. Transl Psychiatry. 2012;2:e107.
69. Forette F, Seux ML, Staessen JA, Thijs L, Babarskiene MR, Babeanu S, et al. The prevention of dementia with antihypertensive treatment: new evidence from the Systolic Hypertension in Europe (Syst-Eur) study. Arch Intern Med. 2002;162(18):2046–52.
70. Ruitenberg A, den Heijer T, Bakker SL, van Swieten JC, Koudstaal PJ, Hofman A, et al. Cerebral hypoperfusion and clinical onset of dementia: the Rotterdam Study. Ann Neurol. 2005;57(6):789–94.
71. Dietschy JM. Central nervous system: cholesterol turnover, brain development and neurodegeneration. Biol Chem. 2009;390(4):287–93.
72. Anstey KJ, Lipnicki DM, Low LF. Cholesterol as a risk factor for dementia and cognitive decline: a systematic review of prospective studies with meta-analysis. Am J Geriatr Psychiatry. 2008;16(5):343–54.
73. Zhou B, Teramukai S, Fukushima M. Prevention and treatment of dementia or Alzheimer's disease by statins: a meta-analysis. Dement Geriatr Cogn Disord. 2007;23(3):194–201.
74. Shepherd J, Blauw GJ, Murphy MB, Bollen EL, Buckley BM, Cobbe SM, et al. Pravastatin in elderly individuals at risk of vascular disease (PROSPER): a randomised controlled trial. Lancet. 2002;360(9346):1623–30.
75. Heart Protection Study Collaborative Group. MRC/BHF Heart Protection Study of cholesterol lowering with simvastatin in 20,536 high-risk individuals: a randomised placebo-controlled trial. Lancet. 2002;360(9326):7–22.
76. Seshadri S, Beiser A, Selhub J, Jacques PF, Rosenberg IH, D'Agostino RB, et al. Plasma homocysteine as a risk factor for dementia and Alzheimer's disease. N Engl J Med. 2002;346(7):476–83.

77. Blasko I, Jellinger K, Kemmler G, Krampla W, Jungwirth S, Wichart I, et al. Conversion from cognitive health to mild cognitive impairment and Alzheimer's disease: prediction by plasma amyloid beta 42, medial temporal lobe atrophy and homocysteine. Neurobiol Aging. 2008;29(1):1–11.
78. Luchsinger JA, Tang MX, Shea S, Miller J, Green R, Mayeux R. Plasma homocysteine levels and risk of Alzheimer disease. Neurology. 2004;62(11):1972–6.
79. O'Brien J, Reisberg B, Erkinjuntti T. Vascular burden of the brain. Int Psychogeriatr. 2003;15 Suppl 1:7–10.
80. Clerici F, Caracciolo B, Cova I, Fusari Imperatori S, Maggiore L, Galimberti D, et al. Does vascular burden contribute to the progression of mild cognitive impairment to dementia? Dement Geriatr Cogn Disord. 2012;34(3–4):235–43.
81. Kivipelto M, Ngandu T, Laatikainen T, Winblad B, Soininen H, Tuomilehto J. Risk score for the prediction of dementia risk in 20 years among middle aged people: a longitudinal, population-based study. Lancet Neurol. 2006;5(9):735–41.
82. Barnes DE, Covinsky KE, Whitmer RA, Kuller LH, Lopez OL, Yaffe K. Predicting risk of dementia in older adults: the late-life dementia risk index. Neurology. 2009;73(3):173–9.
83. Roher AE, Kokjohn TA, Beach TG. An association with great implications: vascular pathology and Alzheimer disease. Alzheimer Dis Assoc Disord. 2006;20(1):73–5.
84. Beydoun MA, Beydoun HA, Wang Y. Obesity and central obesity as risk factors for incident dementia and its subtypes: a systematic review and meta-analysis. Obes Rev. 2008;9(3):204–18.
85. Fitzpatrick AL, Kuller LH, Lopez OL, Diehr P, O'Meara ES, Longstreth Jr WT, et al. Midlife and late-life obesity and the risk of dementia: cardiovascular health study. Arch Neurol. 2009;66(3):336–42.
86. Hughes TF, Borenstein AR, Schofield E, Wu Y, Larson EB. Association between late-life body mass index and dementia: The Kame Project. Neurology. 2009;72(20):1741–6.
87. Atti AR, Palmer K, Volpato S, Winblad B, De Ronchi D, Fratiglioni L. Late-life body mass index and dementia incidence: nine-year follow-up data from the Kungsholmen Project. J Am Geriatr Soc. 2008;56(1):111–6.
88. Fleminger S, Oliver DL, Lovestone S, Rabe-Hesketh S, Giora A. Head injury as a risk factor for Alzheimer's disease: the evidence 10 years on; a partial replication. J Neurol Neurosurg Psychiatry. 2003;74(7):857–62.
89. Yaffe K, Blackwell T, Gore R, Sands L, Reus V, Browner WS. Depressive symptoms and cognitive decline in nondemented elderly women: a prospective study. Arch Gen Psychiatry. 1999;56(5):425–30.
90. Espiritu DA, Rashid H, Mast BT, Fitzgerald J, Steinberg J, Lichtenberg PA. Depression, cognitive impairment and function in Alzheimer's disease. Int J Geriatr Psychiatry. 2001;16(11):1098–103.
91. Ownby RL, Crocco E, Acevedo A, John V, Loewenstein D. Depression and risk for Alzheimer disease: systematic review, meta-analysis, and metaregression analysis. Arch Gen Psychiatry. 2006;63(5):530–8.
92. Rasgon N, Jarvik L. Insulin resistance, affective disorders, and Alzheimer's disease: review and hypothesis. J Gerontol A Biol Sci Med Sci. 2004;59(2):178–83; discussion 184–92.
93. Baldwin RC. Is vascular depression a distinct sub-type of depressive disorder? A review of causal evidence. Int J Geriatr Psychiatry. 2005;20(1):1–11.
94. Lavretsky H, Ercoli L, Siddarth P, Bookheimer S, Miller K, Small G. Apolipoprotein epsilon4 allele status, depressive symptoms, and cognitive decline in middle-aged and elderly persons without dementia. Am J Geriatr Psychiatry. 2003;11(6):667–73.
95. Brookmeyer R, Johnson E, Ziegler-Graham K, Arrighi HM. Forecasting the global burden of Alzheimer's disease. Alzheimers Dement. 2007;3(3):186–91.

Chapter 7
Frontotemporal Lobar Degeneration: Genetics and Clinical Phenotypes

Maria Serpente and Daniela Galimberti

Abstract Frontotemporal lobar degeneration (FTLD) is the most frequent dementia in the presenile population. It presents with different syndromes, including behavioral variant frontotemporal dementia (bvFTD), primary nonfluent aphasia (PNFA), semantic dementia (SD), and logopenic aphasia. In 2011, new diagnostic criteria bvFTD, which include the use of biomarkers, have been published. According to them, bvFTD can be classified into "possible" (clinical features only), "probable" (inclusion of imaging biomarkers), and "definite" (in the presence of a known causal mutation or at autopsy). Motor neuron degeneration often co-occurs with FTLD. In the last few years, different autosomal dominant mutations have been demonstrated to be the cause of the familial aggregation frequently reported in FTLD. Major causal genes so far discovered include microtubule-associated protein tau (*MAPT*), progranulin (*GRN*), and chromosome 9 open reading frame *(C9ORF)72*. Mutations in *MAPT* are generally associated with early onset and with the bvFTD phenotype, whereas mutations in *GRN* and *C9ORF72* are associated with high clinical heterogeneity and age at disease onset. In addition, other genes are linked to rare cases of familial FTLD.

Keywords Frontotemporal lobar degeneration • Tau • Progranulin (*GRN*) • C9ORF72 Genetics • Risk factor

M. Serpente, PhD (✉)
Neurology Unit, Department of Pathophysiology and Transplantation, Centro Dino Ferrari, University of Milan, Fondazione Cà Granda, IRCCS Ospedale Maggiore Policlinico, Via F. Sforza 35, Milan 20122, Italy
e-mail: maria.serpente@policlinico.mi.it

D. Galimberti, PhD
Neurology Unit, Alzheimer Center, University of Milan, Fondazione Ca' Granda, IRCCS Ospedale Policlinico, Milan, Italy

D. Galimberti, E. Scarpini (eds.), *Neurodegenerative Diseases*,
DOI 10.1007/978-1-4471-6380-0_7, © Springer-Verlag London 2014

Introduction

The term *frontotemporal lobar degeneration* (FTLD) encompasses three main clinical syndromes: behavioral variant frontotemporal dementia (bvFTD), progressive nonfluent aphasia (PNFA), and semantic dementia (SD). It represents a common cause of dementia in subjects under 65 years. The age at onset is typically 45–65 years, with a mean average in the 50s, and the prevalence, equal among men and women, is 10–15 individuals out of 100,000. bvFTD is the most frequent FTLD phenotype. It is primarily characterized by behavioral changes and progressive deterioration of personality. Throughout the disease, patients show a wide spectrum of symptoms, including behavioral alterations, such as disinhibition, overeating and impulsiveness, and impairment of cognitive functions, with relative sparing of memory. Changes in social behavior, loss of empathy, and impairment of social insight are early and consistent symptoms of bvFTD. The first histopathological findings were described in 1911 by Alois Alzheimer, who observed ballooned neurons containing tau protein and argyrophilic intracytoplasmic inclusions. He named them "Pick cells" and "Pick bodies," respectively, after Arnold Pick, who reported the first case in 1892.

Both diagnostic criteria for bvFTD and language presentations have been recently revised including neuroimaging and genetic findings [1, 2]. In addition, other phenotypes, such as PSP, corticobasal syndrome (CBS), and bvFTD with MND, are part of the clinical manifestation within the FTLD spectrum. Overall, these different phenotypes well reflect the clinical heterogeneity and the underlying clinical-pathological spectrum of FTLD.

Diagnostic Criteria of bvFTD and Language Presentations

bvFTD is the most frequent FTLD phenotype. It is primarily characterized by a long phase of subclinical behavioral changes and progressive deterioration of personality. Throughout the disease, patients show a wide spectrum of symptoms, including behavioral alterations, such as apathy, disinhibition, overeating and impulsiveness, and impairment of cognitive functions, in particular executive dysfunction, with a relative sparing of memory. This neuropsychological profile mostly corresponds to prefrontal cortex degeneration [3]. Despite recent advances in the characterization of this disorder, the diagnosis is still challenging. The extended preservation of functional activity and of performances in standard neuropsychological tests may lead to misdiagnose subjects as suffering from psychological and psychiatric conditions. Changes in social behavior, loss of empathy, and impairment of social insight are early and consistent symptoms of bvFTD, whose importance and role for the early diagnosis have been emphasized in the new consensus criteria [1]. According to these criteria, the main feature of bvFTD is the progressive deterioration of behavior and/or cognition by observation or history (provided by a knowledgeable informant). If this criterion is satisfied, there are three further levels of certainty for bvFTD: possible, probable, or definite. "Possible" bvFTD requires three out of six clinically discriminating features. "Probable" bvFTD meets the criteria of "possible" bvFTD and (1) a significant functional decline (by caregiver report or evidenced at neuropsychological testing) (2) imaging results consistent with bvFTD, i.e., frontal and/or anterior temporal atrophy on MRI or CT or frontal and/or anterior temporal hypoperfusion or hypometabolism on PET or SPECT. "Definite" bvFTD implies the histopathological evidence of FTLD (post mortem) or the presence of a known pathogenic mutation. These new criteria have a flexible structure to account for the high heterogeneity at initial presentation, i.e., the clinical presentation could be memory impairment, which often leads to an AD clinical diagnosis. These new criteria were significantly more sensitive in cases with early onset (<65 years of age) as compared with late onset. Compared with the former, patients with late onset bvFTD had significantly lower rates of disinhibition, loss of sympathy/empathy, and perseverative/compulsive behaviors. The lower sensitivity of possible bvFTD in older patients may be due to the presence of unusual FTLD-spectrum pathologies or primarily amnestic presentations.

Primary progressive aphasia (PPA) is characterized by early and progressive changes in language functions and includes three subtypes: primary nonfluent aphasia (PNFA), semantic dementia (SD), and logopenic aphasia (LA) [2]. Progressive loss of speech, with hesitant, nonfluent speech output with phonetic/phonological errors and distortions and/or agrammatism, is typical of PNFA [4], whereas loss of knowledge about words and objects, anomia, and single-word comprehension deficits are core features of SD. On the contrary, patients with PNFA usually do not show behavioral changes. Each PPA variant is associated with a distinct pattern of brain atrophy that represents a useful supportive feature for the diagnosis. While the anterior temporal lobe, usually on the left side, characterizes the semantic variant, a focal left-sided perisylvian region involvement is the main neuroradiological feature of PNFA [4]. Logopenic aphasia is characterized by phonological disorders, defective word retrieval and sentence repetition deficits [2]. This PPA subtype seems to be associated, in the majority of cases, with underlying AD pathology [5]. Consistent with the hypothesis of a phonological short-term memory impairment as the main feature of this subtype, left temporoparietal atrophy is an imaging marker of the disease [4].

Genetics: Autosomal Dominant Inherited Mutations and Pathogenic-Related Mechanisms

The majority of FTLD cases are sporadic and likely caused by the interaction between genetic and environmental factors. A number of cases, however, present familial aggregation and are inherited in an autosomal dominant fashion, suggesting a genetic cause [6–8]. Up to 40 % of patients have a family history, suggesting

FTLD in at least one extra family member [7, 9], with a percentage of autosomal dominant cases accounting for 13.4 % of the total according to Goldman et al. [8].

The current knowledge about genetics of FTLD has been recently enlarged by the identification of multiple novel genetic defects and chromosomal loci involved in hereditary forms. At present, many genes have been associated with the FTLD pathology, including *MAPT*, *GRN*, *VCP-1*, *CHMP2B*, and a hexanucleotide expansion in chromosome 9 responsible for familial FTLD-MND cases.

MAPT

The first evidence of a genetic cause for familial FTLD came from the demonstration of a linkage with chromosome 17q21.2 in autosomal dominantly inherited form of FTD with parkinsonism [10], named FTDP-17. The gene responsible for such association, *MAPT*, was discovered few years later [11]. *MAPT* encodes the microtubule-associated protein tau, which is involved in microtubule stabilization and assembly. To date, 44 pathogenic *MAPT* mutations have been described in 134 families (http://www.molgen.vib-ua.be/) and classified according to their position in the gene, their effects on *MAPT* transcription, and their type of tauopathy. *MAPT* mutations include missense mutations, deletions, or intronic mutations located close to the splice donor site of the intron after the alternatively spliced exon 10. They are mainly clustered in exons 9–13, which contain the microtubule-binding regions, except for two mutations in exon 1 [12]. The pathogenic mechanism of each different mutation depends from the type and location of the genetic defect and affects the normal function of tau, i.e., the stabilization of microtubules promoting their assembly by binding tubulin. Some mutations increase the free cytoplasmic portion of the protein promoting tau aggregation, while others lead to an aberrant phosphorylation of tau protein, which damages microtubule stabilization [12]. Alternatively, other mutations affect the alternative splicing, thus producing altered ratios of the different isoforms (3R/4R tau). Grisart and coworkers observed a microduplication on chromosome 17g21.31 that was associated with behavioral problems and skill impairments [13]. The authors suggested that the overexpression of *MAPT* in neurons could contribute to the behavioral changes, and the duplication of the corticotropin-releasing hormone receptor 1 gene *(CRHR1)*, located 59 Kb centromeric from *MAPT*, could explain the impaired motor skills. The presence of structural changes at the *MAPT* locus in presence of behavioral changes led the authors to believe that rearrangements at this locus might be associated with FTLD [13]. Several subsequent studies failed to identify abnormal copy-number variations (CNVs) at the genetic region encompassing *GRN* and *MAPT* [14]. However, in 2009, Rovelet-Lecrux and coworkers identified a heterozygous 17.3 Kb deletion responsible for the removal of exons 6–9 of *MAPT* in one FTD patient [15]. This deletion caused the loss of the first microtubule binding domain and a decrease in the binding abilities of tau to the microtubules. The same group reported a 439-Kb duplication in the region encompassing *CRHR1*, *MAPT*, and saithoin (*STH*) in one patient affected by behavioral and amnestic disorders [16]. This is the first evidence of a possible link between rearrangements at the *MAPT* locus and the FTLD.

Rossi et al. (2013) recently suggested that tau plays a role in genome and chromosome stability that can be ascribed to its function as a microtubule-associated protein as well as a protein protecting chromatin integrity through interaction with DNA [17]. At autopsy, patients with *MAPT* mutations show tau-positive inclusions [12].

The clinical presentation in *MAPT* mutation carriers is mainly consistent with bvFTD, with a mean onset in the 50s. Nevertheless, cases of PNFA have been reported as well, with an onset even in the sixth decade of life [5]. Despite the clinical presentation is heterogeneous in terms of symptoms and age at onset, subjects carrying *MAPT* mutations usually exhibit severe temporal lobe atrophy (medial and lateral regions and temporal pole), mostly on the right side [18].

CHMP2B

Few FTLD families display mutations in the charged multivesicular body protein 2B gene (*CHMP2B*), located on chromosome 3p11.2, which encodes a component of the heteromeric ESCRT III complex, involved in the endosomal trafficking and degradation [19]. In particular, CHMP2B protein is involved in sorting and trafficking surface receptors or proteins into intraluminal vesicles for lysosomal degradation and binding the Vps4 protein responsible for the dissociation of ESCRT components [20]. CHMP2B is a 213 amino acid-long protein that presents a coiled coil domain at the N-terminus, a microtubule-interacting transport (MIT), and a microtubule-interacting region (MIR) at the C-terminus. The first mutation in CHMP2B was identified in one large kindred from Denmark, and it occurs in the splice acceptor site for the 6th and final *CHMP2B* exon, leading to a formation of two novel transcripts termed $CHMP2B^{Intron5}$ and $CHMP2B^{Delta10}$ [21]. To date, 11 different mutations, of which four in five families seem to exert a pathogenic action (http://www.molgen.vib-ua.be/), have been so far described; for this reason *CHMP2B* is considered an extremely rare genetic cause of FTLD pathology. It is important to note that all mutations described (missense and truncation mutations) show a common mechanism of action: the deletion of the C-terminus of the protein [20]. Probably, the loss of the Vsp-4 binding domain located in C-terminus of the protein causes the accumulation of mutated CHMP2B on the endosomal membrane and prevents the recruitment of other proteins necessary for endosomal fusion with lysosome. This phenomenon leads to the impairment of the late endosomal trafficking and contributes to neurodegenerative processes in FTD [21]. This can be observed as enlarged and abnormal endosomal structures in postmortem brain tissue from patients [22]. From a histological point of view, patients with *CHMP2B* mutations present FTLD-U with ubiquitin- and p62-positive but TDP-43-negative neuronal cytoplasmic inclusions [23]. Recently, it was observed in transgenic mice expressing either human $CHMP2B^{intron5}$ or human wild-type protein that only

CHMP2B^{intron5} but not wild-type or CHMP2B knockout mice developed neuropathology consistent with that seen in FTLD patients carrying CHMP2B mutations [24]. These data support the hypothesis that CHMP2B mutations act thought a gain of function mechanism. Moreover, the use of RNA interference approach against mutant *CHMP2B* in primary patient fibroblasts has shown that this treatment reverses the mutant endosomal phenotype. Importantly, this morphological change is also observed in *CHMP2B* mutation brain tissue, suggesting that RNA interference might be a future therapeutic approach for the treatment of FTLD patients with *CHMP2B* mutations [25].

Behavioral and cognitive impairments associated with extrapyramidal and pyramidal signs are the main clinical manifestations in *CHMP2B*. Myoclonus can occur late in the course of the disease, and motor neuron disorders have been described in only two cases [26]. To assess the earliest neuropsychological changes in *CHMP2B* mutation carriers, a longitudinal prospective study spanning over 8 years and including 17 asymptomatic individuals with *CHMP2B* mutations was carried out. Longitudinal analyses showed a gradual decline in psychomotor speed, working memory capacity, and global executive measures in the mutation carrier group compared with controls. This decline starts several years before they fulfill diagnostic criteria for FTD, but the level of cognitive changes over time varied considerably among different individuals [27].

VCP-1

So far, 18 different mutations in valosin-containing protein (*VCP*)*-1* have been described in 48 families (http://www.molgen.vib-ua.be/). *(VCP)-1* gene is located on chromosome 9p13.3 and encodes a monomeric protein composed by 806 amino acids. The VCP hexamer is a member of the AAA-ATPase superfamily that is composed by six monomers, forming a ring around a central pore with two AAA+ protein domains called D1 and D2 domains [28]. VCP is involved in protein homeostasis, maintaining the proper balance between protein synthesis and protein degradation [29].

The phenotype associated with such mutations is characterized not only by bvFTD but also with inclusion body myopathy (IBM) and Paget's disease of the bone [30, 31]. R155C was the first pathogenic missense mutation reported in FTD and is located in the cofactor-binding domain at the N-terminus of the protein; missense mutations associated with IBMFD have been identified in different domains such as N-terminus domain, the linker L1 connecting N-terminus, and D1 domain. One missense mutation was identified in linker L2 and one in D2 domain [30]. Myopathy is the more frequent clinical symptom, present in about 90 % of affected subjects, whereas bvFTD is seen in about 30 %, usually many years after the onset of muscle symptoms. From a histological point of view, brain tissues of patients carrying *VCP* mutations are characterized by tau-negative but ubiquitin-positive inclusions. Moreover, VCP is also present in the inclusions of several diseases including amyotrophic lateral sclerosis (ALS), Parkinson's disease (PD), and Huntington's disease [28]. Very recently, Komatsu et al. (2013) identified a novel mutation (G156S) associated with IBM, Paget disease, and bvFTD [32].

GRN

After the discovery of *MAPT* as causal gene for FTDP-17, there were still numerous autosomal dominant FTLD cases genetically linked to the same chromosomal region of *MAPT* (chr17q21), without any mutation in *MAPT,* in spite of an extensive fine mapping of the gene. A small region rich of genes, localized approximately 6.2 Mb in physical distance to *MAPT* locus, had been recognized as that one containing the gene responsible for the disease in these families. The first identified mutation in *GRN*, identified in 2006, consisted of a 4-bp insertion of *CTGC* between coding nucleotides 90 and 91, causing a frameshift and premature termination in progranulin (C31LfsX34) [33]. In another parallel study, Cruts and coworkers, analyzing other families with an FTLD pathology without *MAPT* mutation, found at the same time another mutation of five base pairs into the intron following the first noncoding exon of *GRN* (IVS1+5G>C) [34]. This mutation causes the splicing out of the intron 0, leading the retention of mRNA within the nucleus and its degradation.

GRN gene encodes for the growth regulation factor named progranulin. It belongs to a family of proteins involved in multiple biological functions, including development, wound repair, and inflammation, by activating signaling cascades that control cell cycle progression and cell motility. Progranulin is a 593-amino acid protein, rich of cysteine with a molecular weight of ~90 kDa, whose role in neuronal survival and function is still unclear. It is subjected to proteolysis by elastase in a process regulated by an SLPI. It is expressed in neurons as well as in activated microglia [35].

Since the original identification of null mutations in FTLD, 69 different mutations have been described so far (http://www.molgen.vib-ua.be/) in 231 families. Most of the known pathogenic *GRN* mutations, particularly frameshift, splice-site, and nonsense mutations, are predicted to result in a premature stop codon. The resulting aberrant mRNA is degraded through the process of nonsense-mediated decay, leading to haploinsufficiency [36].

At the neuropathological examination, *GRN*-mutated FTLD cases displayed ubiquitin-positive, tau-negative inclusions (FTLD-U) similar to the microvacuolar type still observed in a large proportion of apparently sporadic FTLD, which were different from the tau-positive inclusions typical of *MAPT* mutated cases. According to the novel neuropathological classification of FTLD-TDP pathology in FTLD, TDP-43 neuropathological subtype A is consistently found in association with *GRN*-mutated cases [37].

Truncated and hyperphosphorylated isoforms of the TDP-43 were recognized as main components of the ubiquitin-positive inclusions typical of the *GRN*-mutated families, as well as of idiopathic FTLD and of a proportion of ALS cases [38]. Nevertheless, at present, the linkage between progranulin haploinsufficiency and TDP-43 accumulation in the cytoplasm has not been clarified.

GRN mutation accounts for about 5–10 % of all FTD cases, markedly varying depending on the population considered. A collaborative study [39] analyzing *GRN* mutations in 434 FTLD patients, clinically ranging from bvFTD to PNFA, FTLD associated with parkinsonism or MND, estimates a frequency of 6.9 % of all included FTLD-spectrum cases. About 56 % of such cases were represented by FTLD subjects with ubiquitinated inclusions at the neuropathology (FTLD-U) with a positive family history of FTLD. The most common phenotype was bvFTD, but a few patients were diagnosed with PNFA, AD, or CBS. As expected, the majority of *GRN* mutations introduced a premature termination codon, suggesting that their corresponding mRNA has been degraded through nonsense-mediated decay, thus supporting the hypothesis that most *GRN* mutations create a functionally null allele [39].

From a clinical point of view, mutations in *GRN* are associated with extremely heterogeneous phenotypes, including classical FTLD presentations, such as bvFTD, PNFA, or SD, and also AD [40], CBS [41], MCI [42], and Lewy body dementia (LBD) [43]. Age at disease onset is extremely wide, even in the same family, ranging from 47 to 79 years [42]. Although rarely, an overlap between psychiatric disorders and genetically determined FTLD can occur, as shown by Rainero et al. [44], who described a patient with heterosexual pedophilia who was a carrier of a *GRN* mutation and developed bvFTD over time, and by Cerami et al. [45], who reported two clinically different, apparently sporadic FTLD cases sharing the Thr272fs *GRN* mutation, who had had a premorbid bipolar disorder history.

A major contribution to achieve a correct diagnosis independent of the phenotypic presentation is the demonstration that progranulin plasma levels are extremely low in *GRN* mutation carriers, even in asymptomatic subjects [40, 42].

Regarding the function of progranulin, Pickford et al. [46] demonstrated, in an in vitro model, that it has chemotactic properties toward cultured mouse neurons. In addition, progranulin-treated primary neurons secrete a number of cytokines and chemokines, particularly those involved in proliferation (i.e., IL-4), and, importantly, induce microglia to switch from a pro-inflammatory to an anti-inflammatory phenotype [46]. Another recent observation is that progranulin binds the TNFR2, that is expressed specifically in neuronal subtypes and glial cells in the brain, leading to an anti-inflammatory cascade [47].

Yin et al. [48] generated conditional *GRN* knockout mice. They observed that *GRN*-deficient macrophages produced more pro-inflammatory cytokines and chemokines, including CCL2, CXCL1, interleukin (IL)-6, IL-12p40, and TNF, but less anti-inflammatory cytokine IL-10 compared to wild-type (wt) macrophages, when exposed to bacterial lipopolysaccharide. However, *GRN*-deficient mice failed to clear bacterial infection as fast as wt mice and were characterized by an exaggerated inflammatory tissue damage. Immunostaining of brain sections for CD68 revealed greater activation of microglia with age in *GRN*-deficient than wt mice. Moreover, *GRN*-deficient microglia responded to inflammatory stimuli by becoming more cytotoxic than wt microglia, and *GRN*-deficient neurons were more susceptible than wt to damage by activated microglia and by certain cytotoxic stresses, such as depletion of glucose and oxygen. They also showed enhanced hippocampal ubiquitin immunostaining and increased phosphorylation of TDP-43 in the hippocampus and thalamus of old *GRN*-deficient mice. In light of this observation, authors hypothesized that FTLD may arise from the congruence of two independent phenotypes of *GRN* insufficiency: deregulated inflammation and increased neuronal vulnerability to damage [48].

In vivo studies in progranulin heterozygous mice ($Grn^{+/-}$), that mimic progranulin haploinsufficiency, were carried out as well. These mice developed age-dependent social and emotional deficits potentially relevant to bvFTD. Nevertheless, no gliosis or neuroinflammation was observed, suggesting that microglial activation independent from functional deficits, and thus progranulin deficiency, could have effects directly on neurons [49].

C9ORF72

One of the most intriguing discoveries in the genetics of FTLD has been the investigation of FTD/MND families linked to a locus on chromosome 9q21-22. The first evidence of linkage with this locus comes from a study carried out in families with autosomal dominant FTD-MND [50]. Additional data confirmed the linkage to chr9q21-22 in FTD-MND families [51], until, in 2011, two international groups of researchers identified the gene responsible for the disease in this locus, *C9ORF72* [52, 53]. The mutation consists of a large hexanucleotide (GGGGCC) repeat expansion in the first intron of a gene named Open Reading Frame 72 (*C9ORF72),* that segregates with ALS or combined FTD-MND phenotype, and TDP-43 based pathology. Wild-type alleles contain no more than 23–30 repeats, whereas mutated alleles have hundreds to thousand repeats. These studies demonstrated that the *C9ORF72* expansion could represent a major cause of both familial FTD (12 %) and ALS (22.5 %) cases [52], reaching a prevalence of 46 % of all familiar ALS, 21.1 % of sporadic ALS, and 29.3 % of FTD in the Finnish population [53]. In the next few years, a wide number of confirmatory studies were published [54], confirming that this mutation is as frequent as *GRN* and *MAPT* ones in patients with FTLD.

Regarding the clinical phenotype, it was shown that psychosis and obsessive-compulsive disorder were common symptoms at disease onset in patients with FTLD carrying the repeat expansion [55–57]. Moreover, a case showing mystic delusion with visual and auditory hallucinations, in the absence of neurological symptoms and brain atrophy, was recently described [58]. Presentation with memory impairment also occurs quite often (50–65 % according to Mahoney et al. [59], possibly leading to a clinical diagnosis of AD [58, 60]).

A study on a large population confirmed that the expansion is a quite frequent genetic cause of FTLD, and that it is associated with atypical features. In particular, in a population of 651 patients with FTLD, the pathogenic repeat expansion was detected in 39 cases (6 %).

Clinical phenotypes of carriers included 29 patients with bvFTD (5.2 % of all cases diagnosed with bvFTD), 8 with bvFTD/MND (32 % of cases with bvFTD/MND), and 2 with SD (5.9 % of patients with SD). The presentation with late onset psychosis was significantly more frequent in carriers than noncarriers as well as the presence of cognitive impairment at onset [61].

Concerning the product of the *C9ORF72*, Levine et al. [62] suggested, through informatics research, that the gene encodes for a distant homologue of proteins related to *DENN*, which is a GDP/GTP exchange factor that activates Rab-GTPases.

Regarding the function of the *C9ORF72* product and the mechanisms at the basis of the pathogenesis of the disease in the expansion carriers, quite few information are available. It is known that this mutation causes the loss of one alternatively spliced transcript suggesting a potential loss of function. The accumulation of RNA transcript containing the GGGGCC repeat within nuclear foci in the frontal cortex and spinal cord in c9FTD/ALS also suggests a toxic RNA gain of function and multiple disease mechanisms [52]. RNA foci, which lead to the sequestration and altered activity of RNA-binding proteins, have been implicated in several neurodegenerative noncoding expansion disorders [63]. Reddy et al. [64] demonstrated that the r(GGGGCC)n RNA forms extremely Stable G-quadruplex structures, which are known to theoretically affect promoter activity, genetic instability, RNA splicing, translation, and neurite mRNA localization.

Another possible pathogenic mechanism has been proposed by Mori et al. [65],which have demonstrated that the intronic GGGGCC repeat might be aberrantly translated into DPR proteins with an unconventional mechanisms of non-ATG-initiated translation called RAN. Translation via RAN of the GGGGCC repeat originates three DPR aggregates: poly-(Gly-Ala), poly-(Gly-Pro), and poly-(Gly-Arg). This type of translation was first described in 2011 by Zu and coworkers, who reported that RNA translation across expanded CAG repeats occurs in all reading frames to produce homopolymeric proteins of long polyglutamina and polyalanine tracts. It is important to note that these proteins were found accumulated in tissues of patients with spinocerebellar ataxia type 8 and myotonic dystrophy type 1 [66]. Recently, polyclonal antibodies generated against putative GGGGCC repeat RAN-translated peptides (anti-C9RANT) detected high molecular weight, insoluble material in brain homogenates, and neuronal inclusions throughout the central nervous system (CNS) of C9FTD patients. C9RANT immunoreactivity was not found in other neurodegenerative diseases or in peripheral tissue of c9FTD/ASL. This intriguing finding could represent a possible biomarker for this common cause of FTD and ASL [67]. Given that both foci formation and RAN translation in c9FTD/ALS require the synthesis of GGGGCC repeat expansion RNA, therapeutic strategies that target these transcripts and result in their neutralization or degradation could effectively block these two potential pathogenic mechanisms and provide a much needed treatment for c9FTD/ALS.

To date, understanding of pathogenic mechanism has been prevented by the presence of suboptimal cellular and animal models of GGGGCC repeat expansion.

Traditional approaches to disease modeling have a number of limitations, such as a disease gene is often overexpressed. Moreover, long repeat sequences are often unstable posing a significant technical challenge for molecular cloning and disease modeling of C9ORF72 related FTD/ALS in animals. Almeida and coworkers have developed a new C9orf72 cellular model based on the use of induced pluripotent stem cell (iPSC) technology that allows the study of disease's genes in their native genetic context. They generated multiple iPSC lines starting from fibroblast of two patients carrying *GGGGCC* repeat expansion. They observed the presence of RNA foci containing *GGGGCC* repeats in iPSCs, iPSC-derived human neurons, and primary fibroblasts of repeat expansion carriers but not in neurons of healthy subjects or FTD patients without *GGGGCC* expansions. Moreover, RAN translation products were detected in human neurons with *GGGGCC* repeat expansions, and these neurons showed significantly elevated p62 levels and increased sensitivity to cellular stress induced by autophagy inhibitors. For this reason the authors suggested that compromised autophagy pathway could represent a new underlying pathogenic mechanism [68].

New Approaches for FTLD Genetic Comprehension

The genetics of complex diseases such as FTLD, in which multiple genes interact with environmental risk factors to increase risk, has been revolutionized by the genome-wide association study (GWAS) approach. This uses microarray technology to genotype up to ≥1 million SNPs, which span the whole genome. A great advantage of the GWAS approach, in contrast to the candidate gene method, is that it allows associations between completely novel chromosomal loci and disease to be identified.

In 2010, Van Deerlin and coworkers published the first GWAS on 515 FTD patients with TDP-43 pathology; they identified a possible susceptibility locus, which encompasses the transmembrane protein 106b (*TMEM106b*) gene on chromosome 7p21 [69]. In particular, the study identified three associated single nucleotide polymorphisms (SNPs), rs102004, rs6966915, and rs1990622, which are correlated with an increase of *TMEM106b* expression level [69]. Several subsequent studies showed that the highest association with *TMEM106b* locus was found in FTLD-TDP patients with GRN mutations [70, 71]. These results increased our knowledge about the genetics of FTLD-TDP and represent a starting point from where researches can look into a possible new pathogenic pathway. It is also true that these data are specific for a subgroup of FTLD patients, suggesting that the connection between *TMEM106b* and FTLD cannot be extended to the general FTLD population. Therefore, it is important to note that successful GWAS require stringent inclusion criteria and a large well-characterized cohort of subjects in order to confer statistical relevance to the results.

To date, a large part of the genetic cause of FTD remains unknown, and it is particularly evident in families with age at onset between 55 and 70 years in which genes responsible of the disease pathogenesis have not yet been identified. Despite the achievements of GWAS, this approach is limited because it is only able to study relatively common types of variants, those that occur at a frequency of more than 1 %. Several recent studies, indeed, demonstrate that using next-generation sequencing (NGS) technologies (exome and genome sequencing) is possible to identify the genetic causes of different disorders [72, 73]. These results demonstrate that even by using a small number of subjects, it is now possible to uncover genetic variations not only in Mendelian disorders but also in multifactorial diseases. For example, in 2013 Guerreiro and coworkers, using exome sequencing, identified homozygous mutations in the triggering receptor expressed on myeloid cells 2 genes (*TREM2*) in three probands with clinical FTD-like syndrome and members of three different Turkish families. Mutations in this gene have previously been associated with Nasu-Hakola disease, a rare autosomal recessive disease characterized by early-onset dementia and bone cysts. Now, mutations in *TREM2* could be a more frequent cause of dementia than previously considered, even in absence of bone problems [74]. In conclusion, NGS technologies are a new starting point for the discovery of genetic alterations in the neurodegenerative diseases that will have a great utility in the clinical practice for the diagnosis and important implications in the research of novel therapeutic strategies.

Conclusions

The discoveries of the last few years showed that the term "FTLD" actually comprises diseases with a different etiology. It has become clearer and clearer that there are multiple genetic autosomal dominant mutations leading to the development of FTLD. The most frequent are so far *MAPT, GRN*, and *C9ORF72* mutations. The description of peculiar clinical phenotypes showed that there is an overlap among neurodegenerative disorders in terms of symptoms and pathogenic events leading to neurodegeneration. From a clinical point of view, the same genetic defect has been observed in patients with different diseases, i.e., bvFTD, MND, or both, raising the question whether there are additional unknown genetic or environmental factors influencing the phenotype. In addition, *GRN* and *C9ORF72* mutations are associated with a wide range of phenotypes and age at disease onset, including memory and psychosis, making it difficult to predict the presence of a mutation basing on symptoms and/or familial history. Moreover, the situation is even more complex considering the incomplete penetrance of such mutations.

Concerning pathogenic mechanisms related to these mutations, whereas in *MAPT* mutation carriers there is an impaired functioning and the deposition of this protein in the brain, mechanisms at the basis of *GRN* and *C9ORF72* appear more complex. Mutations in *GRN* do not lead to progranulin deposition in the brain but instead to an altered functioning of transcription factors (i.e., TDP-43). Mutations in *C9ORF72* are associated with TDP-43 deposition and DPR aggregates, and yet the function of c9orf72 protein is not known. Therefore, future challenges will be to understand pathways altered in *GRN* and *C9ORF72* carriers, in order to discover novel therapeutic targets.

New findings about genetics and molecular biology of FTLD recently described have some implications for FTLD diagnosis and treatment. First, biomarkers for identifying mutation carriers are needed. So far, given the heterogeneity of age at disease onset and presentation of symptoms, it is not possible to predict the presence of a causal mutation basing on the clinical picture only. In this regard, low plasma progranulin levels are very good predictors of the presence of a *GRN* mutation leading to haploinsufficiency. Second, in view of the availability of future-tailored therapies aimed to modify the course of the disease by acting on pathogenic mechanisms (i.e., replacing progranulin loss or hampering tau deposition), it would be extremely important to develop tools to predict the ongoing pathology (i.e., tau deposition or TDP-43 altered functioning). Lastly, whereas the genetic analysis is becoming part of the diagnostic workup in symptomatic subjects and is included in the new criteria for bvFTD [1], no consensus criteria for a genetic screening in asymptomatic family members are at present available. There is thus a need of such guideline in view of early (even presymptomatic) therapies.

References

1. Rascovsky K, Hodges JR, Knopman D, Mendez MF, Kramer JH, Neuhaus J, et al. Sensitivity of revised diagnostic criteria for the behavioural variant of frontotemporal dementia. Brain. 2011;134:2456–77.
2. Gorno-Tempini ML, Hillis AE, Weintraub S, Kertesz A, Mendez M, Cappa SF, et al. Classification of primary progressive aphasia and its variants. Neurology. 2011;76: 1006–14.
3. Piguet O, Hornberger M, Mioshi E, Hodges JR. Behavioural-variant frontotemporal dementia: diagnosis, clinical staging, and management. Lancet Neurol. 2011;10:162–72.
4. Gorno-Tempini ML, Dronkers NF, Rankin KP, Ogar JM, Phengrasamy L, Rosen HJ, et al. Cognition and anatomy in three variants of primary progressive aphasia. Ann Neurol. 2004;55:335–46.
5. Villa C, Ghezzi L, Pietroboni AM, Fenoglio C, Cortini F, Serpente M, et al. A novel MAPT mutation associated with the clinical phenotype of progressive nonfluent aphasia. J Alzheimers Dis. 2011;26:19–26.
6. Ratnavalli E, Brayne C, Dawson K, Hodges JR. The prevalence of frontotemporal dementia. Neurology. 2002;58:1615–21.
7. Bird T, Knopman D, VanSwieten J, Rosso S, Feldman H, Tanabe H, et al. Epidemiology and genetics of frontotemporal dementia/Pick's disease. Ann Neurol. 2003;54:S29–31.
8. Goldman JS, Farmer JS, Wood EM, Johnson JK, Boxer A, Neuhaus J, et al. Comparison of family histories in FTLD subtypes and related tauopathies. Neurology. 2005;65:1817–9.
9. Pickering-Brown SM. The complex aetiology of frontotemporal lobar degeneration. Exp Neurol. 2007;114:39–47.
10. Wilhelmsen KC, Lynch T, Pavlou E, Higgins M, Nygaard TG. Localization of disinhibition-dementia-parkinsonism-amyotrophy complex to 17q21-22. Am J Hum Genet. 1994;55(6): 1159–65.
11. Hutton M, Lendon CL, Rizzu P, Baker M, Froelich S, Houlden H, et al. Association of missense and 5′-splice-site mutations in tau with the inherited dementia FTDP-17. Nature. 1998;393:702–5.

12. Rademakers R, Cruts M, van Broeckhoven C. The role of tau (MAPT) in frontotemporal dementia and related tauopathies. Hum Mutat. 2004;24:277–95.
13. Grisart B, Willatt L, Destrée A, Fryns JP, Rack K, de Ravel T, et al. 17q21.31 microduplication patients are characterised by behavioural problems and poor social interaction. J Med Genet. 2009;46:524–30.
14. Lladó A, Rodríguez-Santiago B, Antonell A, Sánchez-Valle R, Molinuevo JL, Reñé R, Pérez-Jurado LA. MAPT gene duplications are not a cause of frontotemporal lobar degeneration. Neurosci Lett. 2007;424:61–5.
15. Rovelet-Lecrux A, Lecourtois M, Thomas-Anterion C, Le Ber I, Brice A, Frebourg T, et al. Partial deletion of the MAPT gene: a novel mechanism of FTDP-17. Hum Mutat. 2009; 30:591–602.
16. Rovelet-Lecrux A, Hannequin D, Guillin O, Legallic S, Jurici S, Wallon D, et al. Frontotemporal dementia phenotype associated with MAPT gene duplication. J Alzheimers Dis. 2010;21: 897–902.
17. Rossi G, Conconi D, Panzeri E, Redaelli S, Piccoli E, Paoletta L, et al. Mutations in MAPT gene cause chromosome instability and introduce copy number variations widely in the genome. J Alzheimers. 2013;33:969–82.
18. Van Swieten JC, Heutink P. Mutations in progranulin (GRN) within the spectrum of clinical and pathological phenotypes of frontotemporal dementia. Lancet Neurol. 2008;7:965–74.
19. Skibinski G, Parkinson NJ, Brown JM, Chakrabarti L, Lloyd SL, et al. Mutations in the endosomal ESCRTIII-complex subunit CHMP2B in frontotemporal dementia. Nat Genet. 2005;37: 806–8.
20. Urwin H, Ghazi-Noori S, Collinge J, Isaacs A. The role of CHMP2B in frontotemporal dementia. Biochem Soc Trans. 2009;37:208–12.
21. Lindquist SG, Braedgaard H, Svenstrup K, Isaacs AM, Nielsen JE, FReJA Consortium. Frontotemporal dementia linked to chromosome 3 (FTD-3)-current concepts and the detection of a previously unknown branch of the Danish FTD-3 family. Eur J Neurol. 2008; 15:667–70.
22. Urwin H, Authier A, Nielsen JE, Metcalf D, Powell C, Froud K, et al. Disruption of endocytic trafficking in frontotemporal dementia with CHMP2B mutations. Hum Mol Genet. 2010; 19:2228–38.
23. Isaacs AM, Johannsen P, Holm I, Nielsen JE, FReJA Consortium. Frontotemporal dementia caused by CHMP2B mutations. Curr Alzheimer Res. 2011;8:246–51.
24. Ghazi-Noori S, Froud KE, Mizielinska S, Powell C, Smidak M, Fernandez de Marco M, et al. Progressive neuronal inclusion formation and axonal degeneration in CHMP2B mutant transgenic mice. Brain. 2012;135:819–32.
25. Nielsen TT, Mizielinska S, Hasholt L, Isaacs AM, Nielsen JE, FReJA Consortium. Reversal of pathology in CHMP2B-mediated frontotemporal dementia patient cells using RNA interference. J Gene Med. 2012;14:521–9.
26. Parkinson N, Ince PG, Smith MO, Highley R, Skibinski G, Andersen PM, et al. ALS phenotypes with mutations in CHMP2B (charged multivesicular body protein 2B). Neurology. 2006;67:1074–7.
27. Stokholm J, Teasdale TW, Johannsen P, Nielsen JE, Nielsen TT, Isaacs A, et al. Cognitive impairment in the preclinical stage of dementia in FTD-3 CHMP2B mutation carriers: a longitudinal prospective study. J Neurol Neurosurg Psychiatry. 2013;84:170–6.
28. Weihl CC, Pestronk A, Kimonis VE. Valosin-containing protein disease: inclusion body myopathy with Paget's disease of the bone and fronto-temporal dementia. Neuromuscul Disord. 2009;19:308–15.
29. Forman MS, Mackenzie IR, Cairns NJ, Swanson E, Boyer PJ, Drachman DA, et al. Novel ubiquitin neuropathology in frontotemporal dementia with valosin-containing protein gene mutations. J Neuropathol Exp Neurol. 2006;65:571–81.
30. Watts GD, Wymer J, Kovach MJ, Mehta SG, Mumm S, Darvish D, et al. Inclusion body myopathy associated with Paget disease of bone and frontotemporal dementia is caused by mutant valosin containing protein. Nat Genet. 2004;36:377–81.

31. Kimonis VE, Fulchiero E, Vesa J, Watts G. VCP disease associated with myopathy, Paget disease of bone and frontotemporal dementia: review of a unique disorder. Biochim Biophys Acta. 2008;1782:744–8.
32. Komatsu J, Iwasa K, Yanase D, Yamada M. Inclusion body myopathy with Paget disease of the bone and frontotemporal dementia associated with a novel G156S mutation in the VCP gene. Muscle Nerve. 2013;48(6):995–6.
33. Baker M, Mackenzie IR, Pickering-Brown SM, Gass J, Rademakers R, Lindholm C, et al. Mutations in progranulin cause tau-negative frontotemporal dementia linked to chromosome 17. Nature. 2006;442:916–9.
34. Cruts M, Gijselinck I, van der Zee J, Engelborghs S, Wils H, Pirici D, et al. Null mutations in progranulin cause ubiquitin-positive frontotemporal dementia linked to chromosome 17q21. Nature. 2006;442:920–4.
35. He Z, Bateman A. Progranulin (granulin-epithelin precursor, PC-cell-derived growth factor, acrogranin) mediates tissue repair and tumorigenesis. J Mol Med. 2003;81:600–12.
36. Gass J, Cannon A, Mackenzie IR, Boeve B, Baker M, Adamson J, et al. A harmonized classification system for FTLD-TDP pathology. Acta Neuropathol. 2011;122(1):111–3.
37. Mackenzie IR, Neumann M, Baborie A, Sampathu DM, Du Plessis D, Jaros E, et al. A harmonized classification system for FTLD-TDP pathology. Acta Neuropathol. 2011;122(1):111–3.
38. Neumann M, Sampathu DM, Kwong LK, Truax AC, Micsenyi MC, Chou TT, et al. Ubiquitinated TDP-43 in frontotemporal lobar degeneration and amyotrophic lateral sclerosis. Science. 2006;314:130–3.
39. Yu CE, Bird TD, Bekris LM, Montine TJ, Leverenz JB, Steinbart E, et al. The spectrum of mutations in progranulin: a collaborative study screening 545 cases of neurodegeneration. Arch Neurol. 2010;67:161–70.
40. Carecchio M, Fenoglio C, De Riz M, Guidi I, Comi C, Cortini F, et al. Progranulin plasma levels as potential biomarker for the identification of GRN deletion carriers. A case with atypical onset as clinical amnestic Mild Cognitive Impairment converted to Alzheimer's disease. J Neurol Sci. 2009;287:291–3.
41. Carecchio M, Fenoglio C, Cortini F, Comi C, Benussi L, Ghidoni R, et al. Cerebrospinal fluid biomarkers in Progranulin mutations carriers. J Alzheimers Dis. 2011;27(4):781–90.
42. Pietroboni AM, Fumagalli GG, Ghezzi L, Fenoglio C, Cortini F, Serpente M, et al. Phenotypic heterogeneity of the GRN Asp22fs mutation in a large Italian kindred. J Alzheimers Dis. 2011;24:253–9.
43. Arosio B, Abbate C, Galimberti D, Rossi PD, Inglese S, Fenoglio C, et al. GRN Thr272fs clinical heterogeneity: a case with atypical late onset presenting with a dementia with Lewy bodies phenotype. J Alzheimers Dis. 2013;35:669–74.
44. Rainero I, Rubino E, Negro E, Gallone S, Galimberti D, Gentile S, et al. Heterosexual pedophilia in a frontotemporal dementia patient with a mutation in the progranulin gene. Biol Psychiatry. 2011;70:43–4.
45. Cerami C, Marcone A, Galimberti D, Villa C, Scarpini E, Cappa SF. From genotype to phenotype: two cases of genetic frontotemporal lobar degeneration with premorbid bipolar disorder. J Alzheimers Dis. 2011;27(4):791–7.
46. Pickford F, Marcus J, Camargo LM, Xiao Q, Graham D, Mo JR, et al. Progranulin is a chemoattractant for microglia and stimulates their endocytic activity. Am J Pathol. 2011;178(1): 284–95.
47. Tang W, Lu Y, Tian QY, Zhang Y, Guo FJ, Liu GY, et al. The growth factor progranulin binds to TNF receptors and is therapeutic against inflammatory arthritis in mice. Science. 2011;332(6028):478–84.
48. Yin F, Banerjee R, Thomas B, Zhou P, Qian L, Jia T, et al. Exaggerated inflammation, impaired host defense, and neuropathology in progranulin-deficient mice. J Exp Med. 2010;207(1): 117–28.
49. Filiano AJ, Martens LH, Young AH, Warmus BA, Zhou P, Diaz-Ramirez G, et al. Dissociation of frontotemporal dementia-related deficits and neuroinflammation in progranulin haploinsufficient mice. J Neurosci. 2013;33(12):5352–61.

50. Hosler BA, Siddique T, Sapp PC, Sailor W, Huang MC, Hossain A, et al. Linkage of familial amyotrophic lateral sclerosis with frontotemporal dementia to chromosome 9q21–q22. JAMA. 2000;284:1664–9.
51. Morita M, Al-Chalabi A, Andersen PM, Hosler B, Sapp P, Englund E, et al. A locus on chromosome 9p confers susceptibility to ALS and frontotemporal dementia. Neurology. 2006;66:839–44.
52. DeJesus-Hernandez M, Mackenzie IR, Boeve BF, Boxer AL, Baker M, Rutherford NJ, et al. Expanded GGGGCC hexanucleotide repeat in noncoding region of C9ORF72 causes chromosome 9p-linked FTD and ALS. Neuron. 2011;72:245–56.
53. Renton AE, Majounie E, Waite A, Simón-Sánchez J, Rollinson S, Gibbs JR, et al. A hexanucleotide repeat expansion in C9ORF72 is the cause of chromosome 9p21-linked ALS-FTD. Neuron. 2011;72:257–68.
54. Galimberti D, Scarpini E. Clinical phenotypes and genetic biomarkers of FTLD. J Neural Transm. 2012;119(7):851–60.
55. Snowden JS, Rollinson S, Thompson JC, Harris JM, Stopford CL, Richardson AM, et al. Distinct clinical and pathological characteristics of frontotemporal dementia associated with *C9ORF72* mutations. Brain. 2012;135(Pt 3):693–708.
56. Floris G, Borghero G, Cannas A, Di Stefano F, Costantino E, Murru MR, et al. Frontotemporal dementia with psychosis, parkinsonism, visuo-spatial dysfunction, upper motor neuron involvement associated to expansion of *C9ORF72*: a peculiar phenotype? J Neurol. 2012; 259(8):1749–51.
57. Calvo A, Moglia C, Canosa A, Cistaro A, Valentini C, Carrara G, et al. Amyotrophic lateral sclerosis/frontotemporal dementia with predominant manifestations of obsessive-compulsive disorder associated to GGGGCC expansion of the *c9orf72* gene. J Neurol. 2012;259(12): 2723–5.
58. Arighi A, Fumagalli GG, Jacini F, Fenoglio C, Ghezzi L, Pietroboni AM, et al. Early onset behavioural variant frontotemporal dementia due to the C9ORF72 hexanucleotide repeat expansion: psychiatric clinical presentations. J Alzheimers Dis. 2012;31:447–52.
59. Mahoney CJ, Beck J, Rohrer JD, Lashley T, Mok K, Shakespeare T, et al. Frontotemporal dementia with the *C9ORF72* hexanucleotide repeat expansion: clinical, neuroanatomical and neuropathological features. Brain. 2012;135(Pt 3):736–50.
60. Murray ME, DeJesus-Hernandez M, Rutherford NJ, Baker M, Duara R, Graff-Radford NR, et al. Clinical and neuropathologic heterogeneity of c9FTD/ALS associated with hexanucleotide repeat expansion in C9ORF72. Acta Neuropathol. 2011;122:673–90.
61. Galimberti D, Fenoglio C, Serpente M, Villa C, Bonsi R, Arighi A, et al. Autosomal dominant frontotemporal lobar degeneration due to the C9ORF72 hexanucleotide repeat expansion: late-onset psychotic clinical presentation. Biol Psychiatry. 2013;74:384–91.
62. Levine TP, Daniels RD, Gatta AT, Wong LH, Hayes MJ. The product of C9orf72, a gene strongly implicated in neurodegeneration, is structurally related to DENN Rab-GEFs. Bioinformatics. 2013;29(4):499–503.
63. Renoux AJ, Todd PK. Neurodegeneration the RNA way. Prog Neurobiol. 2012;97:173–89.
64. Reddy K, Zamiri B, Stanley SY, Macgregor RB, Pearson CE. The disease-associated r(GGGGCC)n repeat from the C9orf72 gene forms tract length-dependent uni- and multimolecular RNA G-quadruplex structures. J Biol Chem. 2013;288:9860–6.
65. Mori K, Weng SM, Arzberger T, May S, Rentzsch K, Kremmer E, et al. The C9orf72 GGGGCC repeat is translated into aggregating dipeptide-repeat proteins in FTLD/ALS. Science. 2013;339(6125):1335–8.
66. Zu T, Gibbens B, Doty NS, Gomes-Pereira M, Huguet A, Stone MD, et al. Non-ATG-initiated translation directed by microsatellite expansions. Proc Natl Acad Sci U S A. 2011;108: 260–5.
67. Ash PE, Bieniek KF, Gendron TF, Caulfield T, Lin WL, Dejesus-Hernandez M, et al. Unconventional translation of C9ORF72 GGGGCC expansion generates insoluble polypeptides specific to c9FTD/ALS. Neuron. 2013;77:639–46.

68. Almeida S, Gascon E, Tran H, Chou HJ, Gendron TF, Degroot S, et al. Modeling key pathological features of frontotemporal dementia with C9ORF72 repeat expansion in iPSC-derived human neurons. Acta Neuropathol. 2013;126:385–99.
69. Van Deerlin VM, Sleiman PM, Martinez-Lage M, Chen-Plotkin A, Wang LS, Graff-Radford NR, et al. TMEM106B regulates progranulin levels and the penetrance of FTLD in GRN mutation carriers. Neurology. 2011;76:467–74.
70. Finch N, Carrasquillo MM, Baker M, Rutherford NJ, Coppola G, Dejesus-Hernandez M, et al. http://www.ncbi.nlm.nih.gov/pubmed/21178100. TMEM106B regulates progranulin levels and the penetrance of FTLD in GRN mutation carriers. Neurology. 2011;76(5):467–74.
71. Cruchaga C, Graff C, Chiang HH, Wang J, Hinrichs AL, Spiegel N, et al. Association of TMEM106B gene polymorphism with age at onset in granulin mutation carriers and plasma granulin protein levels. Arch Neurol. 2011;68:581–6.
72. Jin SC, Pastor P, Cooper B, Cervantes S, Benitez BA, Razquin C, et al. Pooled-DNA sequencing identifies novel causative variants in PSEN1, GRN and MAPT in a clinical early-onset and familial Alzheimer's disease Ibero-American cohort. Alzheimers Res Ther. 2012;4(4):34.
73. Lord J, Turton J, Medway C, Shi H, Brown K, Lowe J, et al. Next generation sequencing of CLU, PICALM and CR1: pitfalls and potential solutions. Int J Mol Epidemiol Genet. 2012;3:262–75.
74. Guerreiro RJ, Lohmann E, Brás JM, Gibbs JR, Rohrer JD, Gurunlian N, et al. Using exome sequencing to reveal mutations in TREM2 presenting as a frontotemporal dementia-like syndrome without bone involvement. JAMA Neurol. 2013;70:78–84.

Chapter 8
Alzheimer's Disease and Frontotemporal Lobar Degeneration: Mouse Models

Janet van Eersel, Fabien Delerue, Lars M. Ittner, and Yazi D. Ke

Abstract Genetically modified mouse models have been instrumental in deciphering pathomechanisms in a large variety of human conditions. Similarly, transgenic and knockout mice have contributed to understanding neurodegenerative processes in Alzheimer's disease (AD) and frontotemporal lobar degeneration (FTLD). While the first models for AD and FTLD, based on mutations in APP and tau, respectively, have been generated more than a decade ago, recent years have seen the identification of new genes involved in the disease. This led to the generation of a large number of new transgenic mouse models for FTLD. This chapter provides an overview of APP and tau-based mouse models of AD and FTLD and discusses in detail the more recent FTLD models expressing novel disease genes.

Keywords Mouse model • APP • Tau • TDP-43 • FUS

Different Methods to Genetically Modify Mice

Transgenesis techniques to generate mouse models of disease rely on both gene transfer methods as well as methods to manipulate the early mouse embryo [1]. To date, the most commonly used technique involves microinjection of DNA

J. van Eersel, PhD • Y.D. Ke, PhD
Dementia Research Unit, School of Medical Sciences, University of New South Wales, Sydney, NSW, Australia

F. Delerue
Transgenic Animal Unit, School of Medical Sciences, University of New South Wales, Sydney, NSW, Australia

L.M. Ittner (✉)
Dementia Research Unit and Transgenic Animal Unit, School of Medical Sciences, University of New South Wales, Wallace Wurth Building, Sydney, NSW 2052, Australia
e-mail: l.ittner@unsw.edu.au

D. Galimberti, E. Scarpini (eds.), *Neurodegenerative Diseases*,
DOI 10.1007/978-1-4471-6380-0_8, © Springer-Verlag London 2014

constructs into the pronucleus of a developing zygote, leading to random integration of a transgene into the endogenous DNA [2]. This then produces a "transgenic animal" that has a foreign gene(s) stably incorporated into its genome through human intervention. This integrated recombinant double-stranded DNA is called a "transgene."

Over the years, the development of more sophisticated models has allowed for better control of transgene expression, both temporally and spatially. This includes both inducible and conditional mouse models. Inducible mouse models enable the study of transgene expression in a strictly regulated manner, as they drive transgene expression exclusively upon induction, by either the presence or absence of a drug, in a dose-dependent manner. This allows researchers to overcome some of the problems associated with constitutive transgene expression, such as embryonic lethality. Conditional models involve the generation of mice with altered gene expression in a cell specific manner, through the expression of recombinase enzymes, which are under the control of a selected promoter, that can remove, invert, or translocate DNA segments.

Site-specific manipulation of the genome (gene targeting) allows for the disruption of a specific gene (knockout approach) or the insertion of a transgene in a defined locus (knock-in approach). Very recently, targeted transgenesis has been introduced, which relies on a core technology based on the use of engineered nucleases, such as zinc finger nucleases (ZfN) [3] or transcription activator-like effector nucleases (TALEN) [4]. This new technology enables investigators to manipulate virtually any gene in a diverse range of cell types and organisms with extreme precision (single base pair). Targeted transgenesis, used either for stable overexpression of a transgene or for disruption of endogenous genes, ultimately remains the most powerful tool to understand the mechanisms underlying physiological processes and their pathological counterparts.

Mouse Models of Alzheimer's Disease

The past two decades have seen the generation of a large number of transgenic mouse models of AD, with a focus on amyloid-β (beta) (Aβ [beta])-forming models. These have assisted in a large number of studies investigating mechanisms underlying neuronal dysfunction and neurodegeneration in AD, as well as in developing and testing novel treatments. Aβ (beta)-forming transgenic mouse models have been extensively reviewed before [e.g., [5]]. Therefore, this part of the chapter will provide a rather general overview and highlight only some discoveries made using AD mouse models.

Amyloid-β (Beta) Precursor Protein (APP) Models

Intensive efforts have been made to develop transgenic mouse models that recapitulate the pathology and symptoms of AD over the past decades. While overexpression of human nonmutant APP did not result in plaque formation and memory deficits, it was the identification of pathogenic mutations in APP in familial cases of AD that paved the way for generating the first disease models [6]. Since then, expression of human mutant APP reproduced Aβ (beta) plaque pathology in a large number of transgenic mouse models [5]. In most models, expression of mutant APP results in the production of Aβ (beta) throughout the brain with plaque formation, affecting memory performance of mice in different test paradigms, such as the Morris water maze. APP transgenic models have also been the basis for showing a prion-like transfer of Aβ (beta) pathology between APP transgenic mice in a strain-dependent manner [7].

While initial studies did not report an overt neuronal loss, a limited number of subsequent studies of established lines reported a decrease in numbers of neurons in certain brain areas [8, 9]. However, the absence of pronounced neuronal loss remains a limitation of Aβ (beta)-forming APP transgenic mice.

To determine if loss of APP function contributes to the development of AD, APP knockout mice have been generated. However, their phenotypes are rather mild and possibly due to developmental anomalies [10]. Interestingly, early postnatal death of double knockout mice with deletion of APP and APLP2, the latter belonging to the same protein family, suggests a functional overlap between the family members during development [11]. APP-deficient mice have contributed to the understanding of the possible physiological functions of APP, some of which have implications for the disease [12–14].

In summary, APP transgenic mice have been instrumental in reproducing aspects of AD pathology in vivo and in deciphering mechanisms underlying disease. Over the past decades, APP transgenic mice have become a central in vivo tool in studying pathomechanisms and developing treatments for AD.

Combinatorial Models

In an attempt to accelerate Aβ (beta) pathology onset and progression and to more closely model the human pathology, mutant APP transgenic mice have been crossed with other mutation-harboring mice. For instance, mutations in the presenilin-encoding *PSEN* genes alter the activity of the γ (gamma)-secretase complex which presenilins are part of. Expression of mutant PSEN1 in mice crossed with Aβ (beta)-forming APP transgenic mice resulted in accelerated Aβ (beta) formation and earlier onset of behavioral deficits as well as neuronal loss [15, 16]. Interestingly, the effects of mutant PSEN was even more pronounced in the absence of the murine PSEN, achieved by a mutant human *PSEN1* knock-in approach [17]. Conversely, reduced β (beta)-secretase activity in BACE-deficient mice reduced Aβ (beta) formation and ameliorated behavioral deficits when crossed on an Aβ (beta)-forming APP transgenic strain [18–20], while overexpression of BACE on an APP background increased pathology [21].

Carriers of the APOEε (epsilon)4 allele have a 20-fold increased risk of developing AD, making it the number one risk gene for developing sporadic late-onset AD [22]. In support of a role for ApoE in Aβ (beta) pathology, crossing APP transgenic mice on a ApoE$^{-/-}$ background reduced both Aβ (beta) levels and its deposition [23]. Conversely, expressing human ApoE4 in APP transgenic mice, by viral gene delivery, increased pathology [24].

Aβ (beta)-forming APP mice were used to provide the first in vivo evidence for the amyloid cascade hypothesis that places Aβ (beta) upstream of tau pathology and neurodegeneration in the sequence of pathogenic events. Accordingly, crossing of APP transgenic mice with human mutant tau-expressing mice resulted in increased neurofibrillary tangle (NFT) formation [25]. A similar result has been achieved by injecting synthetic aggregated Aβ (beta)1-42 into brains of P301L mutant tau transgenic pR5 mice [26].

The central role of tau in AD development, particularly in mediating neuronal deficits induced by Aβ (beta), has been shown when APP transgenic mice were crossed on a tau-deficient background [27]. This approach prevented premature mortality and behavioral deficits associated with Aβ (beta) formation, though the levels of Aβ (beta) and numbers of plaques were unchanged. In this context, we showed that tau mediates Aβ (beta)-induced excitotoxicity by controlling Fyn levels at the postsynapse and sensitizing NMDA receptors to become easily hyperexcited [28]. This work provides the first evidence for a non-axonal function of tau in the dendritic compartment of neurons [29], which has since been supported by other studies since [30, 31].

Taken together, combinatorial approaches using APP transgenic mice together with additional mutant strains have provided exciting new insight into the pathogenesis of AD. Although only a selected small number of studies have been presented here, it is reasonable to expect that combinatorial approaches using APP-based AD mouse models will continue to extend our understanding of AD.

Mouse Models of Frontotemporal Lobar Degeneration

Frontotemporal lobar degeneration (FTLD; also referred to as frontotemporal dementia [FTD]) umbrellas a large number of related neurodegenerative conditions with overlapping clinical symptoms. This is paralleled by an increasing number of proteins that have been found to be present in deposits in FTLD brains as well as the identification of more and more genes carrying pathogenic mutations, further distinguishing subforms of FTLD [32]. This chapter will discuss transgenic mouse models generated by expressing or deleting different genes, with an emphasis on more recent models. Tau models, some of which have been around for many years, will be addressed rather generally and by highlighting some of the recent findings in these mice.

Tau Models

While tau deposits in neurons together with the formation of extracellular Aβ (beta) plaques in AD patient brains, tau forms inclusions in the absence of overt Aβ (beta) pathology in FTLD. To model the tau pathology of AD and FTLD in mice, the first transgenic strain was generated to express the longest human isoform of tau without mutations in neurons [33]. These mice presented with accumulation of hyperphosphorylated forms of tau, resembling a pre-tangle state, but they failed to reproduce NFT formation. Interestingly, aged mice of this tau transgenic line developed motor deficits together with a Wallerian degeneration of axonal tracks in the spinal cord, indicating that pre-tangle hyperphosphorylated tau suffices to impair neuronal function and integrity without deposition.

It took close to five more years after the first tau model had been published, until transgenic expression of human tau carrying a pathogenic FTDP-17 mutation, P301L, achieved NFT formation in vivo [34]. These mice are characterized by severe motor and behavioral deficits, axonal degeneration, and early death, resembling aspects of the human disease. Since the generation of this first mutant tau-expressing mouse model, many additional lines have been generated that recapitulated different aspects of the human condition [5]. Interestingly, neuronal loss that characterizes the human disease has not been reproduced in the earlier mutant tau transgenic mice. But eventually, this has been achieved, when mice expressing distinct mutations (N279K [35] or P301S [36, 37]) using conventional neuronal promoters or particularly high levels of P301L mutant human tau using an inducible modified CMV promoter [38] showed pronounced neuronal loss. These lines are characterized by early-onset NFT formation. Neuronal loss has also been achieved in an elegant transgenic model expressing a mutant but truncated tau that is limited to the microtubule-binding repeats and characterized by rapid tau fibril formation and deposition [39]. This model used an inducible modified CMV promoter too and in combination with a complementary model that expresses the same truncated tau variant but with inclusion of two aggregation-preventing point mutations (I277P and I308P) forms an excellent in vivo tool to study tau fibril formation and test anti-aggregation drugs [39].

Since tau pathology in human FTLD is not limited to neurons, transgenic mouse models with non-neuronal mutant tau expression have been generated [40, 41]. Interestingly, both expression in astrocytes and in oligodendrocytes resulted in neuronal dysfunction and axonal degeneration. This is possibly due to impairment of glia in supporting neuronal function and integrity.

Mutant tau transgenic mice have become a highly valuable tool for studying pathomechanisms underlying tau pathology and neurodegeneration in FTLD but also in AD. Accordingly, transgenic mice are currently extensively used to investigate the prion-like disease progression hypothesis for tau, which includes release of distinct tau species from diseased neurons that are then taken up by healthy neurons to form a seed for disease propagation [42]. So far, it has been shown that tau pathology can be transferred from a mutant tau transgenic line with NFT formation to a transgenic strain that expresses nonmutant human tau and does not form NFTs unless inoculated with brain extracts from NFT-forming mice [43] or human patient brains with tau pathology [44] by stereotaxic injection. Furthermore, inducible mutant tau expression limited to a distinct brain area (entorhinal cortex) leads to NFT formation in connected brain areas (hippocampus) as mice age [45].

Mutant tau transgenic mice are also regularly used for preclinical drug development and testing. For instance, more recently, several groups have developed vaccination strategies targeting pathological tau, either by active or passive immunization [46–49]. Each of these studies used different mutant tau transgenic mouse lines to show efficacy and safety of this approach, providing the preclinical evidence needed to further this approach to clinical trials. Similarly, mutant tau transgenic mice have been used to determine the effects of compounds on different aspect of tau pathology [37, 50].

Taken together, it was the generation of mutant tau transgenic mice that provided in vivo evidence that pathogenic FTLD mutations accelerate tau aggregate formation and deposition and drive neuronal dysfunction and loss. Furthermore, mutant tau transgenic mice are important tools for studying pathomechanisms in vivo and to develop and test new therapeutic approaches. Finally, although the pathogenic mutations expressed in these lines originate from FTLD patients, tau transgenic mice are also valuable for studying tau-related aspects of AD, given the similarity of tau pathology in AD and FTLD.

TAR DNA-Binding Protein 43 (TDP-43) Models

In 2006, Neumann and colleagues identified in a groundbreaking publication TDP-43 as the major component of, until then, unidentified ubiquitin-positive deposits in FTLD [51]. Moreover, they showed that similar deposits in amyotrophic lateral sclerosis (ALS) (also referred to as Lou Gehrig's disease or motor neuron disease [MND]) are also made up of TDP-43. TDP-43 is a nuclear protein with two RNA/DNA binding motifs. Consistent with these domains, TDP-43 is involved in RNA/DNA-related processes in cells, including RNA trafficking, RNA splicing, and promoter binding [52]. In disease, TDP-43 accumulates in the cytoplasm due to unknown reasons and undergoes secondary modifications, such as truncation, phosphorylation, and ubiquitination, eventually leading to the formation of aggregates [53].

Similar to tau transgenic mice, the identification of mutations in the TDP-43-encoding *TARDBP* gene has paved the way for the generation of a number of transgenic mouse models with mutant TDP-43 expression. Furthermore, non-disease mutants of TDP-43 with deletion of function domains have been expressed in mice.

The first TDP-43 mouse model published in 2009 expressed human TDP-43 carrying the A315T mutation under the murine prion protein promoter to generate the Prp-TDP43^{A315T} mice [54]. These mice have an approximate threefold expression over endogenous TDP-43 with highest expression present in the brain and spinal cord. Ubiquitination of proteins in layer V neurons of the cortex concomitantly occurs with loss of nuclear staining of TDP-43 in selective neurons in these mice. Reactive gliosis is also present in this region of degenerating neurons.

This initial TDP-43 transgenic line [54] was followed by several new models generated over the past years [55–59]. Wils and colleagues expressed nonmutant human TDP-43 under the neuronal murine Thy1 promoter to generate the

TDP-43WT lines TAR4 and TAR6 [55]. Hemizygous TAR4 and TAR6 have 2.8- and 1.9-fold, and homozygous TAR4/4 and TAR6/6 have 5.1- and 3.8-fold expression over endogenous TDP-43. These mice have nuclear and cytoplasmic inclusions in cortical layer V neurons that are ubiquinated and phosphorylated as well as a marked astrogliosis. The limited neuronal loss observed in these mice correlated with expression levels of TDP-43. In addition, homozygous TAR4 have an accumulation of cytoplasmic full length TDP-43 as well as the 25 kDa and 35 kDa C-terminal fragments. Phenotypically, these mice exhibit complex motor impairments, with hind limb clasping, reduced footstep length, reduced motor performance on the Rota-Rod, as well as reduced survival rate with disease onset and severity dependent on TDP-43 expression levels.

Xu and colleagues expressed nonmutant human TDP-43 under the murine prion protein promoter to generate the TDP-43_{PrP} with a 1.9–2.5-fold expression over endogenous TDP-43 [56]. An increased human TDP-43 mRNA level was observed with a concomitant decrease in mouse TDP-43 mRNA levels. These mice produce ~25 kDa C-terminal TDP-43 fragments, which are urea insoluble, as well as phosphorylated and ubiquinated cytoplasmic inclusions, reactive gliosis, and argyrophilic degenerating neurites and neurons in the spinal cord. Interestingly, these mice also have abnormal clustering and degeneration of mitochondria in their spinal cord neurons. TDP-43_{PrP} mice display lower body weights compared to wild-type littermates at 14 days, together with hindlimb clasping, body tremors, and a "swimming" gait at 21 days. Their survival is limited as they die between 1 and 2 months of age.

Swarup and colleagues generated three TDP-43 transgenic mice (nonmutant human TDP-43, TDP-43^{A315T} and TDP-43^{G348C}) from DNA subcloned from *TARDBP* bacterial artificial chromosomes containing the endogenous Δ4 kB promoter [59]. These mice present with an approximately threefold overexpression of transgenic TDP-43 over the endogenous protein. Significantly more ~25 kDa and 35 kDa C-terminal fragments were observed in TDP-43^{A315T} and TDP-43^{G348C} compared to nonmutant TDP-43 expressing mice. Ubiquitination of cytoplasmic TDP-43 was observed only in the mutant TDP-43 lines. Abnormal aggregates containing peripherin and neurofilament proteins were also present in TDP-43^{G348C} mice. In addition, gliosis and neuroinflammation were observed in all lines. Furthermore, all lines presented with cognitive and motor deficits in the passive avoidance test, Barnes maze test, and Rota-Rod at 7–10 months with these impairments being most severe in the TDP-43^{G348C} line. Interestingly, they revealed that there is significant increase of GFAP promoter activity or astrogliosis before the onset of behavioral impairments.

Igaz and colleagues generated transgenic mice with inducible overexpression of either nonmutant human TDP-43 (hTDP-43 WT) or human TDP-43 with mutated nuclear localization signal (hTDP-43-ΔNLS) [57]. Mutation of the NLS prevents TDP-43 from entering the nucleus, and, hence, it accumulates in the cytoplasm [60]. Neuronal expression was achieved by using a CaMK2α promoter to drive tet-off rTA and a tetracycline responsive promoter to drive hTDP43 expression. hTDP-43 WT mice had an 8- to 9-fold expression over endogenous TDP-43 and hTDP-43-ΔNLS mice 0.4- to 1.7-fold, respectively. Doxycycline treatment started at birth to suppress expression during postnatal brain development was removed at weaning (3 weeks of age) and mice were analyzed at various time points after doxycycline removal. Both models present with urea-insoluble TDP-43 with no concomitant presence of C-terminal fragments. In addition, ubiquitinated and phosphorylated TDP-43 aggregates were found to be present in hTDP-43-ΔNLS mice. Significant neuronal loss was observed in the dentate gyrus of both lines with the hTDP-43-ΔNLS mice having a more acute and severe dentate gyrus degeneration. The presence of axonal loss and gliosis of the corticospinal tract of hTDP-43-ΔNLS mice occur in a time-dependent manner relative to the developments of motor deficits.

Since the abnormal localization of TDP-43 in disease means that the protein is depleted from the nucleus, TDP-43 might not be able to execute its normal functions (=loss of function). To test this in vivo, Kraemer and colleagues employed a gene trap insertion strategy to generate mice lacking TDP-43 [61]. Heterozygous mice are viable in contrast to homozygous mice, which is embryonically lethal. Heterozygous ($Tardbp^{+/-}$) mice have reduced grip strength with no reportable differences in pathology observed.

Progranulin (PGRN) Models

Mutations in the progranulin (PGRN) gene have been shown to cause tau-negative, ubiquitin-positive, and TDP-43-positive FTLD [62, 63]. The majority of these mutations are known to cause messenger RNA (mRNA) instability (resulting in degradation), while other mutations can cause loss of the entire mutant allele [63], cause prematurely truncated protein [63], or result in the generation of mutant PGRN protein that cannot be secreted efficiently [64] or appropriately cleaved [65]. Therefore, through a variety of mechanisms, these mutations all result in either reduced PGRN levels or loss of PGRN function. It is for this reason that PGRN knockout mice have been used to study this particular disorder.

A variety of PGRN knockout strains have been generated [66–70]. Except for one report [71], all of these knockout strains produce offspring with genotypes at an expected Mendelian ratio, suggesting that loss of PGRN does not impair embryonic development and/or survival. One common feature of all of these strains is that aged, homozygote mice all develop severe astrogliosis and microgliosis that increases with age (generally first detected around 12 months of age). Hence, neuroinflammation may play a role in the disease process. Interestingly, PGRN homozygote knockout mice react less efficiently and with more severe inflammation to bacterial listeria infections [67], and both PGRN-deficient microglia and macrophages are more cytotoxic to cultured neurons [67, 69]. In addition to this, hippocampal slices from PGRN homozygote knockout mice show greater neuronal sensitivity to glucose and oxygen starvation [67]. This suggests that FTLD-PGRN may arise from a combination of deregulated inflammation as well as increased neuronal vulnerability to certain stressors.

In all but one strain [68], homozygote PGRN knockout mice have been found to display significantly more ubiquitinated structures in various brain regions by as early as 7 months (ranging from 7 to 18 months), which increase with age. In support of a compromised ubiquitin-proteasome system, increased p62 and cathepsin D (markers of autophagy and lysosomes) were found in addition to increases in neuronal ubiquitin in PGRN knockout mice [70]. These pathological changes are common features of FTLD-TDP but are also associated with aging. Furthermore, in three of the PGRN knockout strains, levels of lipofuscin, a marker of cellular aging, were significantly increased (throughout the brain and also in the liver in one strain) by as early as 8 months. Hence, PGRN knockout mice may undergo accelerated aging, thereby potentially contributing to the disease process. Interestingly, levels of PGRN progressively increased in the brains of aging wild-type animals, suggesting a role for PGRN in aging [71]. However, no neuronal loss or markers of apoptosis have been observed in any of the knockout strains, though some lines have shorter life spans [70, 72].

Although PGRN mutations are associated with TDP-43 neuropathology in humans, it is not clear whether this is also the case in PGRN knockout mice. To date, only some pathologically phosphorylated TDP-43 has been identified in the brains of two strains [67, 70, 73]. It therefore remains unclear what role PGRN mutations play in the development of TDP-43 pathology, though it does not appear that loss of PGRN alone suffices to cause TDP-43 relocalization or aggregation.

The behavioral assessment of different PGRN knockout lines has produced variable results. This could be the result of variation in genetic background or differences in protocols and equipment used. PGRN knockout mice do not have any significant motor impairments (although reduced muscle strength has been reported by Ghoshal and colleagues); however, there have been multiple reports of reduced social engagement and aggression [68, 72, 73] and depression-like behavior and disinhibition [73], which mimics several major behavioral hallmarks of FTLD. In addition, aged PGRN knockout mice show reduced performance during Morris water maze testing [70, 72, 73] and novel object testing [68], suggesting late-onset learning and memory impairments. Although the mechanism by which PGRN deficiency causes these behavioral phenotypes is unclear, Petkau and colleagues (2012) utilized electrophysiological recordings to demonstrate that hippocampal slices from PGRN homozygote knockout mice display reduced postsynaptic responsiveness and occasional LTP dysfunction. Furthermore, CA1 pyramidal neurons showed reduced dendritic length and reduced spine density. Therefore, synaptic dysfunction may play a role in the disease process underlying FTLD-PGRN.

It should be noted that the majority of studies discussed above utilized homozygote PGRN knockout mice, despite the fact that PGRN mutations cause haploinsufficiency in humans. For this reason, it is important to highlight some results obtained from heterozygote PGRN knockout mice [74]. These mice express approximately 50 % less PGRN mRNA and protein (and were maintained on two different genetic backgrounds), but unlike homozygote PGRN knockout mice, they do not develop any significant astrogliosis, microgliosis, and lipofuscinosis or show any electrophysiological changes, nor do they have any motor impairments or memory and learning impairments. Despite this, these animals (regardless of the genetic background) still show social and emotional dysfunction.

In summary, PGRN knockout mice recapitulate a number of hallmark features of FTLD-TDP43, including neuroinflammation, ubiquitinated aggregates, and behavioral impairments. However, the exact role of TDP-43 in this disease and the effects of PGRN haploinsufficiency versus homozygous deficiency remain to be determined.

Valosin-Containing Protein (VCP) Models

Mutations in the valosin-containing protein (VCP) gene are known to cause the multisystem degenerative disorder inclusion body myopathy associated with Paget's disease of the bone and frontotemporal dementia (IBMPFD) [75]. Although muscle weakness and myopathy are the most common clinical features of this disorder, approximately 30 % of patients also develop language and behavioral impairments typical of FTLD [76]. Furthermore, TDP-43- and ubiquitin-positive inclusions are found in both the brain and muscle of IBMPFD patients. Interestingly, some reports also link VCP mutations to amyotrophic lateral sclerosis [77, 78]. Over 20 mutations have been identified in *VCP*, all of which are thought to alter the 3D structure of VCP and thereby perturb the interactions between VCP and its various substrates [79]. Substitution of arginine 155 to histidine (R155H) is the mutation most commonly associated with IBMPFD. It is for this reason that the majority of mouse models utilize this particular mutation.

To develop an animal model of IBMPFD, a number of groups have generated transgenic mice that express mutant VCP [80–84]. Although these strains all express a similar mutant protein, there are a number of inherent differences among the strains. For example, because mouse VCP differs from the human protein by only one amino acid, some groups chose to express human mutant VCP in the mouse model, whereas other models express mutant mouse VCP; various promoters have been used to generate mice that overexpress the mutant protein exclusively in the muscle [81], the brain [80], or ubiquitous expression in all tissues [82]; while other groups generated knock-in mice that express mutant VCP at levels similar to that of the endogenous protein [83, 84].

Despite these inherent differences, however, all mutant VCP mouse strains have been reported to develop VCP-negative, TDP-43-positive, and ubiquitin-positive aggregates. These develop in regions where the mutant protein is expressed, i.e., the muscle, brain, and spinal cord. In heterozygote animals these aggregates appear at around 10–15 months in the muscle and the spinal cord, and at 14–20 months in the brain; and in homozygote mice [84] TDP-43 aggregates were observed as early as 15 days in the muscle, brain, and spinal cord. In some strains, cytoplasmic and nuclear clearance of TDP-43 was observed, as well as insoluble and high molecular weight TDP-43 species [80, 82, 85]. In one particular strain, TDP-43 aggregates were observed to co-localize with the stress granule marker TiA-1, and overall levels of TiA-1 were increased, suggesting an increased stress response, which could potentially alter mRNA transport and translation. Altered stress granule dynamics and/or altered mRNA metabolism may therefore play a role in the disease processes associated with TDP-43 proteinopathies. Despite the presence of TDP-43 aggregates, however, none of the strains show any sign of neurodegeneration in the brain [80, 82, 83], although loss of motor neurons in the spinal cord has been reported [85].

Other pathological features commonly observed in these mice include significant increases in levels of general protein ubiquitination [80, 81, 84, 85] and upregulation of markers of autophagy [83–85] in the muscle, brain, and spinal cord. Combined with the knowledge that VCP is known to play a role in regulating ubiquitin degradation of a number of proteins, this data suggest that dysfunctional protein degradation and accumulation of ubiquitinated proteins may play a role in the development of this disorder. In addition to this, high molecular weight species of TDP-43 were found to pull down with VCP, suggesting a direct interaction between VCP and high molecular weight TDP-43 isoforms in these mice [80]. One possible explanation for this interaction is that VCP may be trying to direct TDP-43 to the proteasome for degradation, and that disruptions to this interaction may cause TDP-43 to accumulate in the cytoplasm and eventually aggregate.

IBMPFD is most commonly characterized by myopathy. In accordance with this, in all the mutant VCP mice strains that express the transgene in muscle tissue, significant pathology is observed. This includes the following: vacuoles, disordered architecture, variation in muscle fiber size, and swollen mitochondria [81–85]. On average, these features were observed at around 6–15 months of age; however, in mice that were bred to homozygosity, muscle abnormalities were already observed after 15 days. Radiographic and biochemical bone deformities consistent with Paget's disease are also commonly observed in IBMPFD. Similar characteristics have been reproduced in the mutant VCP mice, including loss of bone structure, decreased bone density, hypomineralization, and sclerotic lesions at around 13–16 months of age [82–84]. Therefore, these mice recapitulate the wide range of pathological features associated with IBMPFD within the muscle, brain, and bone.

In general, all mutant VCP mouse strains show signs of muscle weakness and reduced Rota-Rod performance, which is in accordance with the clinical presentation in human patients [81–84]. Although some reports show weight loss and reduced survival in certain strains [82, 85], particularly in the homozygote mice which only survive to 14–21 days [84], this has not been observed in all strains. Custer and colleagues (2010) reported increased anxiety in these mice in the elevated zero maze and reduced performance in the novel object test, while other strains did not show any memory deficits [82–84]. Rodriguez-Ortiz and colleagues (2013) used a neuron-specific promoter to overexpress mutant VCP specifically in the forebrain [80]. These mice showed no difference in swim speed and distance in the Morris water maze but showed significant impairment in the probe trial, as well as impairments in object recognition testing, indicating learning and memory impairments. Furthermore, in these studies, higher expressing mutant VCP mice were shown to have greater cognitive deficits than lower expressing mice, and both lines showed greater impairment with age, suggesting that neuronal, mutant VCP expression impairs cognition in an age- and dose-dependent manner in these mice.

In summary, mutant VCP mice develop muscle and brain pathology as well as bone abnormalities that closely match with that observed in human IBMPFD patients. In addition, the spinal cord pathology closely matched that observed in human ALS patients. This therefore raises the question whether inclusion body myopathy, Paget's disease, ALS, and FTLD share a common underlying mechanism. Because these mice develop ubiquitin-positive, TDP-43 aggregates and show relocalization of TDP-43, they can be used not only to study IBMPFD but also mechanisms underlying the development of TDP-43 pathology in general, particularly the neuron-specific expressing mice.

Charged Multivesicular Body Protein 2B (CMBP2B) Models

Although rare, mutations in the charged multivesicular body protein 2B (*CHMP2B)* gene are associated with familial forms of FTLD that display ubiquitin- and p62-positive inclusions that are negative for tau, FUS, and TDP-43 [86]. All mutations identified have been shown to cause a loss of the C-terminus of CHMP2B; therefore, the disease pathogenesis could be caused by either loss of normal CHMP2B function or, more specifically, loss of the CHMP2B C-terminus. To investigate this in greater depth, Ghazi-Noori and colleagues (2012) generated both wild-type (CHMP2B^{wt}) and C-terminally truncated (CHMP2B^{trunc}) CHMP2B transgenic mice, as well as CHMP2B knockout mice [87]. Initially, both the CHMP2B transgenic and knockout mice showed normal survival curves; however, after 500 days the CHMP2B^{trunc} mice showed increased mortality. Interestingly, the CHMP2B^{trunc} mice were shown to develop p62- and ubiquitin-positive inclusions (but TDP-43- and FUS-negative) that were absent in the CHMP2B^{wt} and knockout mice, suggesting that the formation of these inclusions was dependent on the expression of mutant CHMP2B. Furthermore, these inclusions were absent in the knockout mice, this suggests that the pathology is not caused by a loss of function but rather a gain of toxic function. These inclusions were found in a number of brain regions and motor neurons in the spinal cord, as early as 6 months, and were found abundantly by 18 months of age. In addition to the formation of inclusions, the CHMP2B^{trunc} were also shown to develop astrogliosis and microgliosis, which was absent in the CHMP2B^{wt} and knockout mice. Interestingly, there were no signs of astrogliosis in the CHMP2B^{trunc} mice until 12 months of age and thus occurred only after the formation of inclusions, whereas reactive microglia were already present at 6 months of age and therefore coincided with the formation of inclusions. Another feature that was found to develop exclusively in the CHMP2B^{trunc} mice was axonal swellings. These swellings were apparent at 6 months and increased with age and were found to contain mitochondria as well as vesicles from the lysosomal and autophagy degradation pathways. This suggests that axonal dysfunction and impairment, and possibly even axonal transport, may play a role in the disease process underlying FTLD caused by *CMHP2B* mutations.

Fused in Sarcoma (FUS) Models

Mutations in the fused in sarcoma (FUS) gene have been identified not only in rare cases of FTLD [88] but also in a number of familial ALS cases [89, 90]. In contrast to the pathology in ALS, however, FUS-positive inclusions identified in cases of FTLD co-localize with the RNA-binding proteins TAF15 and EWS and are also ubiquitinated. The majority of FUS mutations cluster within the extreme C-terminus of the protein and interfere with the nuclear localization sequence residing in the C-terminus [91]. However, it has been demonstrated that overexpression of nonmutant FUS is sufficient to cause an aggressive phenotype and neuropathology in mice [92] as well as in rats [93].

Mitchell and colleagues (2013) generated both heterozygote ($FUS^{tg/+}$) and homozygote ($FUS^{tg/tg}$) mice overexpressing human nonmutant FUS in the brain, spinal cord, and testis [92]. Although the $FUS^{tg/tg}$ mice expressed higher levels of transgenic human FUS, this was found to decrease endogenous levels of murine FUS. $FUS^{tg/tg}$ mice were found to have a significantly shorter life span that only averaged 82 days, whereas $FUS^{tg/+}$ mice seem to have similar survival to that of nontransgenic littermates. Both the $FUS^{tg/tg}$ and (to a lesser extent) $FUS^{tg/+}$ mice show significant increases in overall levels of nuclear FUS. In addition, the $FUS^{tg/tg}$ mice also show significant increases in levels of cytoplasmic FUS. This matches with the histological finding of numerous perinuclear inclusions throughout the brain and spinal cord and cytoplasmic FUS within cortical neurons of end-stage $FUS^{tg/tg}$ mice, whereas only perinuclear inclusions are found (to a lesser extent) in the brains of $FUS^{tg/+}$ mice. Because nuclear levels of FUS are higher in the $FUS^{tg/tg}$ mice, this suggests that localization of FUS to the cytoplasm is dependent upon the levels of nuclear FUS. Despite the formation of the cytoplasmic FUS aggregates, no nuclear clearance of FUS was observed. In addition to FUS aggregates, significant increases in the levels of ubiquitin were observed in the $FUS^{tg/tg}$ mice, and to a lesser extent in the $FUS^{tg/+}$ mice. However, there was no obvious co-localization between FUS and ubiquitin. Furthermore, these FUS aggregates do not co-localize with EWS and TAF15, as is observed in FTLD. The molecular composition of the FUS aggregates in these mice therefore more closely resembles that observed in ALS rather than in FTLD.

Despite the formation of FUS aggregates, FUS transgenic mice showed no signs of neuronal loss or gliosis in the brain. However, in the spinal cords of $FUS^{tg/tg}$ mice (and to a lesser extent in $FUS^{tg/+}$ mice), a significant decrease was observed in the number of motor neurons, as well as astrogliosis and microgliosis. The $FUS^{tg/tg}$ mice also showed significant muscle atrophy and reduced muscle force, while $FUS^{tg/+}$ mice showed significant disorganization of muscle fibers. This suggests that overexpression of nonmutant FUS is more toxic to motor neurons than cortical neurons, particularly when it aggregates in the cytoplasm.

$FUS^{tg/tg}$ mice develop a severe early-onset motor phenotype, whereas $FUS^{tg/+}$ mice show no signs of motor dysfunction. By 4 weeks the $FUS^{tg/tg}$ mice developed a tremor and stilted gait, and from 4 weeks onward they failed to gain weight and showed significant impairments on the Rota-Rod. By 8 weeks of age, the mice showed significant decreases in general locomotor activity, clenching, and hindlimb paralysis.

In summary, these mice recapitulate various pathological and behavioral features of both ALS and FTLD patients, making them a good model to study these disorders. Exactly how overexpression of FUS causes these features, and whether a similar process occurs in the presence of mutant FUS, and whether the same process occurs in both ALS and FTLD remain to be determined [94].

Conclusion

Genetically modified mouse models are central to in vivo studies in AD and FTLD. Such models have provided insight and in some aspects a detailed understanding of pathological processes. With the identification of new proteins that form intracellular inclusion and novel pathogenic mutations in genes in FTLD, the number of different mouse models has dramatically increased. However, keeping in mind that each of the models reproduces and addresses only certain aspect of the human condition, it is likely that we will see a lot more transgenic models of even long-known candidates such as APP and tau. In addition, many of the new models of FTLD await being used in combination with other genetically modified strains to address complex pathological processes in vivo.

References

1. Brinster RL, Cross PC. Effect of copper on the preimplantation mouse embryo. Nature. 1972;238(5364):398–9.
2. Ittner LM, Gotz J. Pronuclear injection for the production of transgenic mice. Nat Protoc. 2007;2:1206–15.
3. Geurts AM, Cost GJ, Freyvert Y, Zeitler B, Miller JC, Choi VM, et al. Knockout rats via embryo microinjection of zinc-finger nucleases. Science. 2009;325(5939):433.
4. Sung YH, Baek IJ, Kim DH, Jeon J, Lee J, Lee K, et al. Knockout mice created by TALEN-mediated gene targeting. Nat Biotechnol. 2013;31:23–4.
5. Gotz J, Ittner LM. Animal models of Alzheimer's disease and frontotemporal dementia. Nat Rev Neurosci. 2008;9:532–44.
6. Games D, Adams D, Alessandrini R, Barbour R, Berthelette P, Blackwell C, et al. Alzheimer-type neuropathology in transgenic mice overexpressing V717F beta-amyloid precursor protein. Nature. 1995;373(6514):523–7.

7. Meyer-Luehmann M, Coomaraswamy J, Bolmont T, Kaeser S, Schaefer C, Kilger E, et al. Exogenous induction of cerebral beta-amyloidogenesis is governed by agent and host. Science. 2006;313(5794):1781–4.
8. Calhoun ME, Wiederhold KH, Abramowski D, Phinney AL, Probst A, Sturchler-Pierrat C, et al. Neuron loss in APP transgenic mice. Nature. 1998;395(6704):755–6.
9. Wright AL, Zinn R, Hohensinn B, Konen LM, Beynon SB, Tan RP, et al. Neuroinflammation and neuronal loss precede Abeta plaque deposition in the hAPP-J20 mouse model of Alzheimer's disease. PLoS One. 2013;8(4):e59586.
10. Guo Q, Wang Z, Li H, Wiese M, Zheng H. APP physiological and pathophysiological functions: insights from animal models. Cell Res. 2012;22(1):78–89.
11. Wang P, Yang G, Mosier DR, Chang P, Zaidi T, Gong YD, et al. Defective neuromuscular synapses in mice lacking amyloid precursor protein (APP) and APP-Like protein 2. J Neurosci. 2005;25:1219–25.
12. Zheng H, Jiang M, Trumbauer ME, Sirinathsinghji DJ, Hopkins R, Smith DW, et al. beta-Amyloid precursor protein-deficient mice show reactive gliosis and decreased locomotor activity. Cell. 1995;8:525–31.
13. Li ZW, Stark G, Gotz J, Rulicke T, Gschwind M, Huber G, et al. Generation of mice with a 200-kb amyloid precursor protein gene deletion by Cre recombinase-mediated site-specific recombination in embryonic stem cells. Proc Natl Acad Sci U S A. 1996;93:6158–62.
14. Duce JA, Tsatsanis A, Cater MA, James SA, Robb E, Wikhe K, et al. Iron-export ferroxidase activity of beta-amyloid precursor protein is inhibited by zinc in Alzheimer's disease. Cell. 2010;142:857–67.
15. Holcomb L, Gordon MN, McGowan E, Yu X, Benkovic S, Jantzen P, et al. Accelerated Alzheimer-type phenotype in transgenic mice carrying both mutant amyloid precursor protein and presenilin 1 transgenes. Nat Med. 1998;4(1):97–100.
16. Schmitz C, Rutten BP, Pielen A, Schafer S, Wirths O, Tremp G, et al. Hippocampal neuron loss exceeds amyloid plaque load in a transgenic mouse model of Alzheimer's disease. Am J Pathol. 2004;164:1495–502.
17. Wang R, Wang B, He W, Zheng H. Wild-type presenilin 1 protects against Alzheimer disease mutation-induced amyloid pathology. J Biol Chem. 2006;281:15330–6.
18. Ohno M, Sametsky EA, Younkin LH, Oakley H, Younkin SG, Citron M, et al. BACE1 deficiency rescues memory deficits and cholinergic dysfunction in a mouse model of Alzheimer's disease. Neuron. 2004;41:27–33.
19. McConlogue L, Buttini M, Anderson JP, Brigham EF, Chen KS, Freedman SB, et al. Partial reduction of BACE1 has dramatic effects on Alzheimer plaque and synaptic pathology in APP Transgenic Mice. J Biol Chem. 2007;282:26326–34.
20. Ma H, Lesne S, Kotilinek L, Steidl-Nichols JV, Sherman M, Younkin L, et al. Involvement of beta-site APP cleaving enzyme 1 (BACE1) in amyloid precursor protein-mediated enhancement of memory and activity-dependent synaptic plasticity. Proc Natl Acad Sci U S A. 2007;104:8167–72.
21. Willem M, Dewachter I, Smyth N, Van Dooren T, Borghgraef P, Haass C, et al. beta-site amyloid precursor protein cleaving enzyme 1 increases amyloid deposition in brain parenchyma but reduces cerebrovascular amyloid angiopathy in aging BACE x APP[V717I] double-transgenic mice. Am J Pathol. 2004;165:1621–31.
22. Bertram L, Tanzi RE. The genetic epidemiology of neurodegenerative disease. J Clin Invest. 2005;115:1449–57.
23. Bales KR, Verina T, Dodel RC, Du Y, Altstiel L, Bender M, et al. Lack of apolipoprotein E dramatically reduces amyloid beta-peptide deposition. Nat Genet. 1997;17:263–4.
24. Dodart JC, Marr RA, Koistinaho M, Gregersen BM, Malkani S, Verma IM, et al. Gene delivery of human apolipoprotein E alters brain Abeta burden in a mouse model of Alzheimer's disease. Proc Natl Acad Sci U S A. 2005;102:1211–6.
25. Lewis J, Dickson DW, Lin WL, Chisholm L, Corral A, Jones G, et al. Enhanced neurofibrillary degeneration in transgenic mice expressing mutant tau and APP. Science. 2001; 293(5534):1487–91.

26. Gotz J, Chen F, van Dorpe J, Nitsch RM. Formation of neurofibrillary tangles in P301l tau transgenic mice induced by Abeta 42 fibrils. Science. 2001;293(5534):1491–5.
27. Roberson ED, Scearce-Levie K, Palop JJ, Yan F, Cheng IH, Wu T, et al. Reducing endogenous tau ameliorates amyloid beta-induced deficits in an Alzheimer's disease mouse model. Science. 2007;316(5825):750–4.
28. Ittner LM, Ke YD, Delerue F, Bi M, Gladbach A, van Eersel J, et al. Dendritic function of tau mediates amyloid-beta toxicity in Alzheimer's disease mouse models. Cell. 2010;142:387–97.
29. Ittner LM, Gotz J. Amyloid-beta and tau—a toxic pas de deux in Alzheimer's disease. Nat Rev Neurosci. 2011;12:67–72.
30. Mondragon-Rodriguez S, Trillaud-Doppia E, Dudilot A, Bourgeois C, Lauzon M, Leclerc N, et al. Interaction of endogenous tau protein with synaptic proteins is regulated by N-methyl-D-aspartate receptor-dependent tau phosphorylation. J Biol Chem. 2012;287(38):32040–53.
31. Nakanishi N, Ryan SD, Zhang X, Khan A, Holland T, Cho EG, et al. Synaptic protein alpha1-takusan mitigates amyloid-beta-induced synaptic loss via interaction with tau and postsynaptic density-95 at postsynaptic sites. J Neurosci. 2013;33:14170–83.
32. Mackenzie IR, Munoz DG, Kusaka H, Yokota O, Ishihara K, Roeber S, et al. Distinct pathological subtypes of FTLD-FUS. Acta Neuropathol. 2011;121:207–18.
33. Gotz J, Probst A, Spillantini MG, Schafer T, Jakes R, Burki K, et al. Somatodendritic localization and hyperphosphorylation of tau protein in transgenic mice expressing the longest human brain tau isoform. Embo J. 1995;14:1304–13.
34. Lewis J, McGowan E, Rockwood J, Melrose H, Nacharaju P, Van Slegtenhorst M, et al. Neurofibrillary tangles, amyotrophy and progressive motor disturbance in mice expressing mutant (P301L) tau protein. Nat Genet. 2000;25:402–5.
35. Dawson HN, Cantillana V, Chen L, Vitek MP. The tau N279K exon 10 splicing mutation recapitulates frontotemporal dementia and parkinsonism linked to chromosome 17 tauopathy in a mouse model. J Neurosci. 2007;27:9155–68.
36. Allen B, Ingram E, Takao M, Smith MJ, Jakes R, Virdee K, et al. Abundant tau filaments and nonapoptotic neurodegeneration in transgenic mice expressing human P301S tau protein. J Neurosci. 2002;22:9340–51.
37. Yoshiyama Y, Higuchi M, Zhang B, Huang SM, Iwata N, Saido TC, et al. Synapse loss and microglial activation precede tangles in a P301S tauopathy mouse model. Neuron. 2007;53:337–51.
38. Santacruz K, Lewis J, Spires T, Paulson J, Kotilinek L, Ingelsson M, et al. Tau suppression in a neurodegenerative mouse model improves memory function. Science. 2005;309:476–81.
39. Mocanu MM, Nissen A, Eckermann K, Khlistunova I, Biernat J, Drexler D, et al. The potential for beta-structure in the repeat domain of tau protein determines aggregation, synaptic decay, neuronal loss, and coassembly with endogenous Tau in inducible mouse models of tauopathy. J Neurosci. 2008;28:737–48.
40. Forman MS, Lal D, Zhang B, Dabir DV, Swanson E, Lee VM, et al. Transgenic mouse model of tau pathology in astrocytes leading to nervous system degeneration. J Neurosci. 2005;6(25):3539–50.
41. Higuchi M, Zhang B, Forman MS, Yoshiyama Y, Trojanowski JQ, Lee VM. Axonal degeneration induced by targeted expression of mutant human tau in oligodendrocytes of transgenic mice that model glial tauopathies. J Neurosci. 2005;25:9434–43.
42. Aguzzi A, Rajendran L. The transcellular spread of cytosolic amyloids, prions, and prionoids. Neuron. 2009;64:783–90.
43. Clavaguera F, Bolmont T, Crowther RA, Abramowski D, Frank S, Probst A, et al. Transmission and spreading of tauopathy in transgenic mouse brain. Nat Cell Biol. 2009;11:909–13.
44. Clavaguera F, Akatsu H, Fraser G, Crowther RA, Frank S, Hench J, et al. Brain homogenates from human tauopathies induce tau inclusions in mouse brain. Proc Natl Acad Sci U S A. 2013;110:9535–40.
45. Liu L, Drouet V, Wu JW, Witter MP, Small SA, Clelland C, et al. Trans-synaptic spread of tau pathology in vivo. PLoS One. 2012;7(2):e31302.

46. Asuni AA, Boutajangout A, Quartermain D, Sigurdsson EM. Immunotherapy targeting pathological tau conformers in a tangle mouse model reduces brain pathology with associated functional improvements. J Neurosci. 2007;27(34):9115–29.
47. Bi M, Ittner A, Ke YD, Gotz J, Ittner LM. Tau-targeted immunization impedes progression of neurofibrillary histopathology in aged P301L tau transgenic mice. PLoS One. 2011;6(12):e26860. doi:10.1371/journal.pone.0026860. Epub 2011 Dec 8.
48. Boimel M, Grigoriadis N, Lourbopoulos A, Haber E, Abramsky O, Rosenmann H. Efficacy and safety of immunization with phosphorylated tau against neurofibrillary tangles in mice. Exp Neurol. 2010;224:472–85.
49. Chai X, Wu S, Murray TK, Kinley R, Cella CV, Sims H, et al. Passive immunization with anti-Tau antibodies in two transgenic models: reduction of Tau pathology and delay of disease progression. J Biol Chem. 2011;286:34457–67.
50. van Eersel J, Ke YD, Liu X, Delerue F, Kril JJ, Gotz J, et al. Sodium selenate mitigates tau pathology, neurodegeneration, and functional deficits in Alzheimer's disease models. Proc Natl Acad Sci U S A. 2010;107:13888–93.
51. Neumann M, Sampathu DM, Kwong LK, Truax AC, Micsenyi MC, Chou TT, et al. Ubiquitinated TDP-43 in frontotemporal lobar degeneration and amyotrophic lateral sclerosis. Science. 2006;314(5796):130–3.
52. Buratti E, Baralle FE. Multiple roles of TDP-43 in gene expression, splicing regulation, and human disease. Front Biosci. 2008;13:867–78.
53. Liscic RM, Grinberg LT, Zidar J, Gitcho MA, Cairns NJ. ALS and FTLD: two faces of TDP-43 proteinopathy. Eur J Neurol. 2008;15:772–80.
54. Wegorzewska I, Bell S, Cairns NJ, Miller TM, Baloh RH. TDP-43 mutant transgenic mice develop features of ALS and frontotemporal lobar degeneration. Proc Natl Acad Sci U S A. 2009;106:18809–14.
55. Wils H, Kleinberger G, Janssens J, Pereson S, Joris G, Cuijt I, et al. TDP-43 transgenic mice develop spastic paralysis and neuronal inclusions characteristic of ALS and frontotemporal lobar degeneration. Proc Natl Acad Sci U S A. 2010;107:3858–63.
56. Xu YF, Gendron TF, Zhang YJ, Lin WL, D'Alton S, Sheng H, et al. Wild-type human TDP-43 expression causes TDP-43 phosphorylation, mitochondrial aggregation, motor deficits, and early mortality in transgenic mice. J Neurosci. 2010;30:10851–9.
57. Igaz LM, Kwong LK, Lee EB, Chen-Plotkin A, Swanson E, Unger T, et al. Dysregulation of the ALS-associated gene TDP-43 leads to neuronal death and degeneration in mice. J Clin Invest. 2011;121:726–38.
58. Stallings NR, Puttaparthi K, Luther CM, Burns DK, Elliott JL. Progressive motor weakness in transgenic mice expressing human TDP-43. Neurobiol Dis. 2010;40:404–14.
59. Swarup V, Phaneuf D, Bareil C, Robertson J, Rouleau GA, Kriz J, et al. Pathological hallmarks of amyotrophic lateral sclerosis/frontotemporal lobar degeneration in transgenic mice produced with TDP-43 genomic fragments. Brain. 2011;134:2610–26.
60. Winton MJ, Igaz LM, Wong MM, Kwong LK, Trojanowski JQ, Lee VM. Disturbance of nuclear and cytoplasmic TAR DNA-binding protein (TDP-43) induces disease-like redistribution, sequestration, and aggregate formation. J Biol Chem. 2008;283:13302–9.
61. Kraemer BC, Schuck T, Wheeler JM, Robinson LC, Trojanowski JQ, Lee VM, et al. Loss of murine TDP-43 disrupts motor function and plays an essential role in embryogenesis. Acta Neuropathol. 2010;119:409–19.
62. Baker M, Mackenzie IR, Pickering-Brown SM, Gass J, Rademakers R, Lindholm C, et al. Mutations in progranulin cause tau-negative frontotemporal dementia linked to chromosome 17. Nature. 2006;442:916–9.
63. Cruts M, Gijselinck I, van der Zee J, Engelborghs S, Wils H, Pirici D, et al. Null mutations in progranulin cause ubiquitin-positive frontotemporal dementia linked to chromosome 17q21. Nature. 2006;442(7105):920–4.
64. Shankaran SS, Capell A, Hruscha AT, Fellerer K, Neumann M, Schmid B, et al. Missense mutations in the progranulin gene linked to frontotemporal lobar degeneration with ubiquitin-immunoreactive inclusions reduce progranulin production and secretion. J Biol Chem. 2008;283:1744–53.

65. Wang J, Van Damme P, Cruchaga C, Gitcho MA, Vidal JM, Seijo-Martinez M, et al. Pathogenic cysteine mutations affect progranulin function and production of mature granulins. J Neurochem. 2010;112:1305–15.
66. Kayasuga Y, Chiba S, Suzuki M, Kikusui T, Matsuwaki T, Yamanouchi K, et al. Alteration of behavioural phenotype in mice by targeted disruption of the progranulin gene. Behav Brain Res. 2007;185:110–8.
67. Yin F, Banerjee R, Thomas B, Zhou P, Qian L, Jia T, et al. Exaggerated inflammation, impaired host defense, and neuropathology in progranulin-deficient mice. J Exp Med. 2010; 207:117–28.
68. Petkau TL, Neal SJ, Milnerwood A, Mew A, Hill AM, Orban P, et al. Synaptic dysfunction in progranulin-deficient mice. Neurobiol Dis. 2012;45:711–22.
69. Martens LH, Zhang J, Barmada SJ, Zhou P, Kamiya S, Sun B, et al. Progranulin deficiency promotes neuroinflammation and neuron loss following toxin-induced injury. J Clin Invest. 2012;122:3955–9.
70. Wils H, Kleinberger G, Pereson S, Janssens J, Capell A, Van Dam D, et al. Cellular ageing, increased mortality and FTLD-TDP-associated neuropathology in progranulin knockout mice. J Pathol. 2012;228(1):67–76. PubMed PMID: 22733568.
71. Ahmed Z, Sheng H, Xu YF, Lin WL, Innes AE, Gass J, et al. Accelerated lipofuscinosis and ubiquitination in granulin knockout mice suggest a role for progranulin in successful aging. Am J Pathol. 2010;177(1):311–24. PubMed PMID: 20522652. Pubmed Central PMCID: 2893674.
72. Ghoshal N, Dearborn JT, Wozniak DF, Cairns NJ. Core features of frontotemporal dementia recapitulated in progranulin knockout mice. Neurobiol Dis. 2012;45(1):395–408.
73. Yin F, Dumont M, Banerjee R, Ma Y, Li H, Lin MT, et al. Behavioral deficits and progressive neuropathology in progranulin-deficient mice: a mouse model of frontotemporal dementia. FASEB J. 2010;24:4639–47.
74. Filiano AJ, Martens LH, Young AH, Warmus BA, Zhou P, Diaz-Ramirez G, et al. Dissociation of frontotemporal dementia-related deficits and neuroinflammation in progranulin haploinsufficient mice. J Neurosci. 2013;33:5352–61.
75. Watts GD, Wymer J, Kovach MJ, Mehta SG, Mumm S, Darvish D, et al. Inclusion body myopathy associated with Paget disease of bone and frontotemporal dementia is caused by mutant valosin-containing protein. Nat Genet. 2004;36:377–81.
76. Kimonis VE, Mehta SG, Fulchiero EC, Thomasova D, Pasquali M, Boycott K, et al. Clinical studies in familial VCP myopathy associated with Paget disease of bone and frontotemporal dementia. Am J Med Genet A. 2008;146A:745–57.
77. Johnson JO, Mandrioli J, Benatar M, Abramzon Y, Van Deerlin VM, Trojanowski JQ, et al. Exome sequencing reveals VCP mutations as a cause of familial ALS. Neuron. 2010;68:857–64.
78. Abramzon Y, Johnson JO, Scholz SW, Taylor JP, Brunetti M, Calvo A, et al. Valosin-containing protein (VCP) mutations in sporadic amyotrophic lateral sclerosis. Neurobiol Aging. 2012;33:2231 e1–e6.
79. Tang WK, Li D, Li CC, Esser L, Dai R, Guo L, et al. A novel ATP-dependent conformation in p97 N-D1 fragment revealed by crystal structures of disease-related mutants. EMBO J. 2010;29:2217–29.
80. Rodriguez-Ortiz CJ, Hoshino H, Cheng D, Liu-Yescevitz L, Blurton-Jones M, Wolozin B, et al. Neuronal-specific overexpression of a mutant valosin-containing protein associated with IBMPFD promotes aberrant ubiquitin and TDP-43 accumulation and cognitive dysfunction in transgenic mice. Am J Pathol. 2013;183:504–15.
81. Weihl CC, Miller SE, Hanson PI, Pestronk A. Transgenic expression of inclusion body myopathy associated mutant p97/VCP causes weakness and ubiquitinated protein inclusions in mice. Hum Mol Genet. 2007;16:919–28.
82. Custer SK, Neumann M, Lu H, Wright AC, Taylor JP. Transgenic mice expressing mutant forms VCP/p97 recapitulate the full spectrum of IBMPFD including degeneration in muscle, brain and bone. Hum Mol Genet. 2010;19:1741–55.

83. Badadani M, Nalbandian A, Watts GD, Vesa J, Kitazawa M, Su H, et al. VCP associated inclusion body myopathy and Paget disease of bone knock-in mouse model exhibits tissue pathology typical of human disease. PLoS One. 2010;5(10):e13183.
84. Nalbandian A, Llewellyn KJ, Badadani M, Yin HZ, Nguyen C, Katheria V, et al. A progressive translational mouse model of human valosin-containing protein disease: the VCP(R155H/+) mouse. Muscle Nerve. 2013;47(2):260–70.
85. Yin HZ, Nalbandian A, Hsu CI, Li S, Llewellyn KJ, Mozaffar T, et al. Slow development of ALS-like spinal cord pathology in mutant valosin-containing protein gene knock-in mice. Cell Death Dis. 2012;3:e374.
86. Skibinski G, Parkinson NJ, Brown JM, Chakrabarti L, Lloyd SL, Hummerich H, et al. Mutations in the endosomal ESCRTIII-complex subunit CHMP2B in frontotemporal dementia. Nat Genet. 2005;37:806–8.
87. Ghazi-Noori S, Froud KE, Mizielinska S, Powell C, Smidak M, Fernandez de Marco M, et al. Progressive neuronal inclusion formation and axonal degeneration in CHMP2B mutant transgenic mice. Brain. 2012;135(Pt 3):819–32.
88. Van Langenhove T, van der Zee J, Sleegers K, Engelborghs S, Vandenberghe R, Gijselinck I, et al. Genetic contribution of FUS to frontotemporal lobar degeneration. Neurology. 2010;74:366–71.
89. Kwiatkowski Jr TJ, Bosco DA, Leclerc AL, Tamrazian E, Vanderburg CR, Russ C, et al. Mutations in the FUS/TLS gene on chromosome 16 cause familial amyotrophic lateral sclerosis. Science. 2009;323(5918):1205–8.
90. Vance C, Rogelj B, Hortobagyi T, De Vos KJ, Nishimura AL, Sreedharan J, et al. Mutations in FUS, an RNA processing protein, cause familial amyotrophic lateral sclerosis type 6. Science. 2009;323(5918):1208–11.
91. Dormann D, Rodde R, Edbauer D, Bentmann E, Fischer I, Hruscha A, et al. ALS-associated fused in sarcoma (FUS) mutations disrupt Transportin-mediated nuclear import. EMBO J. 2010;29:2841–57.
92. Mitchell JC, McGoldrick P, Vance C, Hortobagyi T, Sreedharan J, Rogelj B, et al. Overexpression of human wild-type FUS causes progressive motor neuron degeneration in an age- and dose-dependent fashion. Acta Neuropathol. 2013;125:273–88.
93. Huang C, Zhou H, Tong J, Chen H, Liu YJ, Wang D, et al. FUS transgenic rats develop the phenotypes of amyotrophic lateral sclerosis and frontotemporal lobar degeneration. PLoS Genet. 2011;7:e1002011.
94. Neumann M, Bentmann E, Dormann D, Jawaid A, DeJesus-Hernandez M, Ansorge O, et al. FET proteins TAF15 and EWS are selective markers that distinguish FTLD with FUS pathology from amyotrophic lateral sclerosis with FUS mutations. Brain. 2011;134(Pt 9):2595–609.

Chapter 9
Cerebrospinal Fluid Biomarkers in Alzheimer's Disease and Frontotemporal Dementia

Niklas Mattsson and Henrik Zetterberg

Abstract Cerebrospinal fluid (CSF) biomarkers, including β-amyloid42, total tau, phosphorylated tau, and neurofilament light, reflect different pathological processes in Alzheimer's disease (AD) and frontotemporal dementia (FTD). These CSF biomarkers have been studied for diagnosis, prognosis, and treatment follow-up and in relation to genetics and neuroimaging. They are now increasingly used in research, drug development, and clinical settings to increase our understanding of AD and FTD and to improve patient management. Key issues for further implementation of CSF biomarkers in research and clinical routine include technical aspects, such as the development of stable, automatized assays, and improved standardization between laboratories, as well as clinical aspects, such as the creation of universally accepted guidelines specifying the role of CSF biomarkers in relation to clinical measures and neuroimaging findings.

Keywords Cerebrospinal fluid • Alzheimer • Frontotemporal lobe dementia • Tau Amyloid • Neurofilament

---

N. Mattsson, MD, PhD (✉)
Department of Veterans Affairs Medical Center,
Center for Imaging of Neurodegenerative Diseases, San Francisco, CA 94121, USA

Clinical Neurochemistry Laboratory, Institute of Neuroscience and Physiology,
Sahlgren's University Hospital/Mölndal, Mölndal, Sweden
e-mail: niklas.mattsson@neuro.gu.se

H. Zetterberg, MD, PhD
Clinical Neurochemistry Laboratory, Institute of Neuroscience and Physiology,
The Sahlgrenska Academy at the University of Gothenburg, Mölndal, Sweden

D. Galimberti, E. Scarpini (eds.), *Neurodegenerative Diseases*,
DOI 10.1007/978-1-4471-6380-0_9, © Springer-Verlag London 2014

Introduction

Cerebrospinal fluid (CSF) is a valuable source for cellular, molecular, and chemical measurements reflecting biological processes in the brain. Such measurements may be called biomarkers, which are defined as characteristics that are objectively measured and evaluated as indicators of normal biological processes, pathogenic processes, or pharmacological responses to therapeutic intervention [1].

CSF biomarkers have been used extensively in neurology since the beginning of the twentieth century when lumbar puncture was introduced by Heinrich Quincke in the management of patients with central nervous system (CNS) infections [2]. Modern technology allows the identification and quantification of several thousand proteins in CSF, which theoretically reflects most cellular compartments in CNS. Although the large dynamic range of protein concentrations in CSF makes it difficult to identify and quantify a large number of proteins simultaneously, rapid technological development allows a more and more detailed characterization of CSF contents. However, a natural limitation of lumbar CSF analyses is that all measurements are done on a mixture of liquid originating from the whole CNS, which precludes anatomical precision. This is likely a disadvantage in dementias, where distinct parts of the brain are affected, especially early in the disease. CSF dynamics (CSF production, transportation, and absorption) may also influence the results, as may damage to the blood–brain barrier, which may lead to different degrees of contamination of the CSF sample with plasma-derived proteins.

During the last three decades, research on CSF biomarkers in dementias has boomed. The database PubMed currently (August 2013) lists more than 5,000 articles for the query "'cerebrospinal fluid' AND biomarker," including thousands of papers on Alzheimer's disease (AD) or frontotemporal dementia (FTD). Here we attempt to summarize key concepts and highlight particularly interesting studies on CSF biomarkers in AD and FTD.

Normal CSF Physiology

Cerebrospinal fluid is a clear liquid which occupies the ventricles and the subarachnoid space around the brain and the spinal cord (see [3] for an extensive review on normal CSF physiology). Histologically, CSF is in close contact with the cells of the CNS and is not separated from the brain tissue by the blood–brain barrier. CSF has several functions in normal physiology, where it (1) creates a condition of neutral buoyancy for the brain, reducing its net weight, which protects blood supply and integrity of neurons, especially in the lower sections of the brain; (2) supplies nutrients, peptides, and hormones to widespread neuronal networks; (3) clears waste products from the normal metabolism into blood stream; (4) provides mechanical protection for the brain, by distributing the impact of an incoming force; and (5) helps to maintain a constant intracranial pressure.

CSF is essentially a highly diluted filtrate of plasma (about 99 % water) which is modified by metabolic processes in the CNS to give it unique biochemical properties. The majority of CSF is produced by the choroid plexus in the ventricles, but CSF is also released from other structures, including the blood vessels and the remaining ventricular ependyma. Driven by arterial pulsations, CSF circulates through a series of compartments, from the ventricles deep inside the brain to the subarachnoid space. CSF is reabsorbed to the venous blood stream through arachnoid granulations. The CSF portion in the caudal lumbar sac is available for sampling by lumbar puncture.

The total CSF volume in an adult human is about 150 mL, with a formation rate of about 0.4 mL per minute, and an overall turnover rate of about 3–4 volumes per day [3]. The normally acquired volume in a lumbar puncture is about 10–20 mL, which is quickly replenished. It is possible that the CSF turnover is reduced due to lower formation rates in normal aging and in dementia, but the impact of this on biomarker measurements is not clear. Lumbar puncture is a relatively easy procedure that can be performed on outpatients [4]. The only known complication is headache, which has an incidence of between 2 and 35 % depending to the largest extent on age (lower incidence in higher age groups) [4–7]. The headache is most often mild, can be symptomatically treated, and resolves by itself within a day or two.

CSF Biomarkers in Alzheimer's Disease

Alzheimer's disease is a continuous process, which begins with an extended preclinical phase that gradually progresses into early and late clinical phases. The overall duration of the disease is several decades, with the majority of time being prior to dementia. CSF biomarker studies have contributed to this understanding, since they have shown that biomarker signs of AD pathology are present several years prior to dementia. The relationship between underlying pathological changes and clinical symptoms in AD is now a major field of research.

Influenced by the idea of AD as a combined preclinical and clinical entity, international working groups published new research criteria for the disease in 2007 [8] and 2010 [9]. These criteria acknowledged that the CSF biomarkers β-amyloid42 (Aβ42), total tau (T-tau), and phosphorylated tau (P-tau) could be used as tools to identify people with AD in different stages of the disease, even prior to onset of symptoms. Another set of criteria were published in 2011 by the National Institute on Aging-Alzheimer's Association workgroup (NIA-AA). This was divided into separate parts for the preclinical stage of AD [10], mild cognitive impairment (MCI) due to AD [11], and dementia due to AD [12] and also incorporated biomarkers in diagnostic algorithms.

The novel criteria rest on the assumption that AD biomarkers may be broadly categorized into markers of Aβ pathology (reduced CSF Aβ42 and increased signal of positron emission tomography [PET] with Aβ tracers) and markers of neuronal

Table 9.1 CSF biomarkers in AD and FTD

CSF biomarker	Change in AD	Change in FTD	Reflects
Aβ1-42	Decreased in preclinical and clinical stages	Unchanged or slightly decreased	Amyloid plaque pathology
BACE1 activity	Unchanged or increased	Unknown	APP processing capacity?
Chitotriosidase	Increased	Unknown	Inflammation/microglial activity
GAP43	Increased	Unknown	Synaptic injury?
Neurogranin	Increased	Unknown	Synaptic injury?
NFL	Unchanged or slightly increased	Increased	Injury to (myelinated) axons
Progranulin	Unchanged	Unchanged or reduced	*GNR* mutations
P-tau	Increased in clinical stages	Unchanged	AD-related axonal injury
sAPP-α and sAPP-β	Unchanged or increased	Unchanged	Overall APP production and processing?
TDP-43	Unchanged	Unchanged or increased	TDP-43 inclusions
T-tau	Increased in clinical stages	Unchanged or slightly increased	Injury to (cortical) axons
YKL-40	Increased	Unknown	Inflammation/microglial activity

The table includes selected CSF biomarkers described in this chapter. The amount of available data on changes in different diseases (or stages of diseases) varies greatly between biomarkers. This summary is based on selected references in the chapter and the authors' clinical experience. Most biomarkers have large overlap between patients and healthy controls, and the disease-related alterations described here are average changes at the group level

injury (increased CSF T-tau and P-tau and decreased signal of PET with fluorodeoxyglucose [FDG], or brain atrophy on magnetic resonance imaging [MRI] or computer tomography [CT]). It is not yet fully resolved to what degree different biomarkers within these categories are interchangeable or provide unique clinically valuable information. Also, several other CSF biomarkers have been studied in AD (Table 9.1), and for most of these, it is not clear how they relate to imaging measurements.

Toward the Optimal CSF AD Biomarker

One reason for the large number of publications on CSF dementia biomarkers is that this is a point of convergence for studies on disease biology, drug development, and clinical applications. Given the many different applications, the ideas about what constitutes an optimal biomarker may differ, but it has been suggested that an ideal dementia biomarker should (1) be linked to fundamental features of the underlying pathology, (2) be validated in neuropathologically confirmed cases, (3) detect the disease early, (4) distinguish the disease from other dementias, (5) be noninvasive, (6) be simple to use, and (7) be inexpensive [13, 14]. We propose that these requirements are now fulfilled in AD for CSF Aβ42 and partly also for CSF T-tau and P-tau. Strong data support that reduced CSF Aβ42 is linked to brain Aβ accumulation, which is a fundamental feature of AD [10, 11], and CSF T-tau and P-tau are related to axonal degeneration and tangle pathology, respectively [15, 16]. Antemortem CSF Aβ42, T-tau, and P-tau for AD diagnosis have been validated in neuropathologically confirmed cases [17] and are altered early in the disease [18–20]. CSF Aβ42 may be reduced already several years prior to symptoms [21, 22]. The combination of CSF Aβ42, T-tau, and P-tau has a relatively unique pattern in AD, enabling differential diagnostics toward other dementias [16]. As detailed above, lumbar puncture is a safe procedure, with benign headache as the most common adverse effect, which is age dependent and very rare at memory clinics [4, 6]. Analysis of CSF Aβ42, T-tau, and P-tau is currently done on small-scale immunoassay systems, which may be partly automated, but there is an ongoing rapid development of larger-scale fully automated systems, which will facilitate the use of CSF biomarkers outside expert centers [23]. The currently available assays have low intra-laboratory variability, and efforts are underway to also reduce inter-laboratory variability [23]. Samples may also easily be sent to expert laboratories. The cost for analysis of CSF Aβ42, T-tau, and P-tau at a public hospital laboratory in Europe is currently around 250 USD. Notably, the alternative method to establish brain Aβ pathology in a living person is a PET scan, which involves the injection of a radioactive ligand into the body, has very low availability outside university hospitals, and currently costs from around 3,000 USD and up.

CSF Aβ Peptides

Aβ peptides are cleaved out from the type I transmembrane protein amyloid precursor protein (APP) by enzymatic activities of α-secretase, β-secretase (BACE1), and γ-secretase and released by neurons in relation to synaptic activity [24]. The isoform with highest concentration in CSF is Aβ1-40, but most studies have focused on the 42 amino acid long isoform Aβ1-42 (Aβ42), since this was found early on to be reduced in AD dementia compared to controls [25]. The main theory explaining the reduced CSF Aβ42 levels is that the peptide is sequestered in plaques and thus has limited access to CSF. Autopsy studies have found inverse correlations between CSF Aβ42 and the amount of amyloid plaques in the brain [15, 26], and the same has been shown in living patients using PET Aβ imaging [14, 27]. Other mechanisms may hypothetically also lower CSF Aβ42, including formation of Aβ oligomers that are not detected by common immunoassays [28], Aβ binding to other proteins that block antibody epitopes [29], and intracellular Aβ accumulation [30].

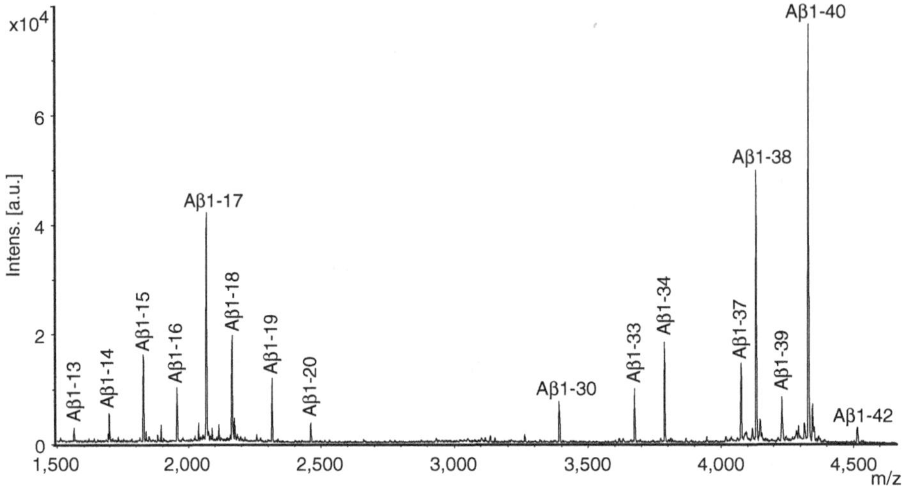

Fig. 9.1 **Aβ** peptides present in CSF. A large number of different Aβ variants are present in CSF, besides the commonly studied Aβ1-40 and Aβ1-42. The figure shows different Aβ isoforms present in normal human CSF, as detected by immunoprecipitation and matrix-assisted laser desorption/ionization time-of-flight mass spectrometry (IP-MALDI-TOF-MS) using the anti-Aβ antibodies 6E10 and 4G8 (Courtesy of Erik Portelius, University of Gothenburg)

Aβ42 is continuously turned over in the CSF, with a clearance and production rate of about 8 % per hour [31]. One study found that patients with sporadic AD have about 30 % reduced clearance rates of Aβ40 and Aβ42 compared to controls, but it is not clear exactly how this relates to plaque accumulation and the reduced CSF Aβ42 in AD, since CSF Aβ40 levels are not reduced in AD and since Aβ42 clearance rates did not correlate to CSF Aβ42 concentrations in that study [32].

Besides Aβ40 and Aβ42, APP processing gives rise to a large number of different soluble Aβ isoforms which are present in plasma and CSF (Fig. 9.1). These include C-terminally truncated long peptides formed by γ-secretase cleavage of APP at different positions (e.g., Aβ1-37 and Aβ1-38 [33]), C-terminally truncated short isoforms formed by combined β-secretase and α-secretase activity (e.g., Aβ1-16 and Aβ1-17 [34, 35]), and N-terminally truncated isoforms formed through other enzymatic activities (e.g., Aβ5-40 [36]).

CSF T-Tau and P-Tau Proteins

The other two extensively studied CSF AD biomarkers are T-tau, which denotes the concentration of tau proteins as measured by unspecific tau assays, and P-tau, which denotes the concentration of tau proteins phosphorylated at a specific threonine or serine amino acid residue. Tau is a neuronal protein, which is mainly found in thin, unmyelinated, cortical axons, where it stabilizes microtubule and facilitates axonal transport mechanisms. Phosphorylation of tau leads to altered properties of the protein and may cause it to aggregate into paired helical filaments and larger assemblies called neurofibrillary tangles. Alternative splicing of exon 10 leads to tau isoforms with three (3R-tau) or four (4R-tau) microtubule-binding repeat domains with only 3R-tau in embryonic brain and comparable levels of 3R- and 4R-tau in normal adult brain [37].

Studies across different neurological diseases have shown that CSF levels of T-tau, but not P-tau, are increased in relation to neuronal injury. The highest CSF T-tau concentrations are seen in conditions with severe injury, such as stroke or Creutzfeldt–Jakob disease. CSF T-tau levels are relatively stable in late clinical stages of AD, suggesting that the concentration is proportional to the rate of neuronal loss rather than the accumulated loss [38, 39].

CSF P-tau levels correlate to the amount of neurofibrillary tangles and hyperphosphorylated tau found in the brain [26, 40, 41]. CSF P-tau is increased in AD, but not in other dementias with neurofibrillary tangles, for reasons that are still unclear. Most likely, we need additional assays for specific tau isoforms or fragments in these conditions. For example, a recent study using 3R-/4R-tau-specific assays revealed selective decreases of 4R-tau in CSF of progressive supranuclear palsy (PSP) and AD patients compared with controls and lower 4R-tau levels in AD compared with Parkinson's disease with dementia [42].

The Dynamic Biomarker Model

The development of pathological CSF biomarkers in AD is believed to follow an ordered sequence, with each biomarker following a specific trajectory. This theory was summarized in 2010 [43] and 2013 [44] as "the dynamic biomarker model." The presumed ordering of biomarkers follows the amyloid cascade hypothesis, which states that the initial pathological event in AD is an abnormal folding and aggregation of Aβ peptides, which secondarily leads to neuroinflammation, synaptic dysfunction, tau pathology, and neuronal degeneration [45]. In line with this, CSF Aβ42 levels are thought to change earlier than CSF T-tau and P-tau, both in autosomal dominant AD [22] and sporadic AD [20, 46]. The second proposition of the dynamic biomarker model is that the biomarkers' trajectories are sigmoid. This is based on several sources, including the finding that CSF Aβ42, T-tau, and P-tau are stable in clinical stages of AD [39], suggesting that they have reached a plateau phase, as well as autopsy studies showing that amyloid accumulation plateaus with increasing disease duration [47]. Ultimately proving that the trajectories are sigmoid requires longitudinal studies with multiple time points per subject, but most data published so far have had short follow-up or been cross-sectional with derived longitudinal measurements based on cognitive scales [48].

CSF Biomarkers in AD Dementia

Since the mid-1990s, many studies have found that AD dementia patients have about 50 % reduced CSF Aβ42, and several times increased CSF T-tau and P-tau levels, compared to cognitively healthy controls [49], with 80–85 % sensitivity and specificity [16]. CSF biomarkers may not only be useful for diagnosis but could also provide prognostic information, since high levels of P-tau or T-tau are associated with rapid clinical progression of AD dementia [50]. During the last years, most attention in CSF AD biomarker research has been focused on the early clinical and preclinical stages of the disease.

CSF Biomarkers in MCI due to AD

MCI is as an objective and clinically relevant cognitive dysfunction which does not cause significant interference with daily functioning. MCI patients have an increased risk of progression to dementia, but from the clinical symptoms alone, it is difficult to predict when or if an individual patient will progress and to determine which underlying disease that causes the symptoms.

CSF Aβ42, T-tau, and P-tau are altered to AD-like levels in MCI patients later progressing to AD dementia, which distinguishes them from patients who develop other dementias or remain cognitively stable, at sensitivities and specificities of 70–90 % [17–19, 51]. Differences between MCI studies may be due to heterogeneity of MCI classifications. Different definitions of MCI exist, and some studies only include "amnestic" MCI patients, while other includes MCI subjects with or without dominating amnestic symptoms. Carefully controlled mono-center studies may achieve very high diagnostic accuracies. For example, in one longitudinal MCI study, the combination of CSF Aβ42 and T-tau had a sensitivity of 95 % and a specificity of 83 % for conversion to AD dementia at a median follow-up of 5 years [19]. When the same study population was evaluated at 9 years follow-up, the ratio of Aβ42 to P-tau at baseline had sensitivities and specificities of 85–90 % for future AD dementia [20]. Among MCI patients who have biomarker evidence of Aβ pathology, high T-tau and P-tau are associated with shorter time to dementia [20, 52], suggesting either that CSF tau levels increase dynamically during the clinical stages of AD or that patients with high CSF tau levels have a more aggressive variant of AD. Most longitudinal studies with serial CSF samples have shown that CSF Aβ42, T-tau, and P-tau are essentially stable during the later disease stage [39, 53, 54].

A major goal for clinical AD research is to establish predictive values of AD biomarkers for people with MCI. The negative predictive values of CSF AD biomarkers are often around 90 %, while the positive predictive values vary between studies, from around 60–90 %. Although diagnostic accuracies decrease with age (mainly due to increased accumulation of Aβ in non-demented subjects), the biomarkers still have stable positive and negative predictive values in older age groups [55].

CSF Biomarkers in Preclinical AD

To some degree, AD can be identified by fluid or imaging measurements prior to any clinical symptom [10]. In autosomal dominant familial AD, the known deterministic relationship between mutations and future clinical disease provides a unique opportunity to investigate preclinical biomarker changes. A cross-sectional study on presymptomatic mutation carriers in a Colombian kindred found that CSF Aβ42 levels were increased in mutation carriers more than two decades prior to expected age of symptom onset [56]. This was in line with experimental data showing that similar mutations resulted in increased Aβ42 production. In a cross-sectional study by the Dominantly Inherited Alzheimer Network (DIAN), CSF Aβ42 levels started to decline about 25 years before expected symptom onset, and this was about 10 years earlier than any other biomarker alteration, including increased CSF tau [22]. CSF Aβ42 initially seemed to be elevated compared to controls, in line with the data from the Colombian cohort. When testing mutation carriers closer to onset of dementia, other studies have found reduced CSF Aβ42 and increased CSF T-tau and P-tau [57–59].

Regarding preclinical sporadic AD, baseline CSF Aβ42 and sometimes CSF tau predict future impairment in people who are cognitively normal [21, 60–62] or who have subjective cognitive impairment [63, 64]. Combinations of pathological Aβ42 and tau levels may be more likely to result in cognitive impairment than individual biomarker positivity [65, 66]. Besides development of cognitive impairment, reduced CSF Aβ42 is also linked to increased brain atrophy rates in cognitively healthy controls [67].

The predictive accuracy of biomarkers to determine future cognitive decline may be increased by also including factors indicating cognitive reserve, such as age, education, and brain volume [68]. Among cognitively normal people with high levels of T-tau or P-tau, long education and large brain volume slow development of cognitive impairment, suggesting that increased tau levels indicate a preclinical disease that results in symptoms later in subjects with protective factors [69].

CSF Aβ42 and Tau in AD Differential Diagnoses

CSF Aβ42, T-tau, and P-tau may be altered in patients with diagnoses other than AD. For example, reduced CSF Aβ42 is seen in subgroups of patients with vascular dementia [70], Lewy body dementia [71], FTD [72], Creutzfeldt–Jakob disease [73], amyotrophic lateral sclerosis [74], and multiple system atrophy [75]. One reason for this may be that CSF biomarkers reflect pathological processes which either are shared between different diseases or are in early stages and not related to the current symptoms (indicating prodromal AD in patients with other diseases). Notably, most studies do not include autopsy validation of clinical diagnoses, which may underestimate the biomarkers' accuracies [76]. To complicate matters, similar biomarker patterns may arise from different pathological processes, since infections and inflammations affect APP metabolism and lower CSF Aβ42 levels without formation of plaques [77–79]. As explained above, CSF T-tau may increase in many neurological conditions with axonal degeneration [80–82]. This motivates analysis of P-tau, since potential AD differential diagnoses often have normal P-tau levels despite T-tau increases. However, P-tau increases have also been reported in some non-AD diseases [83], and CSF P-tau is even elevated during normal brain development [84].

CSF Biomarkers in AD Treatment

Despite intense efforts there is still no recognized disease-modifying treatment for AD. CSF biomarkers have been suggested to facilitate drug development by being implemented for (1) enrichment of participants with underlying AD pathology, (2) stratification according to specific biochemical traits, (3) measurement of pharmacodynamic effects, and (4) monitoring of toxicity and side effects.

CSF Biomarkers to Enrich Study Populations

Traditionally, most AD drug trials have only included patients based on clinical characteristics, which may have resulted in participants without underlying AD pathology. This problem is even more pronounced for studies in early clinical stages and is fundamental for trials in asymptomatic persons. Since the damaging processes in AD may already have reached a point of no return when the patient is demented, trials increasingly recruit participants at early clinical or preclinical stages. In these groups, CSF measurements may identify people at high risk for cognitive decline due to AD, increasing the power or lowering the costs of the study, although all savings are partly offset by prolonged trial duration, since biomarker-based enrichment means that more study subjects must undergo screening. The overall economical benefits of biomarker use is related both to the test costs, the prevalence of the trait used for inclusion, and the anticipated effect size of the drug, which determines the sample size needed to prove efficiency at a certain level of significance. The European Medicines Agency (EMA) supports the use of CSF Aβ42 and T-tau to enrich clinical populations with prodromal AD [85]. At the point of writing, the US counterpart agency Food and Drug Administration has still not released a corresponding statement.

CSF Biomarkers for Patient Stratification

AD is a heterogeneous disease, especially in older age groups, where many patients have multiple pathologies contributing to the overall clinical presentation. CSF biomarkers of different pathologies may be used to stratify patients by treatment effects. For example, it is possible that patients respond differently to an anti-Aβ treatment depending on their degree of Aβ pathology, as reflected by their CSF Aβ42 levels. It is also possible that the ongoing rate of neuronal loss, as reflected by CSF T-tau and P-tau levels, is related to the probability of treatment response at a certain stage of the disease.

CSF Biomarkers of Toxicity and Side Effects

CSF biomarkers may detect signs of drug-induced side effects, including meningoencephalitis, which was a side effect of active Aβ immunotherapy in early trials [86]. Also, CSF profiling at baseline may identify immunoactivities that are present already before treatment (e.g., chronic infection or inflammation), to avoid the risk of misinterpreting inflammatory reactions as adversary effects [87].

CSF Biomarkers of Treatment Effects

CSF Aβ42, T-tau, and P-tau are normally stable or change very little during at least 2–4 years in symptomatic AD, even during treatment with acetylcholine esterase inhibitors [39, 53, 54], which may be useful for identification of drug effects on pathological processes. Biomarkers of drug effects may be classified as primary, secondary, or exploratory pharmacodynamic biomarkers.

Primary biomarkers reflect the intended drug target, for example, CSF measurements of Aβ metabolism for anti-Aβ therapies. Proof-of-concept studies have shown that several classes of therapies directed against Aβ, including aggregation inhibitors [88], BACE1-inhibitors [36, 89], and γ-secretase inhibitors and modulators [90], result in altered CSF (and plasma) levels of different Aβ-related biomarkers. Besides well-studied markers, including CSF Aβ1-40, Aβ1-42, and sAPPβ, many other Aβ peptides are potentially useful as pharmacodynamic biomarkers. For example, γ-secretase inhibition results in increased CSF levels of short Aβ isoforms, such as Aβ1-14, Aβ1-15, and Aβ1-16, and increased levels of long isoforms, from Aβ1-17 and up [91]. Other Aβ peptides (Aβ5-40 and Aβ5-42) may be upregulated during BACE1 inhibition [36]. Measurement of these peptides may be a useful complement to the core biomarkers for specific drug classes.

Secondary pharmacodynamic biomarkers reflect effects on pathological processes downstream of the intended drug target. This includes CSF tau levels for anti-Aβ drugs, since reduced CSF tau levels may indicate lower axonal degeneration after successful blockage of pathological Aβ metabolism. Some Aβ immunotherapy trials have reported reduced CSF tau levels in patients receiving active treatment, suggesting beneficial drug effects on axonal degeneration [92, 93].

Finally, CSF biomarkers may be used as exploratory pharmacodynamic biomarkers, to identify novel drug effects. For example, CSF biomarkers have been measured in patients treated with statins, which was suggested in epidemiological studies to effect AD, although their mechanisms of action are unclear. In presymptomatic carriers of PSEN1 mutations, HMG-CoA reductase inhibitors lowered CSF sAPP-α and sAPP-β levels, without changing CSF Aβ42, P-tau, or T-tau, suggesting that the treatment interfered with APP processing, but not with Aβ plaque pathology or axonal degeneration [94].

Surrogate Biomarkers

The term surrogate biomarker is a regulatory term, indicating a measurement that may serve as a surrogate for a clinical outcome in a specific treatment [95]. The regulatory framework for surrogate markers is very stringent and requires extensive studies of drug effects on both clinical outcome and biomarker response. The extensive studies necessary to qualify a surrogate marker are essentially the same studies that the surrogate was intended to avoid, making the number of surrogate biomarkers in all of medicine very small. CSF biomarkers are unlikely to have broad use as surrogate markers in the regulatory meaning during the foreseeable future. However, if multiple AD drugs show clinical effects coupled to a specific biomarker response, it may result in the qualification of a surrogate biomarker, facilitating the development of next-generation AD drugs.

Technologies for Measurements of Core CSF AD Biomarkers

Several analytical methods are used to measure CSF Aβ42, T-tau, and P-tau. Presently, the most commonly used assays for Aβ42 include an enzyme-linked immunosorbent assay (ELISA) [96], a bead-based multiplex assay for the xMAP platform [97] (both the ELISA and the xMAP assay measure peptides containing the N-terminal 1st amino acid and the C-terminal 42nd amino acid of the Aβ sequence, Aβ1-42), and a plate-based multiplex assay for the Meso Scale Discovery platform [98] (which also detects N-terminal truncated isoforms, AβX-42, although these have minor concentrations relative to Aβ1-42). These assays are believed to measure monomeric Aβ42, rather than aggregated or oligomeric peptides, but concentrations correlate well with the total Aβ42 amount, as measured by a selected reaction monitoring mass-spectrometry method [99]. T-tau and P-tau may also be analyzed by immunoassays, where T-tau assays are constructed to be independent of tau phosphorylation state [100] and P-tau assays are constructed to be specific to tau phosphorylated at amino acid residues 181 or 231 [97, 101, 102]. A popular multiplex xMAP assay simultaneously measures CSF Aβ42, T-tau, and P-tau [97].

Different technologies report different absolute quantifications and may also differ in terms of specific molecules that they actually measure. However, comparisons between ELISA, xMAP, and Meso Scale Discovery show good agreement between the different technologies, especially for T-tau and P-tau [103], and conversion factors may be used to transfer data between technologies. Several other assay formats are currently under development, which may facilitate automation and large-scale clinical use.

Mass-spectrometry-based methods have been used to identify and quantify a large number of different Aβ isoforms (Fig. 9.1), which may be used both for clinical applications and basic research [34, 104]. In combination with mass spectrometry, Stable Isotope Labeling Kinetics (SILK) may be used to measure production and clearance rates of proteins. For this, subjects are administered a stable isotope-labeled amino acid (e.g., $^{13}C_6$ leucine), which becomes incorporated into proteins during normal protein synthesis. Body fluid samples, including CSF, may then be analyzed to compare fractions of labeled versus unlabeled proteins. This technique has been used to determine production and clearance rates of Aβ peptides [31, 32, 105].

Novel Candidate CSF AD Biomarkers

Besides CSF Aβ42, T-tau, and P-tau, several other molecules have been investigated as potential AD biomarkers for diagnosis, prognosis, treatment monitoring, and understanding of basic disease mechanisms [106]. Biomarker discovery may be done by either targeted methods, where a pre-hoc identified molecule is tested for a certain performance, or general methods, where a large number of different molecules are screened and tested simultaneously. Furthermore, identification of novel biomarkers may be done using either clinical information (e.g., comparing biomarker levels between controls, MCI or AD) or a biological trait, for example, Aβ42 pathology or tau pathology, as measured by CSF biomarkers.

Biomarkers of Aβ and APP Metabolism

Several CSF biomarkers related to Aβ metabolism have been explored for AD diagnosis, including different Aβ and APP peptides, and activities of APP processing enzymes. Since Aβ40 levels are unaltered in AD, the ratio between Aβ40 and Aβ42 has been suggested to be a better indicator of AD than Aβ42 alone [107], and this has also been extended to include a large panel of Aβ variants [104]. The enzyme BACE1, which has a rate-limiting function in the formation of Aβ peptides, exists in a soluble form that is measurable in CSF. A few initial studies found increased CSF BACE1 activity in AD or MCI [108–110], but this was not replicated in all studies [111]. However, it is possible that BACE1 levels are increased only early in the disease [111, 112]. A few studies have found increased CSF sAPP-α or sAPP-β in MCI or AD (especially in subjects with pathological CSF Aβ42 or T-tau) [113, 114], but not all studies replicated this [115], and the overlap between groups is likely too large for diagnostic utility. Finally, several studies have measured CSF levels of Aβ oligomers, although these are difficult to quantify and characterize, and results have varied [116–119].

Biomarkers of γ-Secretase Function

γ-Secretase, which liberates the Aβ peptide from the remaining C-terminal APP stub after BACE1-cleavage, is a general proteolytic enzyme residing in the cellular membrane. It has more than 100 known substrates, and several of these are present in CSF. For example, the protein alcadein has been shown to be processed by γ-secretase into several smaller peptides, in a fashion similar to APP, and some of these fragments are present in CSF in concentrations suggesting γ-secretase dysfunction in MCI [120]. Such non-Aβ-related markers of γ-secretase function may be very useful to study the pathogenesis of AD and to evaluate drug effects.

Microglial Markers

Since Aβ plaques are surrounded by activated microglia cells, many CSF markers of inflammatory and microglial activity have been studied in AD [106]. These markers include chitotriosidase enzyme activity [121] and concentrations of YKL-40 [122], which are upregulated in CSF from AD patients.

Synaptic Markers

As synaptic dysfunction is a hallmark of early AD [123], synaptic markers would be a highly interesting class of biomarkers. Several presynaptic and postsynaptic proteins, including rab3A, synaptotagmin, growth-associated protein 43 (GAP43), synaptosomal-associated protein 25 (SNAP25), and neurogranin, have been identified in CSF using protein purification and mass spectrometric techniques [124]. CSF levels of GAP43 and neurogranin are elevated in AD [125, 126], but other markers await better assays to be quantified in CSF in neurodegenerative conditions.

CSF Biomarkers in Relation to AD Genetics

The concept of using CSF biomarkers to enrich genetic studies with patients with AD pathology and to exclude preclinical AD from the controls is supported by a study showing that the odds ratio of *APOE* increased from four to around ten when combining clinical with biomarker data [127]. However, another study failed to show any association between the AD risk genes *BIN1*, *CLU*, *CR1*, and *PICALM* and CSF Aβ42 and P-tau, despite being powered to detect very small effects, suggesting that some AD risk genes mediate risk through Aβ and tau-independent mechanisms [128]. CSF biomarkers have also been used as quantitative traits for genetic analysis to find new risk loci for AD [129].

Relation to Imaging

Combining CSF biomarkers with neuroimaging may help to identify pathological processes underlying biomarker abnormalities and improve the diagnostic performance compared to using individual biomarkers alone. Low CSF Aβ42 is highly correlated with elevated Aβ PET signals [130], which suggests that combining the two modalities provide little benefit. However, some persons may have reduced CSF Aβ42 levels despite low Aβ PET-signal, and the significance of this is still unclear [131].

The classification of controls, MCI and AD dementia, and the prediction of conversion from MCI to AD dementia may be improved by combining CSF and imaging markers (structural MRI [132–134] and functional imaging with FDG PET [135]). In AD patients, CSF Aβ42 and T-tau at baseline are correlated with longitudinal hippocampal atrophy rates [136], and in cognitively healthy elderly, reduced CSF Aβ42 is associated with brain atrophy rates [67].

There is much heterogeneity in biomarker patterns among healthy controls and MCI subjects. For example, healthy controls with MRI gray matter loss indicative of AD are at risk of developing cognitive impairment, but only 60 % of those with an AD-like pattern have reduced CSF Aβ42, compared to 19 % of those without [137]. Considering the dynamic biomarker model, it may be surprising that 40 % of healthy controls with AD-like brain atrophy have nonreduced CSF Aβ42 levels. This suggests that AD-like brain atrophy may develop without concomitant brain Aβ pathology as measurable by current available methods [138] (it remains possible that nondetected Aβ pathology is present).

CSF Biomarkers in FTD

FTD is a clinical term encompassing a heterogeneous group of neurodegenerative disorders, including the behavioral variant (bvFTD), progressive aphasia, semantic dementia (SD), and progressive nonfluent aphasia (PNFA), and these diseases are also associated with other neurological conditions, including corticobasal degeneration (CBD), PSP, and motor neuron disease (MND) [139]. All these entities are linked to molecular pathologies known as frontotemporal lobar degenerations (FTLDs), which is a pathological term. Different FTLDs are classified according to the dominating protein found in pathological inclusions in the disease. In most cases the inclusions consist of tau proteins, transactive response DNA-binding protein of 43 kDa

(TDP-43), or tumor-associated protein fused in sarcoma (FUS). The relationships between the clinical FTD entities and the neuropathological FTLD hallmarks are complex and variable, and the biological mechanisms underlying the different associations are not well understood. For example, bvFTD may be caused by different types of inclusion, while SD cases are dominated by TDP-43 inclusions and PNFA by tau inclusions (see [140] for review). Several CSF biomarkers have been investigated in FTDs, including Aβ42, tau, neurofilament light (NFL), TDP-43, and progranulin.

CSFAβ42 and Tau in FTD

FTD patients generally have higher CSF Aβ42 and lower T-tau and P-tau levels than AD patients, although with some overlap between the dementias, which may be explained by co-occurrence of AD pathology in some FTD patients [141] (most studies lack autopsy verification). FTD patients may have disease-causing mutations in the *MAPT* gene, which encodes for the tau protein, but these still have lower CSF tau levels than AD patients. Recently it was suggested that neurons with *MAPT* mutations have reduced release of extracellular tau, although mutations do not change the concentrations of intracellular tau [142]. This may partly explain why FTD patients with *MAPT* mutations lack prominently increased CSF tau levels.

Neurofilament Light

NFL is one of three neurofilament proteins (the others are the heavy [NFH] and intermediate [NFM] chains), which are important cytoskeletal proteins, predominantly found in large diameter myelinated axons. CSF NFL levels are increased in many conditions with damage to these axons, including normal aging [143], acute cerebral infarctions and vascular dementia [144], atypical parkinsonian disorders (including CBD [145]), white matter disease [146], and amyotrophic lateral sclerosis (ASL [147]).

CSF NFL (and NFH) levels are increased in FTD compared to AD dementia and controls [148]. One study found highest CSF NFL levels in bvFTD and SD, while PNFA patients tended to have lower levels [149]. The same study found that among neuropathologically confirmed cases, CSF NFL was highest in tau-negative pathology and lowest in tau-positive pathology.

Novel CSF FTD Biomarkers

Several studies have explored CSF proteins potentially linked to mutations or pathological inclusions seen in FTD, including CSF TDP-43, which may be increased in FTD and ASL [150, 151]. A large proportion of hereditary FTD cases are caused by mutations in the *GRN* gene, which encodes the secreted protein progranulin [152]. *GNR* mutations have no specificity to any clinical entity in the FTD spectrum, and different diseases may even appear among mutation carriers within the same families. But individuals carrying *GNR* mutations have reduced CSF (and plasma) levels of progranulin, enabling screening tests with high sensitivity and specificity for mutation carriers versus controls or patients with other dementias, including AD [152]. Finally, general proteomics strategies have been used for unbiased detection of novel FTD biomarkers [153, 154], but proteins identified this way need to be verified by independent studies.

Future Challenges

CSF biomarkers are already being used in clinical care in some countries, and with the development of disease-modifying therapies, we foresee that this will increase in the future. However, although the novel AD criteria provide a conceptual framework for biomarker use in AD [8–12], there are still no universally accepted clinical guidelines for use of CSF biomarkers in cognitive investigation. Such guidelines need to cover how CSF analyses shall be weighted toward neuroimaging, management of patients with contradictory biomarker results, and the ethical implications of disclosing diagnoses in early disease stages [155]. We anticipate that the development of such guidelines for AD and other dementias, and the standardization of procedures between centers and laboratories, will be a major topic in clinical dementia research during the next coming years.

Biomarker Trajectories and the Importance of Aβ Pathology

Most studies on CSF biomarkers in preclinical AD have been cross-sectional or have only had a few years follow-up, which should be compared to the two or three decades that it likely takes from the first biomarker signs of pathological Aβ metabolism to dementia. Studies with longer follow-up are needed to clarify exactly how different biomarkers develop over time and to clarify the clinical implication of the brain amyloidosis that is commonly seen in cognitively healthy elderly. This will be crucial if disease-modifying treatment becomes available, since such treatments will raise the question of broad-scale screening, and the biomarkers' performances are heavily impacted by the true disease prevalence [156].

Standardizing Biomarker Measurements

Biomarker measurements vary within and across centers [157], due to many different pre-analytical and analytical confounding factors that affect the biomarker results [158]. This type of variability is not unique to dementia biomarkers, but a general concern in laboratory medicine, and external quality control programs have been initiated to monitor it [159]. The largest of these programs is the Alzheimer's Association Quality Control program, which runs with several rounds every year, and which has reported biomarker variability around 25–30 % across centers [160, 161].

To overcome the variability, there is a need for CSF AD biomarker assays that can stably quantify CSF biomarkers with high analytical precision over long time periods and across centers. For this, several groups are developing mass-spectrometry-based methods, with the aim to allow absolute quantifications in certified, well-characterized, long-term stable reference materials. One such candidate reference method for CSF Aβ42 has been published [99], and work is ongoing for tau markers. Reference materials are being constructed in collaboration with the Institute for Reference Materials and Measurements [23]. These materials will be made available at self-cost for assay vendors to harmonize calibration systems for the same analyte.

Conclusion

CSF biomarkers have increased our understanding of pathological processes in AD and FTD and may be used to diagnose different neurodegenerative conditions. Future challenges include the establishment of universal guidelines for clinical biomarker use, and the development of accurate and stable laboratory methods, to allow standardization of measurements between centers.

References

1. Biomarkers Definitions Working Group. Biomarkers and surrogate endpoints: preferred definitions and conceptual framework. Clin Pharmacol Ther. 2001;69(3):89–95.
2. Quincke H. Die Technik der Lumbalpunction. Berlin: Urban & Schwarzenberg; 1902.
3. Johanson CE, Duncan JA, Klinge PM, Brinker T, Stopa EG, Silverberg GD. Multiplicity of cerebrospinal fluid functions: new challenges in health and disease. Cerebrospinal Fluid Res. 2008;5:10.
4. Zetterberg H, Tullhog K, Hansson O, Minthon L, Londos E, Blennow K. Low incidence of post-lumbar puncture headache in 1,089 consecutive memory clinic patients. Eur Neurol. 2010;63:326–30.
5. Evans RW, Armon C, Frohman EM, Goodin DS. Assessment: prevention of post-lumbar puncture headaches: report of the Therapeutics and Technology Assessment Subcommittee of the American Academy of Neurology. Neurology. 2000;55(7):909–14.
6. Peskind ER, Riekse R, Quinn JF, Kaye J, Clark CM, Farlow MR, et al. Safety and acceptability of the research lumbar puncture. Alzheimer Dis Assoc Disord. 2005;19:220–5.
7. Vilming ST, Kloster R. Post-lumbar puncture headache: clinical features and suggestions for diagnostic criteria. Cephalalgia. 1997;17(7):778–84.

8. Dubois B, Feldman HH, Jacova C, Dekosky ST, Barberger-Gateau P, Cummings J, et al. Research criteria for the diagnosis of Alzheimer's disease: revising the NINCDS-ADRDA criteria. Lancet Neurol. 2007;6(8):734–46.
9. Dubois B, Feldman HH, Jacova C, Cummings JL, Dekosky ST, Barberger-Gateau P, et al. Revising the definition of Alzheimer's disease: a new lexicon. Lancet Neurol. 2010;9: 1118–27.
10. Sperling RA, Aisen PS, Beckett LA, Bennett DA, Craft S, Fagan AM, et al. Toward defining the preclinical stages of Alzheimer's disease: recommendations from the National Institute on Aging-Alzheimer's Association workgroups on diagnostic guidelines for Alzheimer's disease. Alzheimers Dement. 2011;7(3):280–92.
11. Albert MS, DeKosky ST, Dickson D, Dubois B, Feldman HH, Fox NC, et al. The diagnosis of mild cognitive impairment due to Alzheimer's disease: recommendations from the National Institute on Aging-Alzheimer's Association workgroups on diagnostic guidelines for Alzheimer's disease. Alzheimers Dement. 2011;7:270–9.
12. McKhann GM, Knopman DS, Chertkow H, Hyman BT, Jack CR, Kawas CH, et al. The diagnosis of dementia due to Alzheimer's disease: recommendations from the National Institute on Aging-Alzheimer's Association workgroups on diagnostic guidelines for Alzheimer's disease. Alzheimers Dement. 2011;7:263–9.
13. Consensus report of the Working Group on: "Molecular and Biochemical Markers of Alzheimer's Disease". The Ronald and Nancy Reagan Research Institute of the Alzheimer's Association and the National Institute on Aging Working Group. Neurobiol Aging. 1998;19: 109–16.
14. Fagan AM, Mintun MA, Mach RH, Lee SY, Dence CS, Shah AR, et al. Inverse relation between in vivo amyloid imaging load and cerebrospinal fluid Abeta42 in humans. Ann Neurol. 2006;59:512–9.
15. Strozyk D, Blennow K, White LR, Launer LJ. CSF Abeta 42 levels correlate with amyloid-neuropathology in a population-based autopsy study. Neurology. 2003;60:652–6.
16. Blennow K, Hampel H, Weiner M, Zetterberg H. Cerebrospinal fluid and plasma biomarkers in Alzheimer disease. Nat Rev Neurol. 2010;6:131–44.
17. Shaw LM, Vanderstichele H, Knapik-Czajka M, Clark CM, Aisen PS, Petersen RC, et al. Cerebrospinal fluid biomarker signature in Alzheimer's disease neuroimaging initiative subjects. Ann Neurol. 2009;65:403–13.
18. Mattsson N, Zetterberg H, Hansson O, Andreasen N, Parnetti L, Jonsson M, et al. CSF biomarkers and incipient Alzheimer disease in patients with mild cognitive impairment. JAMA. 2009;302:385–93.
19. Hansson O, Zetterberg H, Buchhave P, Londos E, Blennow K, Minthon L. Association between CSF biomarkers and incipient Alzheimer's disease in patients with mild cognitive impairment: a follow-up study. Lancet Neurol. 2006;5:228–34.
20. Buchhave P, Minthon L, Zetterberg H, Wallin AK, Blennow K, Hansson O. Cerebrospinal fluid levels of beta-amyloid 1–42, but not of tau, are fully changed already 5 to 10 years before the onset of Alzheimer dementia. Arch Gen Psychiatry. 2012;69:98–106.
21. Skoog I, Davidsson P, Aevarsson O, Vanderstichele H, Vanmechelen E, Blennow K. Cerebrospinal fluid beta-amyloid 42 is reduced before the onset of sporadic dementia: a population-based study in 85-year-olds. Dement Geriatr Cogn Disord. 2003;15:169–76.
22. Bateman RJ, Xiong C, Benzinger TL, Fagan AM, Goate A, Fox NC, et al. Clinical and biomarker changes in dominantly inherited Alzheimer's disease. N Engl J Med. 2012;367: 795–804.
23. Carrillo MC, Blennow K, Soares H, Lewczuk P, Mattsson N, Oberoi P, et al. Global standardization measurement of cerebral spinal fluid for Alzheimer's disease: an update from the Alzheimer's Association Global Biomarkers Consortium. Alzheimers Dement. 2013;9(2): 137–40.
24. Cirrito JR, Yamada KA, Finn MB, Sloviter RS, Bales KR, May PC, et al. Synaptic activity regulates interstitial fluid amyloid-beta levels in vivo. Neuron. 2005;48:913–22.

25. Motter R, Vigo-Pelfrey C, Kholodenko D, Barbour R, Johnson-Wood K, Galasko D, et al. Reduction of beta-amyloid peptide42 in the cerebrospinal fluid of patients with Alzheimer's disease. Ann Neurol. 1995;38:643–8.
26. Tapiola T, Alafuzoff I, Herukka SK, Parkkinen L, Hartikainen P, Soininen H, et al. Cerebrospinal fluid {beta}-amyloid 42 and tau proteins as biomarkers of Alzheimer-type pathologic changes in the brain. Arch Neurol. 2009;66:382–9.
27. Grimmer T, Riemenschneider M, Förstl H, Henriksen G, Klunk WE, Mathis CA, et al. Beta amyloid in Alzheimer's disease: increased deposition in brain is reflected in reduced concentration in cerebrospinal fluid. Biol Psychiatry. 2009;65(11):927–34.
28. Stenh C, Englund H, Lord A, Johansson AS, Almeida CG, Gellerfors P, et al. Amyloid-beta oligomers are inefficiently measured by enzyme-linked immunosorbent assay. Ann Neurol. 2005;58:147–50.
29. Kanekiyo T, Ban T, Aritake K, Huang ZL, Qu WM, Okazaki I, et al. Lipocalin-type prostaglandin D synthase/beta-trace is a major amyloid beta-chaperone in human cerebrospinal fluid. Proc Natl Acad Sci U S A. 2007;104:6412–7.
30. LaFerla FM, Green KN, Oddo S. Intracellular amyloid-beta in Alzheimer's disease. Nat Rev Neurosci. 2007;8:499–509.
31. Bateman RJ, Munsell LY, Morris JC, Swarm R, Yarasheski KE, Holtzman DM. Human amyloid-beta synthesis and clearance rates as measured in cerebrospinal fluid in vivo. Nat Med. 2006;12:856–61.
32. Mawuenyega KG, Sigurdson W, Ovod V, Munsell L, Kasten T, Morris JC, et al. Decreased clearance of CNS beta-amyloid in Alzheimer's disease. Science. 2010;330(6012):1774.
33. Wiltfang J, Esselmann H, Bibl M, Smirnov A, Otto M, Paul S, et al. Highly conserved and disease-specific patterns of carboxyterminally truncated Abeta peptides 1-37/38/39 in addition to 1-40/42 in Alzheimer's disease and in patients with chronic neuroinflammation. J Neurochem. 2002;81:481–96.
34. Portelius E, Westman-Brinkmalm A, Zetterberg H, Blennow K. Determination of beta-amyloid peptide signatures in cerebrospinal fluid using immunoprecipitation-mass spectrometry. J Proteome Res. 2006;5:1010–6.
35. Portelius E, Zhang B, Gustavsson MK, Brinkmalm G, Westman-Brinkmalm A, Zetterberg H, et al. Effects of gamma-secretase inhibition on the amyloid beta isoform pattern in a mouse model of Alzheimer's disease. Neurodegener Dis. 2009;6:258–62.
36. Mattsson N, Rajendran L, Zetterberg H, Gustavsson M, Andreasson U, Olsson M, et al. BACE1 inhibition induces a specific cerebrospinal fluid beta-amyloid pattern that identifies drug effects in the central nervous system. PLoS One. 2012;7:e31084.
37. Goedert M, Spillantini MG, Jakes R, Rutherford D, Crowther RA. Multiple isoforms of human microtubule-associated protein tau: sequences and localization in neurofibrillary tangles of Alzheimer's disease. Neuron. 1989;3(4):519–26.
38. Buchhave P, Blennow K, Zetterberg H, Stomrud E, Londos E, Andreasen N, et al. Longitudinal study of CSF biomarkers in patients with Alzheimer's disease. PLoS One. 2009;4:e6294.
39. Zetterberg H, Pedersen M, Lind K, Svensson M, Rolstad S, Eckerstrom C, et al. Intra-individual stability of CSF biomarkers for Alzheimer's disease over two years. J Alzheimers Dis. 2007;12:255–60.
40. Buerger K, Ewers M, Pirttila T, Zinkowski R, Alafuzoff I, Teipel SJ, et al. CSF phosphorylated tau protein correlates with neocortical neurofibrillary pathology in Alzheimer's disease. Brain. 2006;129:3035–41.
41. Seppala TT, Nerg O, Koivisto AM, Rummukainen J, Puli L, Zetterberg H, et al. CSF biomarkers for Alzheimer disease correlate with cortical brain biopsy findings. Neurology. 2012;78:1568–75.
42. Luk C, Compta Y, Magdalinou N, Martí MJ, Hondhamuni G, Zetterberg H, et al. Development and assessment of sensitive immuno-PCR assays for the quantification of cerebrospinal fluid three- and four-repeat tau isoforms in tauopathies. J Neurochem. 2012;123(3):396–405.

43. Jack CR, Knopman DS, Jagust WJ, Shaw LM, Aisen PS, Weiner MW, et al. Hypothetical model of dynamic biomarkers of the Alzheimer's pathological cascade. Lancet Neurol. 2010;9:119–28.
44. Jack Jr CR, Knopman DS, Jagust WJ, Petersen RC, Weiner MW, Aisen PS, et al. Tracking pathophysiological processes in Alzheimer's disease: an updated hypothetical model of dynamic biomarkers. Lancet Neurol. 2013;12(2):207–16.
45. Hardy J, Selkoe DJ. The amyloid hypothesis of Alzheimer's disease: progress and problems on the road to therapeutics. Science. 2002;297:353–6.
46. Jack Jr CR, Vemuri P, Wiste HJ, Weigand SD, Aisen PS, Trojanowski JQ, et al. Evidence for ordering of Alzheimer disease biomarkers. Arch Neurol. 2011;68(12):1526–35.
47. Ingelsson M, Fukumoto H, Newell KL, Growdon JH, Hedley-Whyte ET, Frosch MP, et al. Early Abeta accumulation and progressive synaptic loss, gliosis, and tangle formation in AD brain. Neurology. 2004;62(6):925–31.
48. Jack Jr CR, Vemuri P, Wiste HJ, Weigand SD, Lesnick TG, Lowe V, et al. Shapes of the trajectories of 5 major biomarkers of Alzheimer disease. Arch Neurol. 2012;69(7):856–67.
49. Bloudek LM, Spackman DE, Blankenburg M, Sullivan SD. Review and meta-analysis of biomarkers and diagnostic imaging in Alzheimer's disease. J Alzheimers Dis. 2011;26:627–45.
50. Samgard K, Zetterberg H, Blennow K, Hansson O, Minthon L, Londos E. Cerebrospinal fluid total tau as a marker of Alzheimer's disease intensity. Int J Geriatr Psychiatry. 2010;25: 403–10.
51. Visser PJ, Verhey F, Knol DL, Scheltens P, Wahlund LO, Freund-Levi Y, et al. Prevalence and prognostic value of CSF markers of Alzheimer's disease pathology in patients with subjective cognitive impairment or mild cognitive impairment in the DESCRIPA study: a prospective cohort study. Lancet Neurol. 2009;8:619–27.
52. Van Rossum IA, Vos SJB, Burns L, Knol DL, Scheltens P, Soininen H, et al. Injury markers predict time to dementia in subjects with MCI and amyloid pathology. Neurology. 2012;79(17):1809–16.
53. Blennow K, Zetterberg H, Minthon L, Lannfelt L, Strid S, Annas P, et al. Longitudinal stability of CSF biomarkers in Alzheimer's disease. Neurosci Lett. 2007;419:18–22.
54. Mattsson N, Portelius E, Rolstad S, Gustavsson M, Andreasson U, Stridsberg M, et al. Longitudinal cerebrospinal fluid biomarkers over four years in mild cognitive impairment. J Alzheimers Dis. 2012;30:767–78.
55. Mattsson N, Rosen E, Hansson O, Andreasen N, Parnetti L, Jonsson M, et al. Age and diagnostic performance of Alzheimer disease CSF biomarkers. Neurology [Internet]. 2012. Available from: http://www.ncbi.nlm.nih.gov/pubmed/22302554.
56. Reiman EM, Quiroz YT, Fleisher AS, Chen K, Velez-Pardo C, Jimenez-Del-Rio M, et al. Brain imaging and fluid biomarker analysis in young adults at genetic risk for autosomal dominant Alzheimer's disease in the presenilin 1 E280A kindred: a case-control study. Lancet Neurol. 2012;11(12):1048–56.
57. Ringman JM, Coppola G, Elashoff D, Rodriguez-Agudelo Y, Medina LD, Gylys K, et al. Cerebrospinal fluid biomarkers and proximity to diagnosis in preclinical familial Alzheimer's disease. Dement Geriatr Cogn Disord. 2012;33(1):1–5.
58. Moonis M, Swearer JM, Dayaw MPE, St George-Hyslop P, Rogaeva E, Kawarai T, et al. Familial Alzheimer disease: decreases in CSF Abeta42 levels precede cognitive decline. Neurology. 2005;65(2):323–5.
59. Ringman JM, Younkin SG, Pratico D, Seltzer W, Cole GM, Geschwind DH, et al. Biochemical markers in persons with preclinical familial Alzheimer disease. Neurology. 2008;71:85–92.
60. Fagan AM, Roe CM, Xiong C, Mintun MA, Morris JC, Holtzman DM. Cerebrospinal fluid tau/beta-amyloid(42) ratio as a prediction of cognitive decline in nondemented older adults. Arch Neurol. 2007;64:343–9.
61. Gustafson DR, Skoog I, Rosengren L, Zetterberg H, Blennow K. Cerebrospinal fluid beta-amyloid 1–42 concentration may predict cognitive decline in older women. J Neurol Neurosurg Psychiatry. 2007;78:461–4.

62. Li G, Sokal I, Quinn JF, Leverenz JB, Brodey M, Schellenberg GD, et al. CSF tau/Abeta42 ratio for increased risk of mild cognitive impairment: a follow-up study. Neurology. 2007;69:631–9.
63. Stomrud E, Hansson O, Blennow K, Minthon L, Londos E. Cerebrospinal fluid biomarkers predict decline in subjective cognitive function over 3 years in healthy elderly. Dement Geriatr Cogn Disord. 2007;24:118–24.
64. Van Harten AC, Visser PJ, Pijnenburg YAL, Teunissen CE, Blankenstein MA, Scheltens P, et al. Cerebrospinal fluid Aβ42 is the best predictor of clinical progression in patients with subjective complaints. Alzheimers Dement. 2013;9:481–7.
65. Desikan RS, McEvoy LK, Thompson WK, Holland D, Brewer JB, Aisen PS, et al. Amyloid-β–associated clinical decline occurs only in the presence of elevated P-tau. Arch Neurol. 2012;69(6):709–13.
66. Roe CM, Fagan AM, Grant EA, Hassenstab J, Moulder KL, Maue Dreyfus D, et al. Amyloid imaging and CSF biomarkers in predicting cognitive impairment up to 7.5 years later. Neurology. 2013;80(19):1784–91.
67. Schott JM, Bartlett JW, Fox NC, Barnes J, Alzheimer's Disease Neuroimaging Initiative Investigators. Increased brain atrophy rates in cognitively normal older adults with low cerebrospinal fluid Aβ1-42. Ann Neurol. 2010;68(6):825–34.
68. Roe CM, Fagan AM, Williams MM, Ghoshal N, Aeschleman M, Grant EA, et al. Improving CSF biomarker accuracy in predicting prevalent and incident Alzheimer disease. Neurology. 2011;76(6):501–10.
69. Roe CM, Fagan AM, Grant EA, Marcus DS, Benzinger TLS, Mintun MA, et al. Cerebrospinal fluid biomarkers, education, brain volume, and future cognition. Arch Neurol. 2011;68(9):1145–51.
70. Bibl M, Mollenhauer B, Esselmann H, Schneider M, Lewczuk P, Welge V, et al. Cerebrospinal fluid neurochemical phenotypes in vascular dementias: original data and mini-review. Dement Geriatr Cogn Disord. 2008;25(3):256–65.
71. Mollenhauer B, Esselmann H, Trenkwalder C, Schulz-Schaeffer W, Kretzschmar H, Otto M, et al. CSF amyloid-β peptides in neuropathologically diagnosed dementia with Lewy bodies and Alzheimer's disease. J Alzheimers Dis. 2011;24(2):383–91.
72. Bibl M, Gallus M, Welge V, Esselmann H, Wolf S, Rüther E, et al. Cerebrospinal fluid amyloid-β 2–42 is decreased in Alzheimer's, but not in frontotemporal dementia. J Neural Transm. 2012;119(7):805–13.
73. Van Everbroeck B, Green AJE, Pals P, Martin JJ, Cras P. Decreased levels of amyloid-beta 1–42 in cerebrospinal fluid of Creutzfeldt-Jakob disease patients. J Alzheimers Dis. 1999;1(6):419–24.
74. Sjogren M, Davidsson P, Wallin A, Granerus AK, Grundstrom E, Askmark H, et al. Decreased CSF-beta-amyloid 42 in Alzheimer's disease and amyotrophic lateral sclerosis may reflect mismetabolism of beta-amyloid induced by disparate mechanisms. Dement Geriatr Cogn Disord. 2002;13:112–8.
75. Holmberg B, Johnels B, Blennow K, Rosengren L. Cerebrospinal fluid Abeta42 is reduced in multiple system atrophy but normal in Parkinson's disease and progressive supranuclear palsy. Mov Disord. 2003;18:186–90.
76. Toledo JB, Brettschneider J, Grossman M, Arnold SE, Hu WT, Xie SX, et al. CSF biomarkers cutoffs: the importance of coincident neuropathological diseases. Acta Neuropathol. 2012;124:23–35.
77. Augutis K, Axelsson M, Portelius E, Brinkmalm G, Andreasson U, Gustavsson MK, et al. Cerebrospinal fluid biomarkers of β-amyloid metabolism in multiple sclerosis. Mult Scler. 2013;19(5):543–52.
78. Gisslen M, Krut J, Andreasson U, Blennow K, Cinque P, Brew BJ, et al. Amyloid and tau cerebrospinal fluid biomarkers in HIV infection. BMC Neurol. 2009;9:63.
79. Sjogren M, Gisslen M, Vanmechelen E, Blennow K. Low cerebrospinal fluid beta-amyloid 42 in patients with acute bacterial meningitis and normalization after treatment. Neurosci Lett. 2001;314:33–6.

80. Hesse C, Rosengren L, Andreasen N, Davidsson P, Vanderstichele H, Vanmechelen E, et al. Transient increase in total tau but not phospho-tau in human cerebrospinal fluid after acute stroke. Neurosci Lett. 2001;297:187–90.
81. Zetterberg H, Hietala MA, Jonsson M, Andreasen N, Styrud E, Karlsson I, et al. Neurochemical aftermath of amateur boxing. Arch Neurol. 2006;63:1277–80.
82. Otto M, Wiltfang J, Tumani H, Zerr I, Lantsch M, Kornhuber J, et al. Elevated levels of tau-protein in cerebrospinal fluid of patients with Creutzfeldt-Jakob disease. Neurosci Lett. 1997;225:210–2.
83. Krut JJ, Zetterberg H, Blennow K, Cinque P, Hagberg L, Price RW, et al. Cerebrospinal fluid Alzheimer's biomarker profiles in CNS infections. J Neurol. 2013;260(2):620–6.
84. Mattsson N, Savman K, Osterlundh G, Blennow K, Zetterberg H. Converging molecular pathways in human neural development and degeneration. Neurosci Res. 2010;66:330–2.
85. Isaac M, Vamvakas S, Abadie E, Jonsson B, Gispen C, Pani L. Qualification opinion of novel methodologies in the predementia stage of Alzheimer's disease: cerebro-spinal-fluid related biomarkers for drugs affecting amyloid burden–regulatory considerations by European Medicines Agency focusing in improving benefit/risk in regulatory trials. Eur Neuropsychopharmacol. 2011;21:781–8.
86. Orgogozo JM, Gilman S, Dartigues JF, Laurent B, Puel M, Kirby LC, et al. Subacute meningoencephalitis in a subset of patients with AD after Abeta42 immunization. Neurology. 2003;61:46–54.
87. Andreasen N, Blennow K, Zetterberg H. Neuroinflammation screening in immunotherapy trials against Alzheimer's disease. Int J Alzheimers Dis. 2010;2010:638379.
88. Lannfelt L, Blennow K, Zetterberg H, Batsman S, Ames D, Harrison J, et al. Safety, efficacy, and biomarker findings of PBT2 in targeting Abeta as a modifying therapy for Alzheimer's disease: a phase IIa, double-blind, randomised, placebo-controlled trial. Lancet Neurol. 2008;7:779–86.
89. May PC, Dean RA, Lowe SL, Martenyi F, Sheehan SM, Boggs LN, et al. Robust central reduction of amyloid-beta in humans with an orally available, non-peptidic beta-secretase inhibitor. J Neurosci. 2011;31:16507–16.
90. Portelius E, Dean RA, Gustavsson MK, Andreasson U, Zetterberg H, Siemers E, et al. A novel Abeta isoform pattern in CSF reflects gamma-secretase inhibition in Alzheimer disease. Alzheimers Res Ther. 2010;2:7.
91. Portelius E, Price E, Brinkmalm G, Stiteler M, Olsson M, Persson R, et al. A novel pathway for amyloid precursor protein processing. Neurobiol Aging. 2011;32:1090–8.
92. Blennow K, Zetterberg H, Rinne JO, Salloway S, Wei J, Black R, et al. Effect of immunotherapy with bapineuzumab on cerebrospinal fluid biomarker levels in patients with mild to moderate Alzheimer disease. Arch Neurol. 2012;69(8):1002–10.
93. Gilman S, Koller M, Black RS, Jenkins L, Griffith SG, Fox NC, et al. Clinical effects of Abeta immunization (AN1792) in patients with AD in an interrupted trial. Neurology. 2005;64:1553–62.
94. Hinerfeld DA, Moonis M, Swearer JM, Baker SP, Caselli RJ, Rogaeva E, et al. Statins differentially affect amyloid precursor protein metabolism in presymptomatic PS1 and non-PS1 subjects. Arch Neurol. 2007;64(11):1672–3.
95. Katz R. Biomarkers and surrogate markers: an FDA perspective. NeuroRx. 2004;1:189–95.
96. Andreasen N, Hesse C, Davidsson P, Minthon L, Wallin A, Winblad B, et al. Cerebrospinal fluid beta-amyloid(1–42) in Alzheimer disease: differences between early- and late-onset alzheimer disease and stability during the course of disease. Arch Neurol. 1999;56:673–80.
97. Olsson A, Vanderstichele H, Andreasen N, De Meyer G, Wallin A, Holmberg B, et al. Simultaneous measurement of beta-amyloid(1–42), total tau, and phosphorylated tau (Thr181) in cerebrospinal fluid by the xMAP technology. Clin Chem. 2005;51:336–45.
98. Zetterberg H, Andreasson U, Hansson O, Wu G, Sankaranarayanan S, Andersson ME, et al. Elevated cerebrospinal fluid BACE1 activity in incipient Alzheimer disease. Arch Neurol. 2008;65:1102–7.

99. Pannee J, Portelius E, Oppermann M, Atkins A, Hornshaw M, Zegers I, et al. A selected reaction monitoring (SRM)-based method for absolute quantification of Aβ38, Aβ40, and Aβ42 in cerebrospinal fluid of Alzheimer's disease patients and healthy controls. J Alzheimers Dis. 2013;33(4):1021–32.
100. Blennow K, Wallin A, Agren H, Spenger C, Siegfried J, Vanmechelen E. Tau protein in cerebrospinal fluid: a biochemical marker for axonal degeneration in Alzheimer disease? Mol Chem Neuropathol. 1995;26:231–45.
101. Vanmechelen E, Vanderstichele H, Davidsson P, Van Kerschaver E, Van Der Perre B, Sjogren M, et al. Quantification of tau phosphorylated at threonine 181 in human cerebrospinal fluid: a sandwich ELISA with a synthetic phosphopeptide for standardization. Neurosci Lett. 2000;285:49–52.
102. Kohnken R, Buerger K, Zinkowski R, Miller C, Kerkman D, DeBernardis J, et al. Detection of tau phosphorylated at threonine 231 in cerebrospinal fluid of Alzheimer's disease patients. Neurosci Lett. 2000;287:187–90.
103. Wang L-S, Leung YY, Chang S-K, Leight S, Knapik-Czajka M, Baek Y, et al. Comparison of xMAP and ELISA assays for detecting cerebrospinal fluid biomarkers of Alzheimer's disease. J Alzheimers Dis. 2012;31(2):439–45.
104. Portelius E, Zetterberg H, Andreasson U, Brinkmalm G, Andreasen N, Wallin A, et al. An Alzheimer's disease-specific beta-amyloid fragment signature in cerebrospinal fluid. Neurosci Lett. 2006;409:215–9.
105. Bateman RJ, Siemers ER, Mawuenyega KG, Wen G, Browning KR, Sigurdson WC, et al. A gamma-secretase inhibitor decreases amyloid-beta production in the central nervous system. Ann Neurol [Internet]. 2009. Available from: http://www.ncbi.nlm.nih.gov/entrez/query.fcgi?cmd=Retrieve&db=PubMed&dopt=Citation&list_uids=19360898.
106. Fagan AM, Perrin RJ. Upcoming candidate cerebrospinal fluid biomarkers of Alzheimer's disease. Biomark Med. 2012;6:455–76.
107. Schoonenboom NS, Mulder C, Van Kamp GJ, Mehta SP, Scheltens P, Blankenstein MA, et al. Amyloid beta 38, 40, and 42 species in cerebrospinal fluid: more of the same? Ann Neurol. 2005;58:139–42.
108. Holsinger RM, McLean CA, Collins SJ, Masters CL, Evin G. Increased beta-secretase activity in cerebrospinal fluid of Alzheimer's disease subjects. Ann Neurol. 2004;55:898–9.
109. Verheijen JH, Huisman LG, van Lent N, Neumann U, Paganetti P, Hack CE, et al. Detection of a soluble form of BACE-1 in human cerebrospinal fluid by a sensitive activity assay. Clin Chem. 2006;52:1168–74.
110. Holsinger RM, Lee JS, Boyd A, Masters CL, Collins SJ. CSF BACE1 activity is increased in CJD and Alzheimer disease versus [corrected] other dementias. Neurology. 2006;67:710–2.
111. Rosen C, Andreasson U, Mattsson N, Marcusson J, Minthon L, Andreasen N, et al. Cerebrospinal fluid profiles of amyloid beta-related biomarkers in Alzheimer's disease. Neuromolecular Med. 2012;14:65–73.
112. Zhong Z, Ewers M, Teipel S, Burger K, Wallin A, Blennow K, et al. Levels of beta-secretase (BACE1) in cerebrospinal fluid as a predictor of risk in mild cognitive impairment. Arch Gen Psychiatry. 2007;64:718–26.
113. Olsson A, Hoglund K, Sjogren M, Andreasen N, Minthon L, Lannfelt L, et al. Measurement of alpha- and beta-secretase cleaved amyloid precursor protein in cerebrospinal fluid from Alzheimer patients. Exp Neurol. 2003;183:74–80.
114. Gabelle A, Roche S, Gény C, Bennys K, Labauge P, Tholance Y, et al. Correlations between soluble α/β forms of amyloid precursor protein and Aβ38, 40, and 42 in human cerebrospinal fluid. Brain Res. 2010;1357:175–83.
115. Hertze J, Minthon L, Zetterberg H, Vanmechelen E, Blennow K, Hansson O. Evaluation of CSF biomarkers as predictors of Alzheimer's disease: a clinical follow-up study of 4.7 years. J Alzheimers Dis. 2010;21(4):1119–28.
116. Fukumoto H, Tokuda T, Kasai T, Ishigami N, Hidaka H, Kondo M, et al. High-molecular-weight beta-amyloid oligomers are elevated in cerebrospinal fluid of Alzheimer patients. FASEB J. 2010;24(8):2716–26.

117. Gao CM, Yam AY, Wang X, Magdangal E, Salisbury C, Peretz D, et al. Aβ40 oligomers identified as a potential biomarker for the diagnosis of Alzheimer's disease. PLoS One. 2010;5(12):e15725.
118. Pitschke M, Prior R, Haupt M, Riesner D. Detection of single amyloid beta-protein aggregates in the cerebrospinal fluid of Alzheimer's patients by fluorescence correlation spectroscopy. Nat Med. 1998;4:832–4.
119. Santos AN, Torkler S, Nowak D, Schlittig C, Goerdes M, Lauber T, et al. Detection of amyloid-beta oligomers in human cerebrospinal fluid by flow cytometry and fluorescence resonance energy transfer. J Alzheimers Dis. 2007;11:117–25.
120. Hata S, Fujishige S, Araki Y, Taniguchi M, Urakami K, Peskind E, et al. Alternative processing of gamma-secretase substrates in common forms of mild cognitive impairment and Alzheimer's disease: evidence for gamma-secretase dysfunction. Ann Neurol. 2011;69:1026–31.
121. Mattsson N, Tabatabaei S, Johansson P, Hansson O, Andreasson U, Mansson JE, et al. Cerebrospinal fluid microglial markers in Alzheimer's disease: elevated chitotriosidase activity but lack of diagnostic utility. Neuromolecular Med. 2011;13:151–9.
122. Craig-Schapiro R, Perrin RJ, Roe CM, Xiong C, Carter D, Cairns NJ, et al. YKL-40: a novel prognostic fluid biomarker for preclinical Alzheimer's disease. Biol Psychiatry. 2010;68: 903–12.
123. Selkoe DJ. Alzheimer's disease is a synaptic failure. Science. 2002;298:789–91.
124. Davidsson P, Puchades M, Blennow K. Identification of synaptic vesicle, pre- and postsynaptic proteins in human cerebrospinal fluid using liquid-phase isoelectric focusing. Electrophoresis. 1999;20:431–7.
125. Sjogren M, Davidsson P, Gottfries J, Vanderstichele H, Edman A, Vanmechelen E, et al. The cerebrospinal fluid levels of tau, growth-associated protein-43 and soluble amyloid precursor protein correlate in Alzheimer's disease, reflecting a common pathophysiological process. Dement Geriatr Cogn Disord. 2001;12:257–64.
126. Thorsell A, Bjerke M, Gobom J, Brunhage E, Vanmechelen E, Andreasen N, et al. Neurogranin in cerebrospinal fluid as a marker of synaptic degeneration in Alzheimer's disease. Brain Res. 2010;1362:13–22.
127. Andreasson U, Lautner R, Schott JM, Mattsson N, Hansson O, Herukka S-K, et al. CSF biomarkers for Alzheimer's pathology and the effect size of APOE ε4. Mol Psychiatry. 2013. doi:10.1038/mp.2013.18. Epub ahead of print.
128. Kauwe JSK, Cruchaga C, Karch CM, Sadler B, Lee M, Mayo K, et al. Fine mapping of genetic variants in BIN1, CLU, CR1 and PICALM for association with cerebrospinal fluid biomarkers for Alzheimer's disease. PLoS One. 2011;6(2):e15918.
129. Cruchaga C, Kauwe JSK, Mayo K, Spiegel N, Bertelsen S, Nowotny P, et al. SNPs associated with cerebrospinal fluid phospho-tau levels influence rate of decline in Alzheimer's disease. PLoS Genet. 2010;6(9):e1001101.
130. Fagan AM, Mintun MA, Shah AR, Aldea P, Roe CM, Mach RH, et al. Cerebrospinal fluid tau and ptau(181) increase with cortical amyloid deposition in cognitively normal individuals: implications for future clinical trials of Alzheimer's disease. EMBO Mol Med. 2009;1:371–80.
131. Cairns NJ, Ikonomovic MD, Benzinger T, Storandt M, Fagan AM, Shah AR, et al. Absence of Pittsburgh compound B detection of cerebral amyloid beta in a patient with clinical, cognitive, and cerebrospinal fluid markers of Alzheimer disease: a case report. Arch Neurol. 2009;66(12):1557–62.
132. Vos S, van Rossum I, Burns L, Knol D, Scheltens P, Soininen H, et al. Test sequence of CSF and MRI biomarkers for prediction of AD in subjects with MCI. Neurobiol Aging. 2012;33(10):2272–81.
133. Schoonenboom NS, van der Flier WM, Blankenstein MA, Bouwman FH, Van Kamp GJ, Barkhof F, et al. CSF and MRI markers independently contribute to the diagnosis of Alzheimer's disease. Neurobiol Aging. 2008;29:669–75.
134. Brys M, Glodzik L, Mosconi L, Switalski R, De Santi S, Pirraglia E, et al. Magnetic resonance imaging improves cerebrospinal fluid biomarkers in the early detection of Alzheimer's disease. J Alzheimers Dis. 2009;16:351–62.

135. Fellgiebel A, Scheurich A, Bartenstein P, Müller MJ. FDG-PET and CSF phospho-tau for prediction of cognitive decline in mild cognitive impairment. Psychiatry Res. 2007;155(2): 167–71.
136. Wang L, Fagan AM, Shah AR, Beg MF, Csernansky JG, Morris JC, et al. Cerebrospinal fluid proteins predict longitudinal hippocampal degeneration in early-stage dementia of the Alzheimer type. Alzheimer Dis Assoc Disord. 2012;26(4):314–21.
137. Dickerson BC, Wolk DA, Alzheimer's Disease Neuroimaging Initiative. MRI cortical thickness biomarker predicts AD-like CSF and cognitive decline in normal adults. Neurology. 2012;78(2):84–90.
138. Fjell AM, Walhovd KB. Neuroimaging results impose new views on Alzheimer's disease–the role of amyloid revised. Mol Neurobiol. 2012;45(1):153–72.
139. Neary D, Snowden JS, Gustafson L, Passant U, Stuss D, Black S, et al. Frontotemporal lobar degeneration: a consensus on clinical diagnostic criteria. Neurology. 1998;51:1546–54.
140. Josephs KA, Hodges JR, Snowden JS, Mackenzie IR, Neumann M, Mann DM, et al. Neuropathological background of phenotypical variability in frontotemporal dementia. Acta Neuropathol. 2011;122(2):137–53.
141. Irwin DJ, Trojanowski JQ, Grossman M. Cerebrospinal fluid biomarkers for differentiation of frontotemporal lobar degeneration from Alzheimer's disease. Front Aging Neurosci [Internet]. 2013[cited 2013 Aug 3];5. Available from: http://www.ncbi.nlm.nih.gov/pmc/articles/PMC3578350/.
142. Cruchaga C, Kauwe JSK, Harari O, Jin SC, Cai Y, Karch CM, et al. GWAS of cerebrospinal fluid tau levels identifies risk variants for Alzheimer's disease. Neuron. 2013;78(2):256–68.
143. Rosengren LE, Karlsson JE, Karlsson JO, Persson LI, Wikkelso C. Patients with amyotrophic lateral sclerosis and other neurodegenerative diseases have increased levels of neurofilament protein in CSF. J Neurochem. 1996;67:2013–8.
144. Norgren N, Rosengren L, Stigbrand T. Elevated neurofilament levels in neurological diseases. Brain Res. 2003;987:25–31.
145. Bech S, Hjermind LE, Salvesen L, Nielsen JE, Heegaard NH, Jorgensen HL, et al. Amyloid-related biomarkers and axonal damage proteins in parkinsonian syndromes. Parkinsonism Relat Disord. 2012;18:69–72.
146. Sjogren M, Blomberg M, Jonsson M, Wahlund LO, Edman A, Lind K, et al. Neurofilament protein in cerebrospinal fluid: a marker of white matter changes. J Neurosci Res. 2001;66: 510–6.
147. Zetterberg H, Jacobsson J, Rosengren L, Blennow K, Andersen PM. Cerebrospinal fluid neurofilament light levels in amyotrophic lateral sclerosis: impact of SOD1 genotype. Eur J Neurol. 2007;14:1329–33.
148. Petzold A, Keir G, Warren J, Fox N, Rossor MN. A systematic review and meta-analysis of CSF neurofilament protein levels as biomarkers in dementia. Neurodegener Dis. 2007;4:185–94.
149. Landqvist Waldo M, Frizell Santillo A, Passant U, Zetterberg H, Rosengren L, Nilsson C, et al. Cerebrospinal fluid neurofilament light chain protein levels in subtypes of frontotemporal dementia. BMC Neurol. 2013;13:54.
150. Steinacker P, Hendrich C, Sperfeld AD, Jesse S, von Arnim CA, Lehnert S, et al. TDP-43 in cerebrospinal fluid of patients with frontotemporal lobar degeneration and amyotrophic lateral sclerosis. Arch Neurol. 2008;65:1481–7.
151. Kasai T, Tokuda T, Ishigami N, Sasayama H, Foulds P, Mitchell DJ, et al. Increased TDP-43 protein in cerebrospinal fluid of patients with amyotrophic lateral sclerosis. Acta Neuropathol. 2009;117(1):55–62.
152. Ghidoni R, Paterlini A, Benussi L. Circulating progranulin as a biomarker for neurodegenerative diseases. Am J Neurodegener Dis. 2012;1(2):180–90.
153. Hu WT, Chen-Plotkin A, Grossman M, Arnold SE, Clark CM, Shaw LM, et al. Novel CSF biomarkers for frontotemporal lobar degenerations. Neurology. 2010;75(23):2079–86.
154. Mattsson N, Ruetschi U, Pijnenburg YA, Blankenstein MA, Podust VN, Li S, et al. Novel cerebrospinal fluid biomarkers of axonal degeneration in frontotemporal dementia. Mol Med Rep. 2008;1:757–61.

155. Mattsson N, Brax D, Zetterberg H. To know or not to know: ethical issues related to early diagnosis of Alzheimer's disease. Int J Alzheimers Dis. 2010. doi:10.4061/2010/841941. pii: 841941.
156. Mattsson N, Zetterberg H. Future screening for incipient Alzheimer's disease—the influence of prevalence on test performance. Eur Neurol. 2009;62:200–3.
157. Mattsson N, Zetterberg H, Blennow K. Lessons from multicenter studies on CSF biomarkers for Alzheimer's disease. Int J Alzheimers Dis. 2010. doi:10.4061/2010/610613. pii: 610613.
158. Bjerke M, Portelius E, Minthon L, Wallin A, Anckarsater H, Anckarsater R, et al. Confounding factors influencing amyloid Beta concentration in cerebrospinal fluid. Int J Alzheimers Dis. 2010;21:221–8.
159. Mattsson N, Andreasson U, Carrillo MC, Persson S, Shaw LM, Zegers I, et al. Proficiency testing programs for Alzheimer's disease cerebrospinal fluid biomarkers. Biomark Med. 2012;6:401–7.
160. Mattsson N, Andreasson U, Persson S, Carrillo MC, Collins S, Chalbot S, et al. CSF biomarker variability in the Alzheimer's Association quality control program. Alzheimers Dement. 2013;9(3):251–61.
161. Mattsson N, Andreasson U, Persson S, Arai H, Batish SD, Bernardini S, et al. The Alzheimer's Association external quality control program for cerebrospinal fluid biomarkers. Alzheimers Dement. 2011;7:386–395 e6.

Chapter 10
Biomarkers for Alzheimer's Disease and Frontotemporal Lobar Degeneration: Imaging

Marco Bozzali and Laura Serra

Abstract In recent years, our understanding of neurodegenerative dementias has translated into a change in the clinical approach to patients presenting with impairments in cognition and behavior. The diagnosis of different forms of neurodegenerative dementias is currently based not only on their clinical and neuropsychological characterization but also on the use of biomarkers. Advances in neuroimaging techniques, particularly magnetic resonance imaging (MRI), have strongly contributed not only in increasing our understanding of clinical and pathophysiological aspects of dementias but also in improving the diagnostic confidence in clinical settings. MRI, thanks to its ability to image in vivo soft tissues noninvasively and with detailed anatomical resolution, shows high sensitivity in detecting the presence and extension of macroscopic brain abnormalities. In this view, as discussed below, MRI plays the unique role of excluding alternative diagnoses that may mimic a neurodegenerative form of cognitive decline.

Keywords Alzheimer's disease • Amyloid-β • Frontotemporal dementia (FTD) • Frontotemporal lobar degeneration (FTLD) • Imaging

Introduction

In recent years, our understanding of neurodegenerative dementias has translated into a change in the clinical approach to patients presenting with impairments in cognition and behavior. The diagnosis of different forms of neurodegenerative dementias is currently based not only on their clinical and neuropsychological characterization but also on the use of biomarkers. Advances in neuroimaging

M. Bozzali, MD • L. Serra, PhD (✉)
Neuroimaging Laboratory, Santa Lucia Foundation IRCCS,
Via Ardeatina 306, Rome 00179, Italy
e-mail: m.bozzali@hsantalucia.it

D. Galimberti, E. Scarpini (eds.), *Neurodegenerative Diseases*,
DOI 10.1007/978-1-4471-6380-0_10, © Springer-Verlag London 2014

Table 10.1 Neurodegenerative dementia: clinical syndromes

Syndrome	CSF characteristics	Key clinical dysfunction	Early cerebral involvement
Amnestic-Alzheimer's disease	Amyloid-β Hyperphosphorylated tau protein	Episodic memory	Medial temporal lobes
Posterior cortical atrophy	Amyloid-β Hyperphosphorylated tau protein	Visuospatial dysfunctions	Parieto-occipital lobes
Dysexecutive Alzheimer's disease	*No pathological correlations available*	Dysexecutive syndrome	Frontal and temporoparietal lobes
Logopenic PPA	Hyperphosphorylated tau protein	Word retrieval, sentence repetition	Left temporoparietal lobe
Agrammatic PPA	Hyperphosphorylated tau protein	Agrammatism, apraxia of speech	Left posterior frontal lobe and insula
Semantic dementia	Hyperphosphorylated tau protein	Confrontation naming, single-word comprehension	Left temporal pole
bv-FTD	Hyperphosphorylated tau protein	Disinhibition, apathy, sleep disorder, perseverative behavior, dysexecutive syndrome	Frontal, temporal lobes, anterior cingulate, insula

Adapted from McGinnis [1] and Cummings [3]
Abbreviations: *PPA* primary progressive aphasia, *bv-FTD* behavioral variant-frontotemporal lobar degeneration

techniques, particularly magnetic resonance imaging (MRI), have strongly contributed not only in increasing our understanding of clinical and pathophysiological aspects of dementias but also in improving the diagnostic confidence in clinical settings [1]. MRI, thanks to its ability to image in vivo soft tissues noninvasively and with detailed anatomical resolution, shows high sensitivity in detecting the presence and extension of macroscopic brain abnormalities [2]. In this view, as discussed below, MRI plays the unique role of excluding alternative diagnoses that may mimic a neurodegenerative form of cognitive decline.

Neurodegenerative dementias, such as Alzheimer's disease (AD) and frontotemporal dementia (FTD), are typically characterized by an insidious onset which is followed by a gradual progression of symptoms. Especially at early clinical stages, the underlying neurodegenerative processes produce selective cognitive dysfunctions that may correspond to the focal distribution of brain damage [1]. As shown in Table 10.1, the combination of biomarkers' characteristics and neuropsychological profiles improves the potential of a correct and early diagnosis of neurodegenerative dementias [1, 3]. Additionally, as demonstrated by research evidence mostly based on neuroimaging, cognitive and behavioral disabilities in dementias are not only due to focal brain tissue damage but also to disconnection mechanisms. In this context, brain connectivity, as assessed by functional neuroimaging techniques, has revolutionized our understanding of large-scale neuronal networks and clarified the relationship between their disruption/modifications and the clinical evolution of neurodegenerative diseases [4, 5]. Finally, metabolic techniques of brain imaging allow to detect in vivo the anatomical deposition of specific disease biomarkers, such as fibrillar amyloid-β proteins in AD brains.

In the current chapter, we will review, for AD and FTD, respectively, the contribution of neuroimaging in supporting a correct clinical diagnosis and the role of advanced neuroimaging techniques in clarifying and monitoring some pathophysiological aspects of disease.

Alzheimer's Disease

Alzheimer's disease (AD) is the most common cause of dementia in the elderly [6]. Neuropathological studies have identified a sequential accumulation of neurofibrillary tangles and amyloid-β plaques in the brain tissue, as well as the progression of neuronal loss through the cerebral cortex [7]. From a clinical viewpoint, the accumulation of neuropathological abnormalities may precede of many years the clinical onset of AD [8]. In particular, neurofibrillary pathology in the entorhinal cortex, hippocampus, and amygdala is considered as the major neurobiological substrate for episodic memory deficits, which are typically observed in AD since its early stages [8]. In recent years, the increased knowledge on the neuropathological cascade occurring in AD brains and the early cognitive modifications originating from these changes have led to the definition of new diagnostic criteria for preclinical AD [9]. These criteria incorporate several biomarkers, including neuroimaging, to define the presence of AD pathology [9].

Conventional MRI

Conventional MRI has shown the ability to produce brain images with a higher spatial resolution compared to computed tomography (CT), thus showing much more detailed information about macroscopic brain anatomy. Moreover, MRI is particularly helpful in detecting and excluding other neurological conditions mimicking a neurodegenerative form of cognitive decline, such as brain tumors, normal pressure hydrocephalus, subdural hematoma, and cerebrovascular diseases. Some examples of these differential diagnoses are illustrated in Fig. 10.1. After exclusion of secondary causes of dementia, conventional MRI may address a correct diagnosis of AD only in a proportion of cases, mainly based on assessment of regional brain atrophy. The simplest approach to determine regional changes of brain volumes is to use rating scales based on visual examination of T1-weighted MR images [10, 11], such as the "medial temporal lobe atrophy" (MTA) [10] scale. This tool allows a semiquantitative volumetric assessment of the medial temporal lobe structures

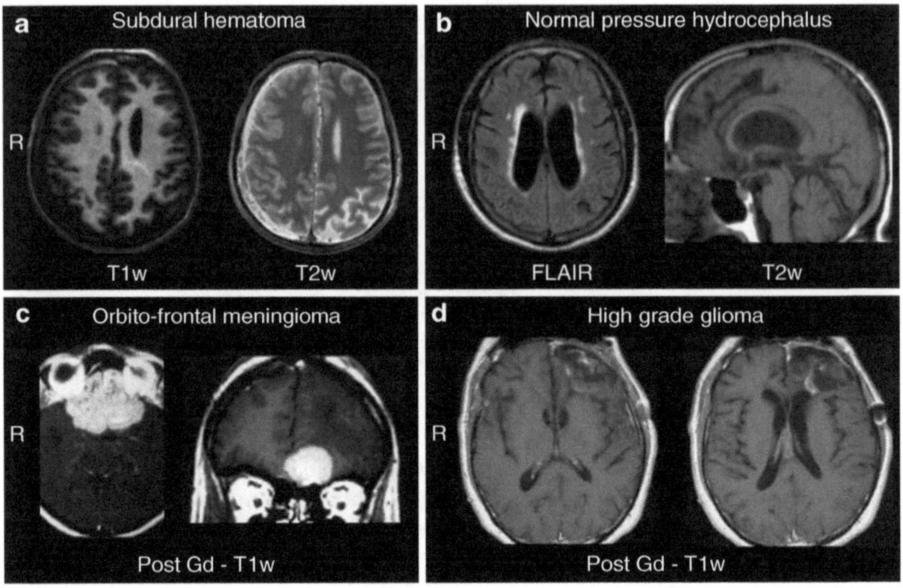

Fig. 10.1 Magnetic resonance imaging to rule out possible secondary causes of cognitive decline. This figure illustrates a few examples of secondary causes of cognitive decline mimicking a degenerative form of dementia. Conventional MR images with different acquisition parameters allow an immediate detection of these causes. (**a**) Subdural hematoma of the right hemisphere, probably due to multiple traumatic events; T1-weighted (T1-w) images highlight a sickle-shaped extracerebral area of hyperintensity (i.e., hemoglobin degradation products in acute-subacute phase) in frontal region, while T2-weighted (T2-w) images highlight a parietotemporal extracerebral hypointensity (hemoglobin degradation products in chronic phase); (**b**) Normal pressure hydrocephalus; both, T2-w and fluid attenuated inversion recovery (FLAIR) images show bilateral ventricle enlargement in the absence of remarkable cortical atrophy. Additionally, FLAIR images demonstrate periventricular hyperintensities which are suggestive for transependymal CSF migration. (**c**) Orbitofrontal meningioma; post-Gd (after i.v. administration of gadolinium) T1-weighted images show a large extraparenchymal enhancing lesion (hyperintense) compressing the brain tissue and causing intraparenchymal edema (hypointense areas). (**d**) High-grade glioma; post-Gd T1-w images show a left frontal nonhomogeneously enhancing lesion that infiltrates the brain tissue

(i.e., hippocampus, dentate gyrus, subiculum, and parahippocampal gyrus) and correspondent enlargements of the temporal horn of the lateral ventricles and choroid fissures (Table 10.2 and Fig. 10.2, top panel). The use of MTA has shown high accuracy in determining the severity of local atrophy in cross-sectional studies that compared AD patients with healthy controls [12]. Conversely, MTA appears to be poorly informative in detecting longitudinal volumetric changes over time [12]. There are specific visual rating scales also to quantify the presence and severity of macroscopic white matter abnormalities. They can be applied to CT images or, with a better definition, to MRI scans (i.e., T2-/proton density [PD]-weighted and/or fluid attenuated inversion recovery [FLAIR] images). The age-related white matter changes (ARWMC) [11] and the Fazekas' (1987) scales [13], whose application criteria are summarized in Table 10.3 and illustrated in Fig. 10.2 (bottom panel),

Table 10.2 Medial temporal atrophy (MTA) visual rating scale

Score	Width of choroid fissure	Width of temporal horn	Height of hippocampal formation
0	Normal	Normal	Normal
1	Mild increase	Normal	Normal
2	Moderate increase	Mild increase	Mild decrease
3	Severe increase	Moderate increase	Moderate decrease
4	Severe increase	Severe increase	Severe decrease

Details on the use of MTA are reported in the text and schematically illustrated in Fig. 10.2

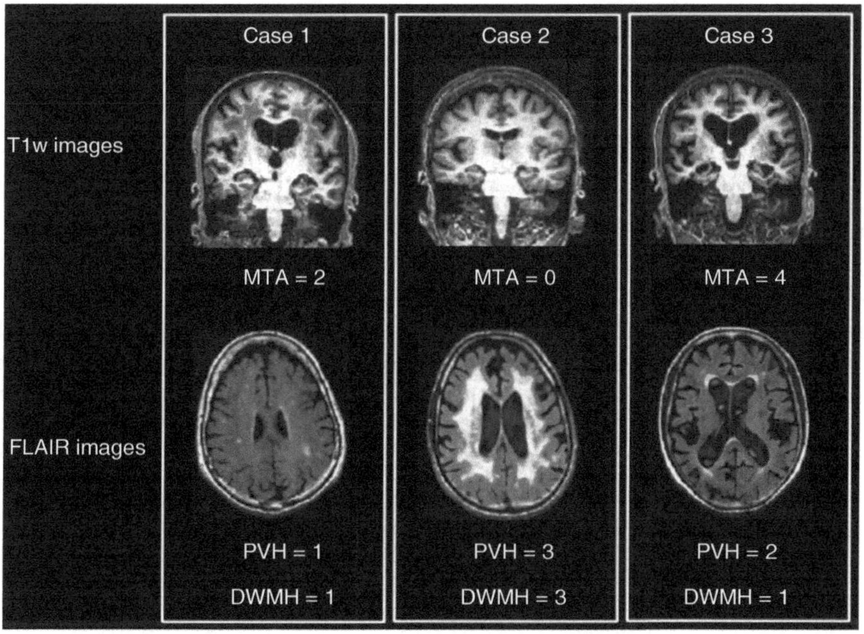

Fig. 10.2 The contribution of conventional magnetic resonance imaging (MRI) in supporting a diagnosis of Alzheimer's disease versus vascular dementia. The *top panel* of the figure shows coronal T1-weighted (T1-w) images used to assess hippocampal atrophy based on the medial temporal lobe atrophy scale (see Table 10.2). *Bottom panel* shows axial fluid attenuated inversion recovery (FLAIR) images used to quantify the severity of macroscopic white matter lesions based on Fazekas' scale (see Table 10.3). Schematically, three cases can be identified. Cases 1 and 2, for which MRI is more informative, indicate, respectively, a relatively pure form of AD and vascular dementia. Case 3 does not allow a clear cutoff between neurodegenerative processes and vascular pathology. *MTA* medial temporal lobe atrophy scale, *PVH* periventricular hyperintensities, *DWMH* deep white matter hyperintensities

allow a simple assessment of white matter macroscopic changes. In the diagnostic suspect of AD, taking altogether the information given by MTA and white matter lesion assessments, three different patterns may schematically be identified (Fig. 10.2): (1) severe MTA and minimal white matter abnormalities, (2) minimal MTA and severe white matter abnormalities, and (3) moderate MTA and moderate/

Table 10.3 Visual rating scales to assess white matter hyperintensities

White matter rating scales	Brain area	White matter lesions
ARWMC (Wahlund et al. [11])	Frontal, parieto-occipital, temporal, and infratentorial	0 = no lesions 1 = focal lesions 2 = beginning confluence of lesions 3 = diffuse involvement of entire regions
	Basal ganglia	0 = no lesions 1 = one focal lesion 2 = more than one focal lesion 3 = confluent lesions
Fazekas scale (Fazekas et al. [13])	Periventricular (PVH) lesions	0 = absent 1 = caps or pencil-thin lining around ventricles 2 = smooth halo around ventricles 3 = irregular PVH extending into DWM
	Deep (DWM) lesions	0 = absent 1 = discrete diffuse lesions 2 = beginning of confluence of foci 3 = large confluent areas

severe white matter abnormalities. In the first two cases, conventional MRI strongly contributes in increasing the diagnostic confidence of degenerative against vascular dementia. In the third case, due to the frequent comorbidity of degenerative and vascular pathology, the contribution of conventional MRI remains limited.

Advanced MRI Techniques

Brain Volumetrics

Several approaches to quantitative brain volumetrics are currently available, and the simplest methods are those based on manual or semiautomatic delineation of brain structures. More recently, the development of sophisticated registration algorithms has made it possible to bring volumetric images from different subjects into a common space and to identify differences between groups (e.g., patients vs. controls) and correlations with clinical/psychometric measures, on a voxel-by-voxel level basis. The most appropriate MR scans for all types of volumetric assessments are the high-resolution T1-weighted volumes, typically obtained using three-dimensional sequences, which provide sufficient anatomical detail, as well as sufficient contrast between gray and white matter tissues.

Manual and Semiautomatic Regional Measurements

Given the relevance of medial temporal lobe atrophy in AD, which corresponds to postmortem evidence of earlier and predominant neurofibrillary degeneration in this region, first attempts to quantify brain damage (i.e., atrophy) employed manual volumetric assessments of the hippocampus on coronal T1-weigthed images [14]. Comparisons between patients with AD and healthy controls have consistently revealed volumetric reductions of the hippocampus of about 40 % [14]. Significant hippocampal reductions have been reported also in patients with MCI [15], thus confirming an involvement of this area as core disease feature since early clinical stages. Interestingly, in clinical follow-up studies, hippocampal volumetrics revealed a more severe atrophy in those MCI patients who convert to AD than in those who remain stable [15]. In terms of potential diagnostic application, hippocampal and entorhinal volumetrics allow a separation of AD and MCI patients from healthy controls with accuracies ranging from 70 % (in early MCI) to 100 % (in AD patients) [14]. Additionally, volumetrics of these brain structures have been reported as predictive for a future conversion from MCI to AD with an accuracy of about 80–85 % [16]. Nevertheless, manual assessments of medial temporal lobe volumes are strongly operator dependent, based on different anatomical landmarks across studies, and time-consuming. So far, these weaknesses have prevented a wide diffusion of manual assessments in clinical settings, despite the recent ongoing efforts of methodological standardization and validation [17].

Automated Methods to Assess Brain Atrophy

For data-driven analyses, voxel-based morphometry (VBM) is one of the most popular techniques to investigate dementias [18]. VBM has proven high reproducibility when using datasets obtained by different MR systems and various optimizations of image processing. This approach is operator independent and does not require any a priori hypothesis on the anatomical localization of the brain tissue loss, as it includes the whole brain (i.e., voxel-wise analysis); for a review see [2]. VBM analysis is particularly suitable for gray matter volume assessments and is based on a series of automatic steps, the main ones including normalization of individual T1-weighted volumes to standard space, brain segmentation and extraction of gray matter maps, and statistical analyses. Different statistical designs can be employed, which allow to perform between-group comparisons as well as correlations between regional distributions of gray matter volumes and clinical, neuropsychological, and behavioral variables. When applied to AD patients at different clinical stages, VBM has demonstrated a widespread pattern of gray matter atrophy, including not only the medial temporal lobe structures but also several other areas of the association cortex [19, 20]. Moreover, in AD and amnestic MCI patients, it has been shown a strict association between cognitive profiles and regional patterns of gray matter atrophy. For instance, hippocampal gray matter loss has been shown to be associated with patients' episodic memory deficits (Fig. 10.3a;[20]), and posterior cortical atrophy has been found associated with constructional apraxia [21]. Associations between regional gray matter atrophy and patients' behavioral features have also been demonstrated in AD and MCI (Fig. 10.3b), suggesting these symptoms to be part of AD pathophysiology [23]. MCI can also be clinically dominated by neuropsychological deficits other than memory (i.e., non-amnestic MCI). Again, VBM has shown the

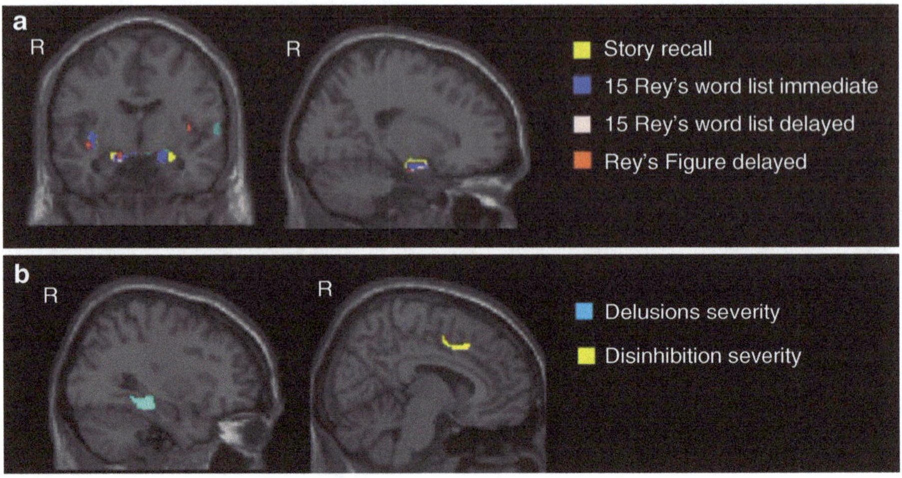

Fig. 10.3 Voxel-based morphometry analysis to assess associations between regional gray matter volumes and higher level dysfunctions in Alzheimer's disease. Panel (**a**) illustrates how hippocampal atrophy accounts for patients' episodic memory impairment as assessed by various neuropsychological tests (Modified by Serra et al. [20]). Panel (**b**) shows associations between regional gray matter atrophy and behavioral and psychological symptoms in patients with AD (Modified by Serra et al. [22]). See text for further details

ability to detect patterns of regional gray matter loss that fit well with the non-amnestic neuropsychological profile, thus allowing a differentiation of MCI patients who are more likely to convert to non-AD forms of dementias [24].

Diffusion Imaging

Diffusion imaging [25] provides, through the measurement of diffusional motion of water molecules into brain cells, unique information to investigate the white matter microarchitecture, connectivity and integrity, documenting the size, shape orientation, and geometry of brain structures (for review see [2]). Neurodegenerative processes, such as those occurring in AD brains, modify tissue integrity and may result in an altered diffusion coefficient, which can be measured in vivo by diffusion MRI. The diffusion of water molecules is facilitated along the principal direction of white matter fibers, and this allows to reconstruct some white matter fiber tracts. The metrics resulting from different steps of diffusion image analysis (e.g., fractional anisotropy, FA; mean diffusivity, MD; radial diffusivity, RD; axial diffusivity, AD) can be statistically analyzed using both automated voxel-wise methods (e.g., by tract-based spatial statistics (TBSS)) [26] or regional approaches (e.g., diffusion tractography reconstruction of white matter tracts) [27].

Diffusion imaging has been widely used in studies investigating MCI and AD patients (for a review see [1]). Some of them have reported a widespread alteration of white matter (WM) tissue integrity in patients with AD at different clinical

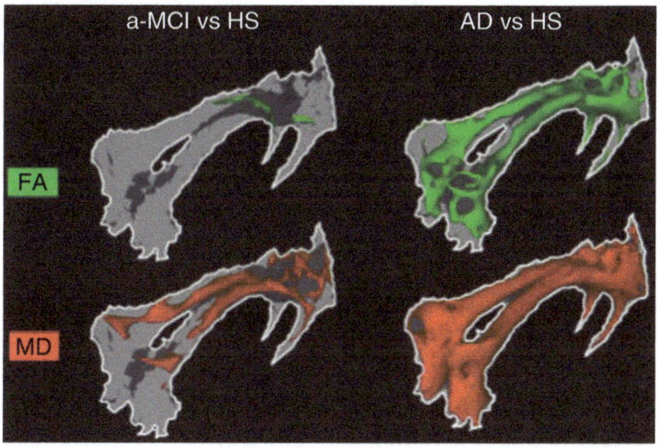

Fig. 10.4 Diffusion imaging-based assessment of structural brain connectivity in Alzheimer's disease. This figure shows the progressive accumulation of microscopic white matter damage (expressed by increases of mean diffusivity (*MD*) and reductions of fractional anisotropy (*FA*) within the cingulum in the transitional stage between normal aging and Alzheimer's disease (*AD*), Healthy Subjects (*HS*) passing through the condition of amnestic mild cognitive impairment. In each subject, using diffusion imaging, MD and FA maps were first calculated. Then, the cingulum was reconstructed using probabilistic tractography and used to quantify MD and FA within the tract. Overall, this figure indicates that structural brain disconnection in specific tracts contributes in determining cognitive disabilities in AD (Modified by Bozzali et al. [30])

stages and using both, a whole brain analysis [20, 28] or focusing on specific WM tracts [29, 30]. For instance, a recent study, based on diffusion tractography of the cingulum (i.e., the main pathway of connection between the limbic areas and the rest of the brain), shows a progressive disruption of this structure over the transitional state from MCI to AD (Fig. 10.4). Interestingly, this white matter damage accounts, in combination with regional gray matter loss, for the cognitive features of preclinical and clinical AD stages. Another interesting tract, implicated in AD pathology, is the uncinate fasciculus. It has been shown how microscopic damage to this tract accounts for cognitive and behavioral aspects, which are typically present at advanced stages of AD.

Finally, a novel method of diffusion imaging analysis, called anatomical connectivity mapping (ACM), has been recently proposed to assess changes in structural brain connectivity across the whole brain [31]. This voxel-wise technique, based on probabilistic tractography, has been able to detect in patients with AD modifications which have been interpreted as phenomena of brain plasticity [32].

Functional MRI

Neuronal activity can be investigated noninvasively, but indirectly, through blood oxygenation level-dependent (BOLD) functional MRI (fMRI). fMRI can be used to assess changes of brain activation in response to patients' performance at cognitive tasks involving specific higher level functions (e.g., memory, visuospatial, executive functions, emotion processing, etc.). On the other hand, fMRI can also be used at rest to record coherent fluctuations of brain activity over time (resting-state fMRI). In this latter case, fMRI provides information on functional brain connectivity within specific networks, some of whom have been associated with specific higher level functions. When using fMRI with active tasks, patients' cooperation is essential, and findings obtained in patients with fully developed AD remain rather controversial. Investigations based on episodic memory tasks have reported, in AD patients, reductions of functional activity in the hippocampus and other temporal areas and increased activity in the parietal association cortex [33]. In contrast, other studies have reported a decrease of functional activity (during memory tasks) not only in the temporal lobe but also in parietal and frontal regions [34]. On the other hand, studies involving patients with MCI have generally reported increased activations in brain areas related to the administered tasks (for a review see [35]). There is some evidence that these increases of functional activity might represent compensatory mechanisms against the incipient occurrence of brain tissue loss. In a group of patients with amnestic MCI single domain, it has been shown increased brain activation in a set of tasks exploring memory and visuospatial attention domains, in the presence of a maintained performance during task execution [36].

Resting-state fMRI has been largely used to investigate neurodegenerative diseases. This technique does not require any active performance of tasks and allows to record spontaneous brain activity fluctuations when subjects lie in the scanner at rest. Therefore, resting-state fMRI provides information on the integrity of functional brain connectivity [37] and permits to identify different brain networks and to assess the strength of connectivity within them [37]. Among all brain networks, the default-mode network (DMN) has been extensively investigated in patients with dementia. This network includes the posterior cingulated cortex, the inferior parietal, and the medial prefrontal cortex. These regions are particularly interesting as they are believed to be similarly modulated by cognitive tasks [37]. Several studies have been performed in patients with AD at different stages and documented a selective disruption within the DMN nodes [4, 38]. Recently, a study [4] involving patients with AD, patients with a-MCI, and healthy controls has assessed both GM atrophy and functional connectivity into the DMN nodes. This study showed that functional disconnection precedes the occurrence of gray matter atrophy in the posterior cingulate cortex (Fig. 10.5), thus supporting the hypothesis that gray matter atrophy in some specific regions of AD brains is likely due to brain disconnection mechanisms [4]. Furthermore, this study indicates that accumulation of gray matter loss in the PCC parallels the conversion from a-MCI to AD [4].

Metabolic Imaging

Positron emission tomography (PET) is a sensitive molecular imaging technique for the in vivo quantification of radiotracer concentrations in a picomolar range. PET scanning allows a noninvasive assessment of molecular processes at their sites of

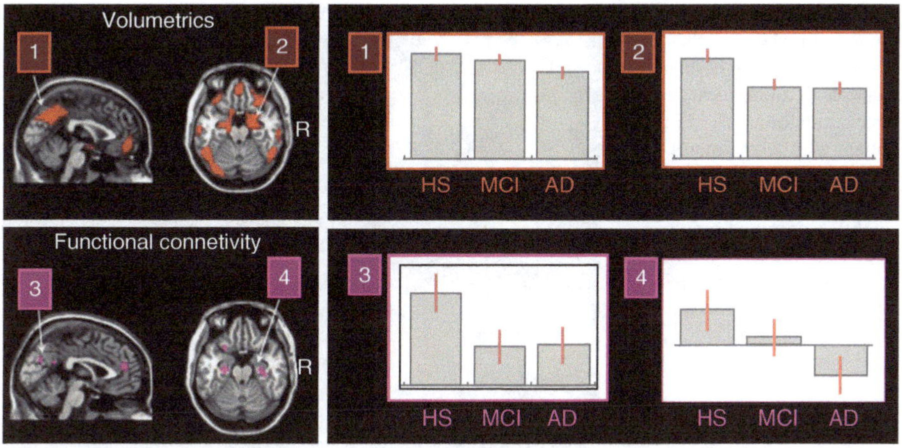

Fig. 10.5 Evolution of functional brain disconnection and gray matter loss in patients with Alzheimer's disease at different stages. In the *top panel* of the figure, red areas illustrate the pattern of gray matter atrophy observed in Alzheimer's disease (*AD*) patients as compared to healthy controls, which includes the medial temporal lobes and prefrontal cortex and the posterior cingulate cortex (PCC). Looking at the three groups of subjects (see plot *2* for AD, mild cognitive impairment (*MCI*), and healthy subjects) local atrophy was equally present in the hippocampi of both patients' groups (AD and MCI). Conversely, the PCC (see plot *1*) was atrophic in AD but not in MCI patients. On the other hand, when looking at functional connectivity (*bottom panel*; *pink areas*), the hippocampi (see plot *4*), as well as the PCC (see plot *3*), revealed a loss of functional cognitivity at both clinical stages, MCI and AD. These findings suggest that atrophy of the hippocampi (present since the earliest AD stages) causes disconnection between these structures and the PCC, and that, at least partially due to this disconnection, the PCC becomes atrophic at later clinical stages of AD (Modified by Gili et al. [4])

action and is in principle capable of detecting disease processes when there is no evidence of structural changes [39]. ^{18}Fluorodeoxyglucose (^{18}FDG-PET) is a widely available PET tracer [40] that reflects the local glucose metabolism as a proxy index for neuronal activity [40]. Typical ^{18}FDG-PET finding in patients with AD is a pattern of reduced glucose uptake in temporoparietal association areas, including the precuneus and the posterior cingulate cortex [40]. ^{18}FDG-PET has demonstrated a high specificity in discriminating between patients with AD and healthy subjects (ranging from 70 to 90 %) [40] and between patients with AD and those with other forms of degenerative dementia (specificity of 87 %) [40]. On the other hand, the ability of ^{18}FDG-PET to identify patients at preclinical AD stages still remains controversial [41].

Another useful application in clinical practice is the use of single-photon emission computed tomography (SPECT) after administration of dopamine-transporter (DAT) ligands (e.g., [^{123}I]FP-CIT, [^{123}I]β-CIT, [^{99m}Tc]-TRODAT-1), the so-called DAT-scan technique. DAT-scan allows the detection of striatal dopaminergic dysfunction, which is typically present in patients with Parkinson-related disorders [42] and not in patients with AD.

Available evidence supports the position that an abnormal processing of amyloid-β (Aβ) peptides is the initiating event of AD pathophysiology, which

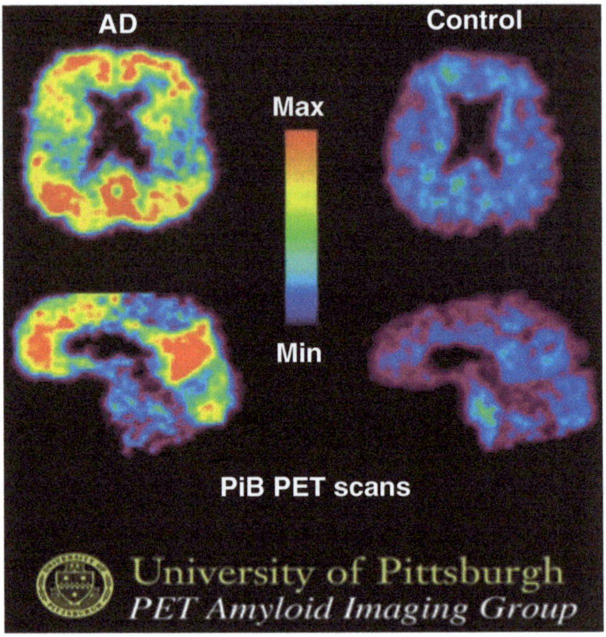

Fig. 10.6 Amyloid PET imaging in Alzheimer's disease. It is given here an example of anatomical distribution of amyloid in a patient with Alzheimer's disease (*left*) in comparison with a healthy control (*right*). Maxima of amyloid tracer binding (*red-yellow areas*) were in the posterior cingulate and prefrontal areas

eventually leads to accumulation of Aβ plaques in the brain tissue [43]. This process occurs when individuals are still cognitively intact, many years before the clinical onset of AD. In this picture, amyloid PET imaging has been proposed as a tool for early detection of AD pathology in vivo, and for the differential diagnosis of dementia [44]. In AD, PET amyloid-β imaging has shown increased tracer binding in areas known to have high concentrations of amyloid plaques, such as the medial and orbitofrontal regions, the lateral parietal and temporal cortex, the precuneus, and the posterior cingulate [44]. An example of PET amyloid-β imaging in a patient with AD as compared to a healthy control is shown in Fig. 10.6. Advances in biomarkers for AD pathology have recently led to proposals for more definitive diagnoses in patients with MCI as individuals at a prodromal AD stage (International Working Group for New Research Criteria for Diagnosis of AD) [45] or MCI as due to AD (National Institute on Aging and Alzheimer's Association Workgroup) [9]. In the latter case, MCI can be defined as due to AD with "high likelihood" whenever both an amyloid and a neurodegenerative biomarker are positive, with "intermediate likelihood" when one biomarker only is positive, and "low likelihood" when both biomarkers are negative for AD pathology. In this perspective, several pharmacological approaches aimed at reducing Aβ levels in the brain tissue are being developed and tested, and many efforts have been focused on generating radiotracers for imaging Aβ in vivo [46]. Currently, the [^{11}C] Pittsburgh compound-B (PIB) is the most popular radiotracer used in AD patients, due to its high affinity and selectivity for fibrillar Aβ in plaques and other Aβ-containing lesions [47]. Most importantly, there are available studies showing that the PIB cortical retention primarily reflects Aβ-related cerebral amyloidosis rather than Lewy bodies or neurofibrillary tangles [48]. This indicates that PIB can be particularly useful for patients' diagnostic definition since early clinical stages [9, 45].

Frontotemporal Dementias

Frontotemporal dementia (FTD) is the second most common neurodegenerative disease, especially in patients with a presenile clinical onset (age <65 years) [49]. FTD can be defined as a heterogeneous cluster of disorders including two major clinical conditions, one characterized by predominant deficits of language functions (primary progressive aphasia) and one characterized by prominent behavioral symptoms (bv-FTD). Nevertheless, several other cognitive deficits (such as impairment in problem-solving, reasoning, planning, attention, and decision-making) can be present in both clinical syndromes, and behavioral disorders can be observed in all FTD clinical variants [50]. In the suspect of FTD, after exclusion of severe macroscopic white matter damage (see also section on "Conventional MRI"), specific patterns of regional brain atrophy may be present and fit with specific clinical/neuropsychological syndromes [49], thus increasing the diagnostic confidence of a correct diagnosis. Some examples of the neuroradiological presentations of the major FTD variants are illustrated in Fig. 10.7.

Histopathological and Genetic Aspects

Over the last few years, many different classifications of FTD variants have been introduced, mainly based on genetic and neuropsychological profiles. Neuroimaging has been used to identify anatomo-functional substrates for these classifications. In all FTD variants, neurodegeneration is mainly due to neuronal loss and gliosis [51], despite a large variety of underling pathophysiology. Specific protein abnormalities have been identified in a heterogeneous group of diseases, namely, the frontotemporal lobar degeneration (FTLD), which are in most cases associated with a FTD syndrome [51].

The first classification of FTLD has identified two main categories, one associated with the deposition of microtubule-associated protein tau (FTLD-tau) and one associated with deposition of ubiquitin-only immunoreactivity (FTLD-U) (see [52] for a review). In some cases of this latter group of disorders, patients show also motor neuron degeneration (FTLD-MND) [52]. More recently, additional subclassifications have been introduced in the FTLD-U group on the basis of specific molecular features. The presence/absence of transactive response DNA-binding protein of 43 kDa has identified the FTLD-TDP against the aFTLD-U form [53]. In turn, within the aFTLD-U group, another subclassification has been introduced

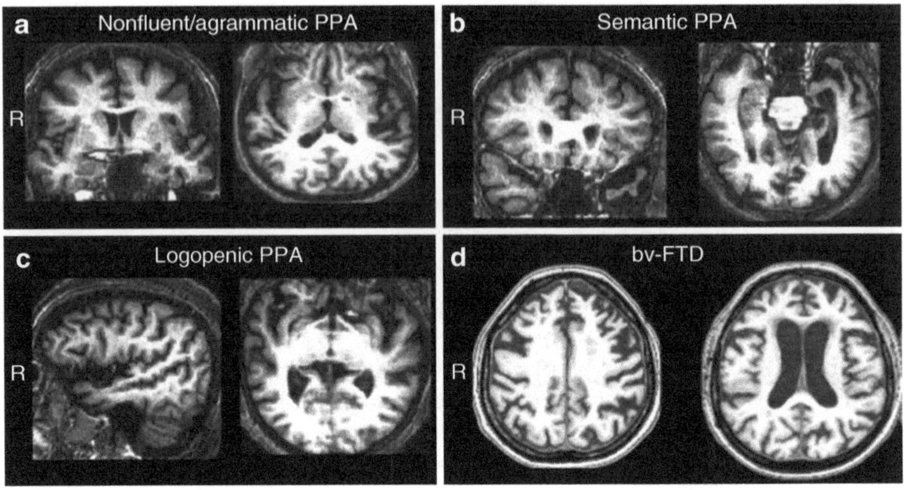

Fig. 10.7 Patterns of anatomical distribution of brain atrophy suggestive for FTD. Panels (**a**–**c**) show three typical cases of neuroradiological presentation of primary progressive aphasia (*PPA*), while panel (**d**) shows a case of behavioral variant of FTD (*bv-FTD*). (**a**) Agrammatic nonfluent PPA, T1-w images show a pattern of bilateral frontotemporal insular atrophy, predominantly involving the left hemisphere; (**b**) Semantic PPA, T1-w images show a selective atrophy of the left temporal pole; (**c**) Logopenic PPA, T1-w images show a focal pattern of brain atrophy in the left perisylvian area; (**d**) bv-FTD, T1-w images show a pattern of bilateral prefrontal atrophy

based on the presence/absence of ubiquitinated protein fused in sarcoma (FUS) [54]. In summary, FTLD currently includes the following forms: FTLD-tau, FTLD-TDP, and FTLD-FUS [51].

In most patients with different FTLD forms, several genetic varieties have been identified, the main ones including the following gene mutations: (1) microtubule-associated protein tau (MAPT) [55], (2) progranulin (GRN) [56], and (3) C9ORF72 [57]. MAPT mutations are commonly observed in the FTLD-tau form [57], while GRN and C9ORF72 mutations are commonly associated to the FTLD-TDP form [57]. Moreover, MAPT mutations have been found in patients with progressive supranuclear palsy, corticobasal syndrome, and progressive nonfluent aphasia (PNFA), while GRN mutations [57] have been found as associated with semantic dementia (SD), behavioral variant of FTD (bv-FTD), and FTLD-MND.

For the purpose of this chapter, which is focused on the neuroimaging, we will limit our description to the clinical classification of FTD as linguistic and behavioral variants.

Clinical Aspects

The Linguistic Variants of FTD

A progressive disorder of language associated with atrophy of the frontal and temporal regions of the left hemisphere was first described in the 1890s by Pick [58]. In the last century, several attempts have been made to further classify the FTD language variant in more specific subtypes. In 1982, Mesulam [59] reported a series of cases with "slowly progressive aphasia" and defined them as primary progressive aphasia (PPA) [60]. In the 1990s, the progressive nonfluent aphasia (PNFA) and the semantic dementia [61, 62] have been characterized. Each FTD language variety presents with a well-defined pattern of cognitive deficits and brain abnormalities [63]. According to the most recent diagnostic criteria, we can recognize the following PPA subtypes: nonfluent/agrammatic, semantic, and logopenic.

The Behavioral Variant of FTD

The behavioral variant is the most frequent clinical presentation of FTD [64] and is characterized by predominant changes in personality and several behavioral symptoms, including disinhibition, apathy, eating and sleep disorders, lack of empathy, and obsessive-compulsive disorders. A progressive deterioration of social behavior and cognition is also considered as a common feature.

Structural MRI in FTD

Linguistic Variants

The nonfluent/agrammatic PPA is clinically characterized by a prominent impairment in speech production dominated by agrammatism and apraxia [63]. Other typical aphasic deficits include comprehension impairment for sentences, anomia, and phonemic errors [63]. Tau but not TDP pathology has been mostly associated to the presence of prominent apraxia of speech [65].

Several studies on the nonfluent/agrammatic variant of PPA have shown a typical pattern of atrophy localized in the left insula and perisylvian area (Fig. 10.7, panel a) [66, 67]. However, an involvement of the left opercular region, Broca's area, and motor/premotor cortex has also been reported [66]. Other studies have shown subcortical gray matter atrophy in the thalamus, in the basal ganglia, and in the amygdala [66]. A recent study based on VBM and diffusion imaging [68] has compared different variants of FTD patients against healthy controls to assess both gray and white matter damage. In patients with the nonfluent/agrammatic variant of PPA, gray matter (GM) and WM loss were found in frontal and temporal language areas, with a selective microscopic damage in the left superior longitudinal/arcuate fasciculus [68]. Patients with the semantic variant of PPA (SD) clinically show fluent aphasia, characterized by anomia, single-word comprehension deficits, impaired object knowledge, surface dyslexia, and dysgraphia. In contrast, repetition and speech production are relatively spared (without agrammatism or apraxia of speech [63]). SD is in most cases associated to FTLD-TDP pathology [69], with a pattern of brain atrophy mainly involving the anterior part of the left temporal lobe and, specifically, the temporal pole (Fig. 10.7, panel b) [70]. When considering the white matter tissue, microstructural damage has been found confined to the uncinate fasciculus bilaterally and to the anterior part of the left inferior longitudinal fasciculus [68].

The logopenic variant of PPA is clinically characterized by impairment in single-word retrieval in spontaneous speech and deficits in sentence repetition [63]. In some cases, phonologic errors during spontaneous speech can also be present. Conversely, single-word comprehension, object knowledge, and motor speech are typically preserved in the absence of agrammatism [63]. This PPA variant is more likely associated to AD pathology [69]. Some studies have reported an asymmetric left-side atrophy involving the perisylvian areas, the posterior part of the superior temporal cortex, and the inferior parietal lobes (Fig. 10.7, panel c) [63, 65]. Additionally, atrophy can also be found in the PCC/precuneus and in the medial temporal lobe [65].

Behavioral Variants

Neuroimaging studies on bv-FTD have demonstrated a widespread pattern of brain atrophy including several frontal areas, such as the anterior medial portion of frontal lobe, the gyrus rectus, the superior frontal gyrus, and the anterior cingulate [71]. An example of neuroradiological presentation of bv-FTD is illustrated in Fig. 10.7, panel d. The insula and the thalamus can also be affected by atrophy [71]. This pattern of tissue loss is typically bilateral with a mild right-side prevalence [71]. The earliest changes are detectable in the anterior cingulate, orbitofrontal, and frontoinsular cortices, and atrophy in these regions may help in distinguishing FTD from AD patients [49].

Nevertheless, different subtypes of bv-FTD can be identified in terms of neuronal networks disrupted. In this view, Whitwell and coworkers (2009) [72] have reported four principal variants of bv-FTD: frontal dominant, temporal dominant, frontotemporal, and temporo-fronto-parietal. The only variant showing significant correlations between brain atrophy and pathological subtype is the temporal dominant, which is associated with MAPT genetic mutations [72].

Within the bv-FTD spectrum, damage to specific brain areas has been associated to specific clinical symptoms. The presence of dorsomedial frontal atrophy is more likely associated a clinical phenotype dominated by apathy and aberrant motor behavior. In contrast, atrophy in orbitofrontal regions is more frequently associated with disinhibition [73]. Finally, consistent with the bv-FTD clinical phenotype, white matter atrophy is predominantly distributed to the anterior part of the corpus callosum [68].

Functional MRI in FTD

Abnormalities in brain connectivity have been reported within specific networks in patients with FTD. A peculiar disruption of functional connectivity (as assessed by resting-state fMRI) has been consistently found in the salience network of bv-FTD patients, which includes the frontal and insular cortex and the anterior cingulate gyrus [67]. A recent study shows that loss of functional connectivity within the salience network is associated with disease severity [74]. Additionally, alterations in insular or fronto-limbic areas have been found to predict patients' behavioral worsening (i.e., increasing of apathy scores) in both phenotypes, bv-FTD and semantic PPA [75].

Recently, Borroni and coworkers [5] have investigated not only patients with sporadic FTD but also patients with a genetic progranulin variant of FTD and a group of preclinical mutation carriers. Interestingly, increased functional connectivity was found in the salience network of presymptomatic mutation carriers, interpreted as a compensatory mechanism preceding the clinical onset of disease. Conversely, in symptomatic patients, the salience network was found to be disrupted, while the DMN, which is typically targeted by AD pathology, revealed here an increase of connectivity [5]. Also in this case, increase in functional connectivity was interpreted as a compensatory mechanism. Conversely, in another study including patients with bv-FTD, the increases of functional connectivity found in the DMN and in the dorsal attention network (DAN) were interpreted as changes associated to patients' clinical features, such as apathy symptoms and executive dysfunctions [76].

References

1. McGinnis SM. Neuroimaging in neurodegenerative dementias. Semin Neurol. 2012;32(4): 347–60.
2. Bozzali M, Padovani A, Caltagirone C, Borroni B. Regional grey matter loss and brain disconnection across Alzheimer disease evolution. Curr Med Chem. 2011;18(16):2452–8.
3. Cummings JL. Toward a molecular neuropsychiatry of neurodegenerative diseases. Ann Neurol. 2003;54(2):147–54.
4. Gili T, Cercignani M, Serra L, Perri R, Giove F, Maraviglia B, Caltagirone C, Bozzali M. Regional brain atrophy and functional disconnection across Alzheimer's disease evolution. J Neurol Neurosurg Psychiatry. 2011;82(1):58–66.
5. Borroni B, Alberici A, Cercignani M, Premi E, Serra L, Cerini C, et al. Granulin mutation drives brain damage and reorganization from preclinical to symptomatic FTLD. Neurobiol Aging. 2012;33(10):2506–20.
6. Sosa-Ortiz AL, Acosta-Castillo I, Prince MJ. Epidemiology of dementias and Alzheimer's disease. Arch Med Res. 2012;43(8):600–8.
7. Braak H, Braak E. Staging of Alzheimer's disease-related neurofibrillary changes. Neurobiol Aging. 1995;16(3):271–8.
8. Markesbery WR. Neuropathologic alterations in mild cognitive impairment: a review. J Alzheimers Dis. 2010;19(1):221–8.
9. Albert MS, DeKosky ST, Dickson D, Dubois B, Feldman HH, Fox NC, Gamst A, Holtzman DM, Jagust WJ, Petersen RC, Snyder PJ, Carrillo MC, Thies B, Phelps CH. The diagnosis of mild cognitive impairment due to Alzheimer's disease: recommendations from the National Institute on Aging-Alzheimer's Association workgroups on diagnostic guidelines for Alzheimer's disease. Alzheimers Dement. 2011;7(3):270–9.
10. Scheltens P, Leys D, Barkhof F, Huglo D, Weinstein HC, Vermersch P, Kuiper M, Steinling M, Wolters EC, Valk J. Atrophy of medial temporal lobes on MRI in "probable" Alzheimer's disease and normal ageing: diagnostic value and neuropsychological correlates. J Neurol Neurosurg Psychiatry. 1992;55(10):967–72.

11. Wahlund LO, Barkhof F, Fazekas F, Bronge L, Augustin M, Sjögren M, et al.; European Task Force on Age-Related White Matter Changes. A new rating scale for age-related white matter changes applicable to MRI and CT. Stroke. 2001;32(6):1318–22.
12. Ridha BH, Barnes J, van de Pol LA, Schott JM, Boyes RG, Siddique MM, et al. Application of automated medial temporal lobe atrophy scale to Alzheimer disease. Arch Neurol. 2007;64(6):849–54.
13. Fazekas F, Chawluk JB, Alvavi A, Hurtig HI, Zimmerman RA. MR signal abnormalities at 1.5T in Alzheimer's disease and normal aging. AJR Am J Roentgenol. 1987;8:421–6.
14. Seab JP, Jagust WJ, Wong ST, Roos MS, Reed BR, Budinger TF. Quantitative NMR measurements of hippocampal atrophy in Alzheimer's disease. Magn Reson Med. 1988;8(2):200–8.
15. Convit A, De Leon MJ, Tarshish C, De Santi S, Tsui W, Rusinek H, George A. Specific hippocampal volume reductions in individuals at risk for Alzheimer's disease. Neurobiol Aging. 1997;18(2):131–8.
16. Jack Jr CR, Petersen RC, Xu YC, O'Brien PC, Smith GE, Ivnik RJ, Boeve BF, Waring SC, Tangalos EG, Kokmen E. Prediction of AD with MRI-based hippocampal volume in mild cognitive impairment. Neurology. 1999;52(7):1397–403.
17. Frisoni GB, Jack CR. Harmonization of magnetic resonance-based manual hippocampal segmentation: a mandatory step for wide clinical use. Alzheimers Dement. 2011;7(2):171–4.
18. Ashburner J, Friston KJ. Voxel-based morphometry–the methods. Neuroimage. 2000;11(6): 805–21.
19. Bozzali M, Filippi M, Magnani G, Cercignani M, Franceschi M, Schiatti E, et al. The contribution of voxel-based morphometry in staging patients with mild cognitive impairment. Neurology. 2006;67(3):453–60.
20. Serra L, Cercignani M, Lenzi D, Perri R, Fadda L, Caltagirone C, et al. Grey and white matter changes at different stages of Alzheimer's disease. J Alzheimers Dis. 2010;19(1):147–59.
21. Serra L, Fadda L, Perri R, Spanò B, Marra C, Castelli D, et al. Constructional apraxia as a distinctive cognitive and structural brain feature of pre-senile Alzheimer's disease. J Alzheimers Dis. 2014;38:391–402.
22. Serra L, Perri R, Cercignani M, Spanò B, Fadda L, Marra C, et al. Are the behavioral symptoms of Alzheimer's disease directly associated with neurodegeneration? J Alzheimers Dis. 2010;21(2):627–39.
23. Serra L, Cercignani M, Petrosini L, Basile B, Perri R, Fadda L, et al. Neuroanatomical correlates of cognitive reserve in Alzheimer disease. Rejuvenation Res. 2011;14(2):143–51.
24. Serra L, Giulietti G, Cercignani M, Spanò B, Torso M, Castelli D, et al. Mild cognitive impairment: same identity for different entities. J Alzheimers Dis. 2013;33(4):1157–65.
25. Basser PJ, Jones DK. Diffusion-tensor MRI: theory, experimental design and data analysis - a technical review. NMR Biomed. 2002;15(7–8):456–67.
26. Smith SM, Jenkinson M, Johansen-Berg H, Rueckert D, Nichols TE, Mackay CE, et al. Tract-based spatial statistics: voxelwise analysis of multi-subject diffusion data. Neuroimage. 2006;31(4):1487–505.
27. Jones DK. Studying connections in the living human brain with diffusion MRI. Cortex. 2008;44(8):936–52.
28. Liu Y, Spulber G, Lehtimäki KK, Könönen M, Hallikainen I, Gröhn H, et al. Diffusion tensor imaging and tract-based spatial statistics in Alzheimer's disease and mild cognitive impairment. Neurobiol Aging. 2011;32(9):1558–71.
29. Serra L, Cercignani M, Basile B, Spanò B, Perri R, Fadda L, et al. White matter damage along the uncinate fasciculus contributes to cognitive decline in AD and DLB. Curr Alzheimer Res. 2012;9(3):326–33.
30. Bozzali M, Giulietti G, Basile B, Serra L, Spanò B, Perri R, et al. Damage to the cingulum contributes to Alzheimer's disease pathophysiology by deafferentation mechanism. Hum Brain Mapp. 2012;33(6):1295–308.
31. Bozzali M, Parker GJ, Serra L, Embleton K, Gili T, Perri R, et al. Anatomical connectivity mapping: a new tool to assess brain disconnection in Alzheimer's disease. Neuroimage. 2011;54(3):2045–51.

32. Bozzali M, Parker GJ, Spanò B, Serra L, Giulietti G, Perri R, et al. Brain tissue modifications induced by cholinergic therapy in Alzheimer's disease. Hum Brain Mapp. 2013;34(12):3158–67.
33. Peters F, Collette F, Degueldre C, Sterpenich V, Majerus S, Salmon E. The neural correlates of verbal short-term memory in Alzheimer's disease: an fMRI study. Brain. 2009;132(7):1833–46.
34. Golby A, Silverberg G, Race E, Gabrieli S, O'Shea J, Knierim K, et al. Memory encoding in Alzheimer's disease: an fMRI study of explicit and implicit memory. Brain. 2005;128(4):773–87.
35. Pihlajamäki M, Jauhiainen AM, Soininen H. Structural and functional MRI in mild cognitive impairment. Curr Alzheimer Res. 2009;6(2):179–85.
36. Lenzi D, Serra L, Perri R, Pantano P, Lenzi GL, Paulesu E, et al. Single domain amnestic MCI: a multiple cognitive domains fMRI investigation. Neurobiol Aging. 2011;32(9):1542–57.
37. Greicius MD, Krasnow B, Reiss AL, Menon V. Functional connectivity in the resting brain: a network analysis of the default mode hypothesis. Proc Natl Acad Sci U S A. 2003; 100(1):253–8.
38. Greicius MD, Supekar K, Menon V, Dougherty RF. Resting-state functional connectivity reflects structural connectivity in the default mode network. Cereb Cortex. 2009;19(1):72–8.
39. Phelps ME. PET: the merging of biology and imaging into molecular imaging. J Nucl Med. 2000;41(4):661–81.
40. Bohnen NI, Djang DS, Herholz K, Anzai Y, Minoshima S. Effectiveness and safety of 18F-FDG PET in the evaluation of dementia: a review of the recent literature. J Nucl Med. 2012;53(1):59–71.
41. Knopman DS. Diagnostic tests for Alzheimer disease: FDG-PET imaging is a player in search of a role. Neurol Clin Pract. 2012;2(2):151–3.
42. Scherfler C, Schwarz J, Antonini A, Grosset D, Valldeoriola F, Marek K, et al. Role of DAT-SPECT in the diagnostic work up of parkinsonism. Mov Disord. 2007;22(9):1229–38.
43. Hardy J, Selkoe DJ. The amyloid hypothesis of Alzheimer's disease: progress and problems on the road to therapeutics. Science. 2002;297(5580):353–6.
44. Rowe CC, Villemagne VL. Amyloid imaging with PET in early Alzheimer disease diagnosis. Med Clin North Am. 2013;97(3):377–98.
45. Visser PJ, Vos S, van Rossum I, Scheltens P. Comparison of International Working Group criteria and National Institute on Aging-Alzheimer's Association criteria for Alzheimer's disease. Alzheimers Dement. 2012;8(6):560–3.
46. Villemagne VL, Rowe CC, Macfarlane S, Novakovic KE, Masters CL. Imaginem oblivionis: the prospects of neuroimaging for early detection of Alzheimer's disease. J Clin Neurosci. 2005;12(3):221–30.
47. Cohen AD, Rabinovici GD, Mathis CA, Jagust WJ, Klunk WE, Ikonomovic MD. Using Pittsburgh Compound B for in vivo PET imaging of fibrillar amyloid-beta. Adv Pharmacol. 2012;64:27–81.
48. Ikonomovic MD, Klunk WE, Abrahamson EE, Mathis CA, Price JC, Tsopelas ND, et al. Post-mortem correlates of in vivo PiB-PET amyloid imaging in a typical case of Alzheimer's disease. Brain. 2008;131(6):1630–45.
49. Rabinovici GD, Miller BL. Frontotemporal lobar degeneration: epidemiology, pathophysiology, diagnosis and management. CNS Drugs. 2010;24(5):375–98.
50. Whitwell JL, Josephs KA, Avula R, Tosakulwong N, Weigand SD, Senjem ML, et al. Altered functional connectivity in asymptomatic MAPT subjects: a comparison to bvFTD. Neurology. 2011;77(9):866–74.
51. Josephs KA, Hodges JR, Snowden JS, Mackenzie IR, Neumann M, Mann DM, Dickson DW. Neuropathological background of phenotypical variability in frontotemporal dementia. Acta Neuropathol. 2011;122(2):137–53.
52. Whitwell JL, Josephs KA. Neuroimaging in frontotemporal lobar degeneration–predicting molecular pathology. Nat Rev Neurol. 2012;8(3):131–42.
53. Neumann M, Sampathu DM, Kwong LK, Truax AC, Micsenyi MC, Chou TT, et al. Ubiquitinated TDP-43 in frontotemporal lobar degeneration and amyotrophic lateral sclerosis. Science. 2006;314(5796):130–3.

54. Neumann M, Rademakers R, Roeber S, Baker M, Kretzschmar HA, Mackenzie IR. A new subtype of frontotemporal lobar degeneration with FUS pathology. Brain. 2009;132(11):2922–31.
55. Hutton M, Lendon CL, Rizzu P, Baker M, Froelich S, Houlden H, et al. Association of missense and 5′-splice-site mutations in tau with the inherited dementia FTDP-17. Nature. 1998;393(6686):702–5.
56. Cruts M, Gijselinck I, van der Zee J, Engelborghs S, Wils H, Pirici D, et al. Null mutations in progranulin cause ubiquitin-positive frontotemporal dementia linked to chromosome 17q21. Nature. 2006;442(7105):920–4.
57. Renton AE, Majounie E, Waite A, Simón-Sánchez J, Rollinson S, Gibbs JR, et al. A hexanucleotide repeat expansion in C9ORF72 is the cause of chromosome 9p21-linked ALS-FTD. Neuron. 2011;72(2):257–68.
58. Pick A. Ubeer die Beziehungen der senile Hirnatrophie zur aphasie. Prager Med Wochenschr. 1892;17:165–7.
59. Mesulam MM. Slowly progressive aphasia without generalized dementia. Ann Neurol. 1982;11(6):592–8.
60. Mesulam MM, Weintraub S. Spectrum of primary progressive aphasia. Baillieres Clin Neurol. 1992;1(3):583–609.
61. Warrington EK. The selective impairment of semantic memory. Q J Exp Psychol. 1975;27(4):635–57.
62. Snowden JS, Goulding PJ, Neary D. Semantic dementia: a form of circumscribed cerebral atrophy. Behav Neurol. 1989;2:167–82.
63. Gorno-Tempini ML, Hillis AE, Weintraub S, Kertesz A, Mendez M, Cappa SF, et al. Classification of primary progressive aphasia and its variants. Neurology. 2011;76(11):1006–14.
64. Rascovsky K, Hodges JR, Knopman D, Mendez MF, Kramer JH, Neuhaus J, et al. Sensitivity of revised diagnostic criteria for the behavioural variant of frontotemporal dementia. Brain. 2011;134(9):2456–77.
65. Rohrer JD, Paviour D, Bronstein AM, O'Sullivan SS, Lees A, Warren JD. Progressive supranuclear palsy syndrome presenting as progressive nonfluent aphasia: a neuropsychological and neuroimaging analysis. Mov Disord. 2010;25(2):179–88.
66. Gorno-Tempini ML, Dronkers NF, Rankin KP, Ogar JM, Phengrasamy L, Rosen HJ, et al. Cognition and anatomy in three variants of primary progressive aphasia. Ann Neurol. 2004;55(3):335–46.
67. Seeley WW, Crawford RK, Zhou J, Miller BL, Greicius MD. Neurodegenerative diseases target large-scale human brain networks. Neuron. 2009;62(1):42–52.
68. Zhang Y, Tartaglia MC, Schuff N, Chiang GC, Ching C, Rosen HJ, et al. MRI signatures of brain macrostructural atrophy and microstructural degradation in frontotemporal lobar degeneration subtypes. J Alzheimers Dis. 2013;33(2):431–44.
69. Rohrer JD. Structural brain imaging in frontotemporal dementia. Biochim Biophys Acta. 2012;1822(3):325–32.
70. Hodges JR, Patterson K. Semantic dementia: a unique clinicopathological syndrome. Lancet Neurol. 2007;6(11):1004–14.
71. Schroeter ML, Raczka K, Neumann J, von Cramon DY. Neural networks in frontotemporal dementia–a meta-analysis. Neurobiol Aging. 2008;29(3):418–26.
72. Whitwell JL, Przybelski SA, Weigand SD, Ivnik RJ, Vemuri P, Gunter JL, et al. Distinct anatomical subtypes of the behavioural variant of frontotemporal dementia: a cluster analysis study. Brain. 2009;132(Pt 11):2932–46.
73. Rosen HJ, Allison SC, Schauer GF, Gomo-Tempini ML, Weiner MW, Miller BL. Neuroanatomical correlates of behavioural disorders in dementia. Brain. 2005;128:2612–25.
74. Zhou J, Greicius MD, Gennatas ED, Growdon ME, Jang JY, Rabinovici GD, et al. Divergent network connectivity changes in behavioural variant frontotemporal dementia and Alzheimer's disease. Brain. 2010;133(5):1352–67.
75. Day GS, Farb NA, Tang-Wai DF, Masellis M, Black SE, Freedman M, et al. Salience network resting-state activity: prediction of frontotemporal dementia progression. JAMA Neurol. 2013;70:1249–53.
76. Rytty R, Nikkinen J, Paavola L, Abou Elseoud A, Moilanen V, Visuri A, et al. GroupICA dual regression analysis of resting state networks in a behavioral variant of frontotemporal dementia. Front Hum Neurosci. 2013;7:461.

Chapter 11
Amyotrophic Lateral Sclerosis: Genotypes and Phenotypes

Nicola Ticozzi and Vincenzo Silani

Abstract Mendelian forms of amyotrophic lateral sclerosis (ALS) account for nearly 10 % of all cases. To date, 19 disease genes, usually but not exclusively inherited with an autosomal dominant pattern, have been reported to be associated with ALS or with atypical motor neuron diseases with or without associated frontotemporal dementia (ALS-FTD). Often, it is possible to draw correlations between distinct ALS-associated mutations and specific clinical phenotypes. This information is essential for biologists and clinicians alike, providing at the same time an unparalleled insight into the pathogenesis of the disease and invaluable tools for genetic counseling, diagnosis, and development of preventive strategies and treatments for ALS.

Keywords Genetic epidemiology • Genotype-phenotype correlation • *SOD1* • *TARDBP* • *FUS* • *C9orf72*

N. Ticozzi, MD
Department of Neurology and Laboratory of Neuroscience, IRCCS Istituto Auxologico Italiano, P.le Brescia 20, Milan 20149, Italy

Department of Pathophysiology and Transplantation, Dino Ferrari Center, Università degli Studi di Milano, Milan, Italy

V. Silani, MD (✉)
Department of Neurology and Laboratory of Neuroscience, IRCCS Istituto Auxologico Italiano, P.le Brescia 20, Milan 20149, Italy

Department of Pathophysiology and Transplantation, Dino Ferrari Center, Università degli Studi di Milano, Milan, Italy
e-mail: vincenzo.silani@unimi.it

D. Galimberti, E. Scarpini (eds.), *Neurodegenerative Diseases*,
DOI 10.1007/978-1-4471-6380-0_11,

Introduction

Amyotrophic lateral sclerosis (ALS) is an adult-onset, rapidly progressive neurodegenerative disorder, traditionally interpreted as caused by the selective loss of upper and lower motor neurons in the cerebral cortex, brainstem, and spinal cord. Neuronal degeneration leads to weakness, muscular atrophy, and spasticity that evolve to paralysis. It is today also quite indisputable that neurons in the prefrontal and temporal cortex are affected in ALS as well, but to varying degrees. Degeneration of such neurons results in frontal executive dysfunction in many patients with this disorder and concomitant frontotemporal dementia (FTD) in about 15 % of patients. This is known as ALS with frontotemporal lobe degeneration (FTLD). ALS and FTD are therefore at opposite ends of the spectrum of a single disease. The typical age at onset is between 50 and 60 years, and the global incidence is 1–2 new cases per 100,000 individuals every year. The disease is fatal within 2–5 years of onset, generally due to respiratory failure. As the cause of ALS is still unknown, there is presently no effective treatment for it [1]. Although the majority of ALS cases are sporadic (sporadic ALS [SALS]), 10 % of patients have a positive familial history for motor neuron disease, generally with an autosomal dominant inheritance pattern, although recessive pedigrees have been described (familial ALS [FALS]) [2]. The clinical phenotype of FALS cases is usually indistinguishable from SALS. However, in comparison to SALS, FALS is characterized by an equal male:female sex ratio, an earlier age at onset, and generally a longer disease duration. The first signs of the disease often occur in the lumbosacral segment, and atypical symptoms may be present at onset [2–5]. To date, 19 disease genes have been reported to be associated with typical ALS or atypical motor neuron diseases with or without associated FTD (ALS-FTD) (Table 11.1). Mutations in four major genes, namely, *SOD1*, *TARDBP*, *FUS*, and *C9orf72*, account for ~50 % of all FALS cases and ~10 % of SALS. Given the relatively high mutational frequency, a robust genotype-phenotype correlation can be drawn for these genes. Conversely, pathogenic mutations in the other 15 ALS-associated genes (*ALS2*, *SETX*, *SPG11*, *VAPB*, *ANG*, *FIG4*, *OPTN*, *ATXN2*, *VCP*, *UBQLN2*, *SIGMAR1*, *CHMP2B*, *PFN1*, *ERBB4*, and *HNRNPA1*) are collectively responsible for less than 5 % of cases. Usually, those variants are found in isolated pedigrees, often with atypical ALS phenotypes [6–13], and are private mutations, thus making a clear genotype-phenotype correlation extremely difficult. In this chapter, we will describe the main clinical features of *SOD1*-, *TARDBP*-, *FUS*-, and *C9orf72*-associated ALS, briefly outline the atypical phenotypes of minor genes, and discuss the genetic overlap between ALS and FTD.

SOD1-associated ALS (ALS1)

The linkage of ALS1 to chromosome 21q22.1 was initially described in 1991 [14]. Two years later, 11 disease-associated mutations in the *SOD1* gene, encoding for the Cu/Zn superoxide dismutase, a cytoplasmic enzyme responsible for the catabolism

Table 11.1 Disease genes reported to be associated with typical ALS or atypical motor neuron diseases with or without associated FTD

ALS type	Onset	Inheritance	Locus	Gene	Protein
ALS1	Adult	AD[a]	21q22.1	*SOD1*	Cu/Zn superoxide dismutase
ALS2	Juvenile	AR	2q33-35	*ALS2*	Alsin
ALS3	Adult	AD	18q21	Unknown	Unknown
ALS4	Juvenile	AD	9q34	*SETX*	Senataxin
ALS5	Juvenile	AR	15q15-21	*SPG11*	Spatacsin
ALS6	Adult	AD[b]	16p11.2	*FUS*	*FUS*ed in sarcoma
ALS7	Adult	AD	20p13	Unknown	Unknown
ALS8	Adult	AD	20q13.33	*VAPB*	VAMP-associated protein B
ALS9	Adult	AD	14q11	*ANG*	Angiogenin
ALS10	Adult	AD	1q36	*TARDBP*	TAR DNA-binding protein
ALS11	Adult	AD	6q21	*FIG4*	PI(3,5) P(2)5-phosphatase
ALS12	Adult	AR/AD	10p15-p14	*OPTN*	Optineurin
ALS13	Adult	Susceptibility	12q13.12	*ATXN2*	Ataxin-2
ALS14	Adult	AD	9p13.3	*VCP*	Valosin-containing protein
ALS15	Adult	XD	Xp11.21	*UBQLN2*	Ubiquilin-2
ALS16	Juvenile	AR	9p13.3	*SIGMAR1*	Sigma nonopioid receptor 1
ALS17	Adult	AD	3p11.2	*CHMP2B*	Charged multivesicular body protein 2B
ALS18	Adult	AD	17p13.2	*PFN1*	Profilin-1
ALS19	Adult	AD	2q34	*ERBB4*	V-erb-b2 avian erythroblastic leukemia viral oncogene homolog 4
ALS20	Adult	AD	12q13.3	*HNRNPA1*	Heterogeneous nuclear ribonucleoprotein A1
FTDALS	Adult	AD	9p21.2	*C9orf72*	*C9orf72*

[a]With the exception of the D90A mutation which is inherited with a recessive pattern in the majority of affected families and with a dominant pattern in a minority of FALS

[b]With the exception of the H517Q mutation which has been found in homozygosity in a family with apparent autosomal recessive inheritance

of superoxide radicals to hydrogen peroxide and molecular oxygen, were identified [15]. The gene spans 9.3 kb, is composed of five exons, and encodes for a 153 residues long protein. *SOD1* is a 32 kDa omodimeric metalloenzyme, ubiquitously expressed, and highly conserved, and represents ~1 % of all cytoplasmic proteins. To date, more than 166 different *SOD1* mutations have been reported (http://alsod.iop.kcl.ac.uk), the vast majority of which are missense substitutions distributed throughout the five exons of the gene. Eight frameshift deletions and five insertions, all clustered in exons 4 and 5, which lead to a premature truncation of the protein, have also been described. Collectively, *SOD1* mutations are found in ~20 % of all FALS patients and in ~3 % of SALS cases [16]. *SOD1* mutations are characterized by a considerable interfamilial and intrafamilial variability of the phenotype with regard to the age, site of onset, and disease duration [17]. A notable exception is represented by A4V, the mutation most frequently observed in ALS1 pedigrees, which is consistently associated with a high penetrance, younger age at onset, prevalence of lower motor neuron signs, and a very rapid disease course, usually less than 12 months [17, 18]. Atypical symptoms, such as external ophthalmoplegia, hyperacusis, and neuralgic pain, have occasionally been reported in A4V long survivors on artificial ventilatory support. Interestingly, the same aggressive phenotype is shared by other less common mutations in the same region of exon 1, such as A4T, C6F, C6G, and G10V [19–22]. Conversely, other mutations, such as G41D, H46R, and G93D, display a very mild phenotype, often with carriers surviving more than 20 years after the onset of the disease [17, 23, 24]. Atypical lower motor neuron phenotypes, such as cramp-fasciculation syndrome, flail limb, and/or pseudopolyneuritic (Patrikios) forms of ALS, are also often observed in patients with less aggressive *SOD1* mutations. The penetrance of mutations is also variable, being almost complete for A4V, and less than 30 % at 70 years for I113T [25]. It must be noticed, however, that the majority of the *SOD1* variants described so far are private mutations. Thus, conclusive genotype-phenotype correlation can be safely drawn only for a handful of them. Moreover, although *SOD1* mutations have never been observed in the general population, the pathogenic role of private variants has recently started to be questioned. For instance, Felbecker et al. [26] described four families in which the E100K and D90A mutations are present in some affected individuals, but do not segregate with the disease within the pedigree. These findings must be taken into serious consideration with regard to genetic testing and counseling in the clinical practice. All *SOD1* mutations are inherited as dominant traits, with the exception of the D90A variant, that is observed both in recessive pedigrees in Scandinavia and in dominant pedigrees in the rest of the world [27, 28]. D90A homozygous families display a milder phenotype, characterized by an asymmetrical, slowly progressive, ascending paraparesis with upper motor neuron signs, compared to heterozygous individuals that develop classic ALS. That and the finding that recessive families share a common ancestor suggest the existence of a protective genetic factor in linkage with D90A in the Swedish population [29]: this is also relevant in the Mediterranean area where the Normans migrated and homozygous D90A patients have been identified [30].

TARDBP-associated ALS (ALS10)

TDP-43 is the major proteinaceous component of ubiquitinated cytoplasmic inclusions in ALS. Within aggregates, the protein is hyperphosphorylated and cleaved to generate abnormal C-terminal fragments. Moreover, while in unaffected neurons

TDP-43 localizes in the cell nucleus, it is absent from the nuclei of inclusion-bearing neurons, suggesting a nucleocytoplasmic redistribution of the protein. TDP-43 is a 414 amino acid multifunctional DNA-/RNA-binding protein that plays a role in several biological processes, including gene transcription, splicing regulation, and transport and stabilization of mRNA molecules [31].

Mutations in the *TARDBP* gene, encoding for TDP-43, have been identified in several populations of different geographic origin [32–41] with a proposed mutational frequency of ~5 % for FALS and 2 % for SALS. To date, more than 30 different *TARDBP* mutations have been described, all of which are missense substitutions. With few exceptions, they are clustered in the C-terminal glycine-rich region encoded by exon 6. As a general rule, ALS10 patients display a classic ALS phenotype; although the disease onset appears to be anticipated compared to nonmutated cases, the upper limbs are most often involved at onset, and the duration is longer [42]. The penetrance of *TARDBP* mutations appears to be somewhat lower than that of other ALS-causing genes, thus explaining in part their relatively high occurrence in sporadic cases. However, since most *TARDBP* mutations are private, it is extremely difficult to establish clear genotype-phenotype correlation for each of them. It has been suggested that A382T, which is the variant most commonly observed in the Mediterranean area, may be associated with a low-penetrance (<20 % at age 70), predominantly lower motor neuron disease with an asymmetrical onset in the distal muscles of the limbs, subsequently spreading to proximal muscles, with relative sparing of the bulbar muscles [20]. Interestingly, A382T has occasionally been reported in ALS patients displaying parkinsonian and/or cerebellar features [43, 44]. This mutation is extremely common in Sardinia, accounting for one third of all ALS cases, being also present in ~1 % of healthy controls [45, 46].

FUS-associated ALS (ALS6)

The ALS6 locus on chromosome 16p12.1-q21 has been independently reported by three studies on six European and North American pedigrees with autosomal dominant classic ALS [47–49], and variants in the *FUS* gene have been subsequently identified as the disease-causing mutations in ALS6 families (2009) [50]. Similar to *TARDBP*, the *FUS* gene also encodes for a DNA-/RNA-binding protein involved in several cellular pathways, including transcriptional regulation, maintenance of genomic stability, and splicing, nucleocytoplasmic shuttling, transport, and maturation of mRNAs [51]. Following the original reports, several other groups identified additional variants in ALS cohorts of different ethnicities, proposing an overall mutational frequency of ~4 % in FALS and ~1 % in SALS [28, 38, 40, 52–62]. To date, more than 30 different mutations have been described, the vast majority of which are missense substitutions and the rest are frameshift or nonsense mutations. Although genotype-phenotype correlations are not possible for the majority of *FUS* variants identified so far, it has been suggested that mutations within the C-terminal nuclear localization signal may result in an uncommon phenotype characterized by a symmetrical, proximal spinal onset with early involvement of the scapular and/or pelvic girdle. Respiratory muscles and the axial muscles of the neck and the trunk are also prominently affected, with head drop being a common phenotypic feature. Compared to individuals harboring mutations in other ALS-associated genes, *FUS* patients often show an early onset during their third or fourth decade and an accelerated disease course [63]. ALS6 thus ranks among the most aggressive forms of genetic ALS.

C9orf72-associated ALS-FTD

For several years, the existence of a major ALS locus on chromosome 9p21 was repeatedly suggested by linkage analysis on informative pedigrees [64–68] and genome-wide association studies [69–71]. A major breakthrough came out when two independent groups identified in this region the presence of an expanded hexanucleotide repeat in the first intron of the *C9orf72* gene [46, 72]. This repeat (GGGGCC) is highly polymorphic in the normal population (2–23 units) but is expanded both in ALS and FTD patients up to 4,400 units [73]. In the inbred Finnish population, the mutational frequency of *C9orf72* was reported to be as high as 46 % in FALS and 21 % in SALS [46]. In other populations of European descent, the mutational frequencies ranged from 23 to 47 % in FALS, 4 to 5 % in SALS, 12 to 29 % in FTD, and 6 to 86 % in ALS-FTD patients [2, 20, 28, 69, 72, 74, 75].

The motor phenotype of *C9orf72*-positive patients is often characterized by an early age at onset and shorter survival time compared to nonmutated individuals, possibly due to a very early bulbar involvement in a majority of cases [20, 28, 69]. Similar to *TARDBP*, and unlike *SOD1*, the upper limbs are more frequently affected at onset compared to the lower limbs. It has also been suggested that *C9orf72*-positive patients display a predominance of upper motor neuron signs, although no patients with pure primary lateral sclerosis (PLS) have been diagnosed so far [20].

It has consistently been reported that patients with concurrent ALS and FTD or with a family history of dementia or motor neuron disease have a higher risk of harboring *C9orf72* RE (33–86 %), further indicating that the two diseases belong to the same pathogenic continuum [69].

The cognitive deficit of *C9orf72*-positive patients is usually consistent with a diagnosis of behavioral variant of frontotemporal dementia (bvFTD) and characterized by socially inappropriate, impulsive behavior and general deterioration in ability to perform routine daily tasks [76], and/or apathy, social isolation, and emotional lability [68]. Patients often display prominent psychiatric features such as visual hallucinations, paranoid behavior, persecutory delusions, aggressive behavior, and/or suicidal thoughts [2, 20, 28, 69, 74].

The heterogeneity of the clinical phenotype associated to *C9orf72* repeat expansions is further complicated by the occasional observations of ALS-FTD patients with concurrent extrapyramidal and/or cerebellar signs [77]. These atypical features can be so prominent in some patients that a diagnosis of concurrent corticobasal syndrome [77, 78], progressive supranuclear palsy, cerebellar ataxia [46], or olivopontocerebellar degeneration [78] can be made, suggesting that *C9orf72* may contribute to the pathogenesis of a broad spectrum of neurodegenerative diseases beyond ALS and FTD. An alternative and possibly more appealing explanation for these findings is that *C9orf72*-positive patients display the same distribution of TDP-43 pathology within the central nervous system as nonmutated cases, although with a greater regional burden of lesions [79]. An increased lesion load in extramotor areas, including the basal ganglia, the cerebellum, and/or brainstem nuclei, may cross a critical threshold, thus causing in *C9orf72* carriers the early appearance of extrapyramidal features otherwise extremely uncommon in nonmutated ALS patients.

Mutations in other ALS-associated Genes and FTD

The recent discovery of *C9orf72* gene as the main cause of ALS and FTD definitively consolidated the hypothesis that the two diseases belong to the same phenotypical, neuropathological, and genetic spectrum. Even before this momentous discovery, however, similar neuropathological features and common mutations in several other genes have been described in both diseases. TDP-43 immunoreactive ubiquitinated inclusions are in fact present in >50 % of all FTD cases [80], while a *FUS* pathology similar to ALS6 is also observed in atypical FTD, basophilic inclusion body dementia, and neuronal intermediate filament inclusion disease [81–83]. Not surprisingly, *TARDBP* and *FUS* mutations, originally identified in ALS cases, have been subsequently found in bvFTD patients with or without motor neuron signs [62, 84]. The concurrence of behavior and executive dysfunction has also been reported for ALS patients carrying mutations in the *UBQLN2* gene [85].

Conversely, other genes initially identified in FTD pedigrees have subsequently been associated to ALS, with or without dementia. Amyotrophy of the limbs has been described in patients with FTD-parkinsonism carrying mutations in the *MAPT* gene [86, 87]. *CHMP2B* mutations have been described in bvFTD patients showing an insidious change in personality and behavior, memory loss, apathy, aggressiveness, stereotyped behavior, disinhibition, dysgraphia, and dyscalculia, as well as in ALS cases [88, 89]. Lastly, mutations in the *VCP* gene, originally identified as causative for an uncommon type of FTD associated with inclusion body myopathy and Paget's disease of the bone [90] have been later found in ALS patients [91].

Uncommon ALS Phenotypes

Mutations in minor ALS genes are usually linked to uncommon clinical phenotypes, often overlapping with hereditary spastic paraparesis and/or hereditary neuropathies. For instance, frameshift deletions or nonsense mutations in *ALS2*, encoding for the protein alsin, are consistently associated either to a juvenile-onset motor neuron disease characterized by distal muscle atrophy and weakness, and spasticity progressively ascending from the lower limbs to the cervical and bulbar segments [92], or to the milder phenotypes of juvenile PLS and infantile-onset ascending hereditary spastic paraparesis [8, 12, 93–97]. Similarly, homozygous or heterozygous mutations in *SPG11* have been identified in a juvenile-onset motor neuron disease (ALS5) characterized by distal muscle atrophy and weakness with pyramidal signs and involvement of bulbar muscles [11]. ALS5 is allelic to a second neurodegenerative disorder named ARHSP-TCC (autosomal recessive hereditary spastic paraplegia with thin corpus callosum) and may belong to the same phenotypic spectrum [98]. Mutations in *SETX*, which encodes for the DNA helicase senataxin, have been also associated with an autosomal dominant, juvenile-onset, slowly progressive, predominantly upper motor neuron disease with consistent sparing of the bulbar muscles (ALS4) [99, 100]. Interestingly, ALS4 is also allelic with another neurodegenerative disease, namely, ataxia-ocular apraxia type 2 (AOA2) [101].

Other genes have been associated with adult-onset ALS with predominantly upper motor neuron signs. Among the patients carrying heterozygous mutations in the *FIG4* gene identified so far, two had been diagnosed with PLS, while the majority of the others had prominent signs of corticospinal tract degeneration. Quite surprisingly, this phenotype is strikingly different from that of Charcot-Marie-Tooth disease type 4 J, an autosomal recessive demyelinating neuropathy characterized by an infantile onset and rapid progression, which is found in compound heterozygous patients [102]. A single patient with PLS carrying a heterozygous mutation in the *OPTN* gene has also been described.

Pure lower motor neuron syndromes have also been described. Patients carrying mutations in *VAPB* often display isolated weakness and wasting of the limb muscles, without concomitant upper neuron signs. Postural tremor, fasciculations, and painful cramps are common features of the clinical phenotype, and in some cases the disease manifested with a very slow-progressive, late-onset spinal muscular atrophy [10, 103, 104]. A spinal onset, predominantly lower motor neuron disease with little or no involvement of bulbar muscles, has also been observed in most individuals carrying *PFN1* mutations [105].

It must be noticed, however, that the clinical information available in the literature about mutated patients is so scant that no safe genotype-phenotype correlations can be drawn for minor ALS-associated genes.

Conclusions

The information provided by genetic studies toward the understanding of the pathogenesis of both FALS and SALS has been invaluable. It has led over the years to the identification of novel cellular pathways involved in motor neuron degeneration, has provided potential therapeutic targets, and made possible the engineering of animal models of ALS, providing essential tools for validating drugs' efficacy.

The ongoing effort aimed at correlating mutations in each ALS-associated gene to distinct clinical phenotypes is equally important. The ALS-FTD spectrum represents today one of the most difficult challenge among neurodegenerative diseases for nosologists, clinicians, and molecular biologists, which have often been confounded by the simultaneous pleiotropism (i.e., the multiple end effects of mutations in a single gene) and genetic heterogeneity (i.e., the existence of mutations in several causative genes leading to the same clinical phenotype) of the two diseases. In fact, the two phenomena should not be mutually exclusive. By emphasizing the pleiotropism of ALS and FTD ("lumping" approach), clinical neurologists were able to recognize the existence of a broad spectrum of neurodegenerative diseases encompassing both motor neuron disorders and dementias, for which molecular biologists demonstrated a common pathological background. Conversely, by focusing on the genetic heterogeneity of the two diseases ("splitting" approach), geneticists were able to provide invaluable information into the etiology and pathomechanisms of ALS and FTD. The application of this splitting approach to the clinical practice has its own merits too. In fact, a more comprehensive understanding of ALS genetic epidemiology among different populations and an extensive genotype-phenotype correlation are essential prerequisites for a successful genetic counseling for patients and their families, for development of preventive strategies and treatments, for reducing diagnostic delays and costs, and, lastly, for optimizing the design of clinical trials.

References

1. Rowland LP, Shneider NA. Amyotrophic lateral sclerosis. N Engl J Med. 2001;344:1688–700.
2. Simon-Sanchez J, Dopper EG, Cohn-Hokke PE, Hukema RK, Nicolaou N, Seelaar H, et al. The clinical and pathological phenotype of C9ORF72 hexanucleotide repeat expansions. Brain. 2012;135:723–5.
3. de Belleroche J, Orrell R, King A. Familial amyotrophic lateral sclerosis/motor neurone disease (FALS): a review of current developments. J Med Genet. 1995;32:841–7.
4. Dejesus-Hernandez M, Rayaprolu S, Soto-Ortolaza AI, Rutherford NJ, Heckman MG, Traynor S, et al. Analysis of the C9orf72 repeat in Parkinson's disease, essential tremor and restless legs syndrome. Parkinsonism Relat Disord. 2013;19(2):198–201.
5. Li TM, Alberman E, Swash M. Comparison of sporadic and familial disease amongst 580 cases of motor neuron disease. J Neurol Neurosurg Psychiatry. 1988;51:778–84.
6. Chow CY, Landers JE, Bergren SK, Sapp PC, Grant AE, Jones JM, et al. Deleterious variants of FIG4, a phosphoinositide phosphatase, in patients with ALS. Am J Hum Genet. 2009;84:85–8.
7. Greenway MJ, Andersen PM, Russ C, Ennis S, Cashman S, Donaghy C, et al. ANG mutations segregate with familial and 'sporadic' amyotrophic lateral sclerosis. Nat Genet. 2006;38:411–3.
8. Hadano S, Hand CK, Osuga H, Yanagisawa Y, Otomo A, Devon RS, et al. A gene encoding a putative GTPase regulator is mutated in familial amyotrophic lateral sclerosis 2. Nat Genet. 2001;29:166–73.

9. Maruyama H, Morino H, Ito H, Izumi Y, Kato H, Watanabe Y, et al. Mutations of optineurin in amyotrophic lateral sclerosis. Nature. 2010;465:223–6.
10. Nishimura AL, Mitne-Neto M, Silva HC, Richieri-Costa A, Middleton S, Cascio D, et al. A mutation in the vesicle-trafficking protein VAPB causes late-onset spinal muscular atrophy and amyotrophic lateral sclerosis. Am J Hum Genet. 2004;75:822–31.
11. Orlacchio A, Babalini C, Borreca A, Patrono C, Massa R, Basaran S, et al. SPATACSIN mutations cause autosomal recessive juvenile amyotrophic lateral sclerosis. Brain. 2010;133:591–8.
12. Yang Y, Hentati A, Deng HX, Dabbagh O, Sasaki T, Hirano M, et al. The gene encoding alsin, a protein with three guanine-nucleotide exchange factor domains, is mutated in a form of recessive amyotrophic lateral sclerosis. Nat Genet. 2001;29:160–5.
13. Yeh TH, Lai SC, Weng YH, Kuo HC, Wu-Chou YH, Huang CL. Screening for C9orf72 repeat expansions in parkinsonian syndromes. Neurobiol Aging. 2012;34:1311.e3–4.
14. Siddique T, Figlewicz DA, Pericak-Vance MA, Haines JL, Rouleau G, Jeffers AJ, et al. Linkage of a gene causing familial amyotrophic lateral sclerosis to chromosome 21 and evidence of genetic-locus heterogeneity. N Engl J Med. 1991;324:1381–4.
15. McCord JM, Fridovich I. Superoxide dismutase. An enzymic function for erythrocuprein (hemocuprein). J Biol Chem. 1969;244:6049–55.
16. Andersen PM, Sims KB, Xin WW, Kiely R, O'Neill G, Ravits J, et al. Sixteen novel mutations in the Cu/Zn superoxide dismutase gene in amyotrophic lateral sclerosis: a decade of discoveries, defects and disputes. Amyotroph Lateral Scler Other Motor Neuron Disord. 2003;4:62–73.
17. Cudkowicz ME, McKenna-Yasek D, Sapp PE, Chin W, Geller B, Hayden DL, et al. Epidemiology of mutations in superoxide dismutase in amyotrophic lateral sclerosis. Ann Neurol. 1997;41:210–21.
18. Juneja T, Pericak-Vance MA, Laing NG, Dave S, Siddique T. Prognosis in familial amyotrophic lateral sclerosis: progression and survival in patients with glu100gly and ala4val mutations in Cu, Zn superoxide dismutase. Neurology. 1997;48:55–7.
19. Akimoto C, Forsgren L, Linder J, Birve A, Backlund I, Andersson J, et al. No GGGGCC-hexanucleotide repeat expansion in C9ORF72 in parkinsonism patients in Sweden. Amyotroph Lateral Scler Frontotemporal Degener. 2013;14:26–9.
20. Ratti A, Corrado L, Castellotti B, Del Bo R, Fogh I, Cereda C. C9ORF72 repeat expansion in a large Italian ALS cohort: evidence of a founder effect. Neurobiol Aging. 2012;33:2528.e7–14.
21. Kohno S, Takahashi Y, Miyajima H, Serizawa M, Mizoguchi K. A novel mutation (Cys6Gly) in the Cu/Zn superoxide dismutase gene associated with rapidly progressive familial amyotrophic lateral sclerosis. Neurosci Lett. 1999;276:135–7.
22. Morita M, Aoki M, Abe K, Hasegawa T, Sakuma R, Onodera Y, et al. A novel two-base mutation in the Cu/Zn superoxide dismutase gene associated with familial amyotrophic lateral sclerosis in Japan. Neurosci Lett. 1996;205:79–82.
23. Aoki M, Ogasawara M, Matsubara Y, Narisawa K, Nakamura S, et al. Mild ALS in Japan associated with novel SOD mutation. Nat Genet. 1993;5:323–4.
24. Luigetti M, Madia F, Conte A, Marangi G, Zollino M, Del Grande A, et al. SOD1 G93D mutation presenting as paucisymptomatic amyotrophic lateral sclerosis. Amyotroph Lateral Scler. 2009;10(5–6):479–82.
25. Orrell RW, Habgood JJ, Malaspina A, Mitchell J, Greenwood J, Lane RJ, deBelleroche JS. Clinical characteristics of SOD1 gene mutations in UK families with ALS. J Neurol Sci. 1999;169:56–60.
26. Felbecker A, Camu W, Valdmanis PN, Sperfeld AD, Waibel S, Steinbach P, et al. Four familial ALS pedigrees discordant for two SOD1 mutations: are all SOD1 mutations pathogenic? J Neurol Neurosurg Psychiatry. 2010;81:572–7.
27. Andersen PM, Forsgren L, Binzer M, Nilsson P, Ala-Hurula V, Keranen ML. Autosomal recessive adult-onset amyotrophic lateral sclerosis associated with homozygosity for Asp90Ala CuZn-superoxide dismutase mutation. A clinical and genealogical study of 36 patients. Brain. 1996;119(Pt 4):1153–72.
28. Gijselinck I, Van Langenhove T, van der Zee J, Sleegers K, Philtjens S, Kleinberger G, et al. A C9orf72 promoter repeat expansion in a Flanders-Belgian cohort with disorders of the

frontotemporal lobar degeneration-amyotrophic lateral sclerosis spectrum: a gene identification study. Lancet Neurol. 2011;11:54–65.
29. Al-Chalabi A, Andersen PM, Chioza B, Shaw C, Sham PC, Robberecht W, et al. Recessive amyotrophic lateral sclerosis families with the D90A SOD1 mutation share a common founder: evidence for a linked protective factor. Hum Mol Genet. 1998;7:2045–50.
30. Gellera C, Castellotti B, Riggio MC, Silani V, Morandi L, et al. Superoxide dismutase gene mutations in Italian patients with familial and sporadic amyotrophic lateral sclerosis: identification of three novel missense mutations. Neurol Disord. 2001;11:404–10.
31. Buratti E, Baralle FE. Multiple roles of TDP-43 in gene expression, splicing regulation, and human disease. Front Biosci. 2008;13:867–78.
32. Gitcho MA, Baloh RH, Chakraverty S, Mayo K, Norton JB, Levitch D, et al. TDP-43 A315T mutation in familial motor neuron disease. Ann Neurol. 2008;63:535–8.
33. Iida A, Kamei T, Sano M, Oshima S, Tokuda T, Nakamura Y, Ikegawa S. Large-scale screening of TARDBP mutation in amyotrophic lateral sclerosis in Japanese. Neurobiol Aging. 2012;33(4):786–90.
34. Kabashi E, Valdmanis PN, Dion P, Spiegelman D, McConkey BJ, Vande Velde C, et al. TARDBP mutations in individuals with sporadic and familial amyotrophic lateral sclerosis. Nat Genet. 2008;40:572–4.
35. Kamada M, Maruyama H, Tanaka E, Morino H, Wate R, Ito H, et al. Screening for TARDBP mutations in Japanese familial amyotrophic lateral sclerosis. J Neurol Sci. 2009; 284:69–71.
36. Kuhnlein P, Sperfeld AD, Vanmassenhove B, Van Deerlin V, Lee VM, Trojanowski JQ, et al. Two German kindreds with familial amyotrophic lateral sclerosis due to TARDBP mutations. Arch Neurol. 2008;65:1185–9.
37. Lemmens R, Race V, Hersmus N, Matthijs G, Van Den Bosch L, Van Damme P, et al. TDP-43 M311V mutation in familial amyotrophic lateral sclerosis. J Neurol Neurosurg Psychiatry. 2009;80:354–5.
38. Millecamps S, Salachas F, Cazeneuve C, Gordon P, Bricka B, Camuzat A, et al. SOD1, ANG, VAPB, TARDBP, and FUS mutations in familial amyotrophic lateral sclerosis: genotype-phenotype correlations. J Med Genet. 2010;47:554–60.
39. Sreedharan J, Blair IP, Tripathi VB, Hu X, Vance C, Rogelj B, et al. TDP-43 mutations in familial and sporadic amyotrophic lateral sclerosis. Science. 2008;319:1668–72.
40. Tsai CP, Soong BW, Lin KP, Tu PH, Lin JL, Lee YC. FUS, TARDBP, and SOD1 mutations in a Taiwanese cohort with familial ALS. Neurobiol Aging. 2010;32(3):553.e13–21.
41. Van Deerlin VM, Leverenz JB, Bekris LM, Bird TD, Yuan W, Elman LB, et al. TARDBP mutations in amyotrophic lateral sclerosis with TDP-43 neuropathology: a genetic and histopathological analysis. Lancet Neurol. 2008;7:409–16.
42. Corcia P, Valdmanis P, Millecamps S, Lionnet C, Blasco H, Mouzat K, et al. Phenotype and genotype analysis in amyotrophic lateral sclerosis with TARDBP gene mutations. Neurology. 2012;78:1519–26.
43. Borghero G, Floris G, Cannas A, Marrosu MG, Murru MR, Costantino E. A patient carrying a homozygous p.A382T TARDBP missense mutation shows a syndrome including ALS, extrapyramidal symptoms, and FTD. Neurobiol Aging. 2012;32:2327 e1–5.
44. Ticozzi N, Tiloca C, Mencacci NE, Morelli C, Doretti A, Rusconi D, et al. Oligoclonal bands in the cerebrospinal fluid of amyotrophic lateral sclerosis patients with disease-associated mutations. J Neurol. 2012;260:85–92.
45. Orru S, Manolakos E, Orru N, Kokotas H, Mascia V, Carcassi C, Petersen MB. High frequency of the TARDBP p.Ala382Thr mutation in Sardinian patients with amyotrophic lateral sclerosis. Clin Genet. 2011;81:172–8.
46. Renton AE, Majounie E, Waite A, Simon-Sanchez J, Rollinson S, Gibbs JR, et al. A hexanucleotide repeat expansion in C9ORF72 is the cause of chromosome 9p21-linked ALS-FTD. Neuron. 2011;72:257–68.
47. Abalkhail H, Mitchell J, Habgood J, Orrell R, de Belleroche J. A new familial amyotrophic lateral sclerosis locus on chromosome 16q12.1-16q12.2. Am J Hum Genet. 2003;73:383–9.

48. Ruddy DM, Parton MJ, Al-Chalabi A, Lewis CM, Vance C, Smith BN. Two families with familial amyotrophic lateral sclerosis are linked to a novel locus on chromosome 16q. Am J Hum Genet. 2003;73:390–6.
49. Sapp PC, Hosler BA, McKenna-Yasek D, Chin W, Gann A, Genise H, et al. Identification of two novel loci for dominantly inherited familial amyotrophic lateral sclerosis. Am J Hum Genet. 2003;73:397–403.
50. Vance C, Rogelj B, Hortobagyi T, De Vos KJ, Nishimura AL, Sreedharan J, et al. Mutations in FUS, an RNA processing protein, cause familial amyotrophic lateral sclerosis type 6. Science. 2009;323:1208–11.
51. Law WJ, Cann KL, Hicks GG. TLS, EWS and TAF15: a model for transcriptional integration of gene expression. Brief Funct Genomic Proteomic. 2006;5:8–14.
52. Belzil VV, Valdmanis PN, Dion PA, Daoud H, Kabashi E, Noreau A, et al. Mutations in FUS cause FALS and SALS in French and French Canadian populations. Neurology. 2009;73:1176–9.
53. Blair IP, Williams KL, Warraich ST, Durnall JC, Thoeng AD, Manavis J, et al. FUS mutations in amyotrophic lateral sclerosis: clinical, pathological, neurophysiological and genetic analysis. J Neurol Neurosurg Psychiatry. 2009;81:639–45.
54. Chio A, Restagno G, Brunetti M, Ossola I, Calvo A, Mora G, et al. Two Italian kindreds with familial amyotrophic lateral sclerosis due to FUS mutation. Neurobiol Aging. 2009;30:1272–5.
55. Cooper-Knock J, Hewitt C, Highley JR, Brockington A, Milano A, Man S, et al. Clinico-pathological features in amyotrophic lateral sclerosis with expansions in C9ORF72. Brain. 2012;135:751–64.
56. Van Damme P, Goris A, Race V, Hersmus N, Dubois B, Bosch LV, et al. The occurrence of mutations in FUS in a Belgian cohort of patients with familial ALS. Eur J Neurol. 2009;17:754–6.
57. Drepper C, Herrmann T, Wessig C, Beck M, Sendtner M. C-terminal FUS/TLS mutations in familial and sporadic ALS in Germany. Neurobiol Aging. 2011;32(3):548.e1–4.
58. Rademakers R, Stewart H, Dejesus-Hernandez M, Krieger C, Graff-Radford N, Fabros M, et al. Fus gene mutations in familial and sporadic amyotrophic lateral sclerosis. Muscle Nerve. 2010;42:170–6.
59. Robertson J, Bilbao J, Zinman L, Hazrati LN, Tokuhiro S, Sato C. A novel double mutation in FUS gene causing sporadic ALS. Neurobiol Aging. 2011;3:553.e27–30.
60. Waibel S, Neumann M, Rabe M, Meyer T, Ludolph AC. Novel missense and truncating mutations in FUS/TLS in familial ALS. Neurology. 2010;75(9):815–7.
61. Yamamoto-Watanabe Y, Watanabe M, Okamoto K, Fujita Y, Jackson M, Ikeda M, et al. A Japanese ALS6 family with mutation R521C in the FUS/TLS gene: A clinical, pathological and genetic report. J Neurol Sci. 2010;296:59–63.
62. Yan J, Deng HX, Siddique N, Fecto F, Chen W, Yang Y, et al. Frameshift and novel mutations in FUS in familial amyotrophic lateral sclerosis and ALS/dementia. Neurology. 2010;75:807–14.
63. Lattante S, Conte A, Zollino M, Luigetti M, Del Grande A, Marangi G, et al. Contribution of major amyotrophic lateral sclerosis genes to the etiology of sporadic disease. Neurology. 2012;79:66–72.
64. Le Ber I, Camuzat A, Berger E, Hannequin D, Laquerriere A, Golfier V. Chromosome 9p-linked families with frontotemporal dementia associated with motor neuron disease. Neurology. 2009;72:1669–76.
65. Luty AA, Kwok JB, Thompson EM, Blumbergs P, Brooks WS, Loy CT, et al. Pedigree with frontotemporal lobar degeneration–motor neuron disease and Tar DNA binding protein-43 positive neuropathology: genetic linkage to chromosome 9. BMC Neurol. 2008;8:32.
66. Morita M, Al-Chalabi A, Andersen PM, Hosler B, Sapp P, Englund E, et al. A locus on chromosome 9p confers susceptibility to ALS and frontotemporal dementia. Neurology. 2006;66:839–44.
67. Valdmanis PN, Dupre N, Bouchard JP, Camu W, Salachas F, Meininger V, et al. Three families with amyotrophic lateral sclerosis and frontotemporal dementia with evidence of linkage to chromosome 9p. Arch Neurol. 2007;64:240–5.

68. Vance C, Al-Chalabi A, Ruddy D, Smith BN, Hu X, Sreedharan J, et al. Familial amyotrophic lateral sclerosis with frontotemporal dementia is linked to a locus on chromosome 9p13.2-21.3. Brain. 2006;129:868–76.
69. Byrne S, Elamin M, Bede P, Shatunov A, Walsh C, Corr B, et al. Cognitive and clinical characteristics of patients with amyotrophic lateral sclerosis carrying a C9orf72 repeat expansion: a population-based cohort study. Lancet Neurol. 2012;11:232–40.
70. Laaksovirta H, Peuralinna T, Schymick JC, Scholz SW, Lai SL, Myllykangas L, et al. Chromosome 9p21 in amyotrophic lateral sclerosis in Finland: a genome-wide association study. Lancet Neurol. 2010;9:978–85.
71. van Es MA, Veldink JH, Saris CG, Blauw HM, van Vught PW, Birve A, et al. Genome-wide association study identifies 19p13.3 (UNC13A) and 9p21.2 as susceptibility loci for sporadic amyotrophic lateral sclerosis. Nat Genet. 2009;41:1083–7.
72. DeJesus-Hernandez M, Mackenzie IR, Boeve BF, Boxer AL, Baker M, Rutherford NJ, et al. Expanded GGGGCC hexanucleotide repeat in noncoding region of C9ORF72 causes chromosome 9p-linked FTD and ALS. Neuron. 2011;72:245–56.
73. Beck J, Poulter M, Hensman D, Rohrer JD, Mahoney CJ, Adamson G, et al. Large C9orf72 hexanucleotide repeat expansions are seen in multiple neurodegenerative syndromes and are more frequent than expected in the UK population. Am J Hum Genet. 2013;92:345–53.
74. Snowden JS, Rollinson S, Thompson JC, Harris JM, Stopford CL, Richardson AM, et al. Distinct clinical and pathological characteristics of frontotemporal dementia associated with C9ORF72 mutations. Brain. 2012;135(Pt 3):693–708.
75. Stewart H, Rutherford NJ, Briemberg H, Krieger C, Cashman N, Fabros M, et al. Clinical and pathological features of amyotrophic lateral sclerosis caused by mutation in the C9ORF72 gene on chromosome 9p. Acta Neuropathol. 2012;123:409–17.
76. Hosler BA, Siddique T, Sapp PC, Sailor W, Huang MC, Hossain A. Linkage of familial amyotrophic lateral sclerosis with frontotemporal dementia to chromosome 9q21-q22. JAMA. 2000;284:1664–9.
77. Boxer AL, Mackenzie IR, Boeve BF, Baker M, Seeley WW, Crook R, et al. Clinical, neuroimaging and neuropathological features of a new chromosome 9p-linked FTD-ALS family. J Neurol Neurosurg Psychiatry. 2011;82:196–203.
78. Lindquist SG, Duno M, Batbayli M, Puschmann A, Braendgaard H, Mardosiene S, et al. Corticobasal and ataxia syndromes widen the spectrum of C9ORF72 hexanucleotide expansion disease. Clin Genet. 2013;83:279–83.
79. Brettschneider J, Tredici KD, Toledo JB, Robinson JL, Irwin DJ, Grossman M, et al. Stages of pTDP-43 pathology in amyotrophic lateral sclerosis. Ann Neurol. 2013;74(1):20–38.
80. Neumann M, Sampathu DM, Kwong LK, Truax AC, Micsenyi MC, Chou TT, et al. Ubiquitinated TDP-43 in frontotemporal lobar degeneration and amyotrophic lateral sclerosis. Science. 2006;314:130–3.
81. Neumann M, Rademakers R, Roeber S, Baker M, Kretzschmar HA, Mackenzie IR. Frontotemporal lobar degeneration with FUS pathology. Brain. 2009;132(Pt 11):2922–31.
82. Seelaar H, Klijnsma KY, de Koning I, van der Lugt A, Chiu WZ, Azmani A, et al. Frequency of ubiquitin and FUS-positive, TDP-43-negative frontotemporal lobar degeneration. J Neurol. 2010;257(5):747–53.
83. Youngsaye W, Dockendorff C, Vincent B, Hartland CL, Bittker JA, Dandapani S, et al. Overcoming fluconazole resistance in Candida albicans clinical isolates with tetracyclic indoles. Bioorg Med Chem Lett. 2012;22:3362–5.
84. Borroni B, Bonvicini C, Alberici A, Buratti E, Agosti C, Archetti S, et al. Mutation within TARDBP leads to frontotemporal dementia without motor neuron disease. Hum Mutat. 2009;30:E974–83.
85. Deng HX, Chen W, Hong ST, Boycott KM, Gorrie GH, Siddique N, et al. Mutations in UBQLN2 cause dominant X-linked juvenile and adult-onset alS and ALS/dementia. Nature. 2011;477:211–5.
86. Wilhelmsen KC, Lynch T, Pavlou E, Higgins M, Nygaard TG. Localization of disinhibition-dementia-parkinsonism-amyotrophy complex to 17q21-22. Am J Hum Genet. 1994;55:1159–65.

87. Zarranz JJ, Ferrer I, Lezcano E, Forcadas MI, Eizaguirre B, Atares B, et al. A novel mutation (K317M) in the MAPT gene causes FTDP and motor neuron disease. Neurology. 2005;64:1578–85.
88. Parkinson N, Ince PG, Smith MO, Highley R, Skibinski G, Andersen PM, et al. ALS phenotypes with mutations in CHMP2B (charged multivesicular body protein 2B). Neurology. 2006;67:1074–7.
89. Skibinski G, Parkinson NJ, Brown JM, Chakrabarti L, Lloyd SL, Hummerich H, et al. Mutations in the endosomal ESCRTIII-complex subunit CHMP2B in frontotemporal dementia. Nat Genet. 2005;37:806–8.
90. Watts GD, Wymer J, Kovach MJ, Mehta SG, Mumm S, Darvish D, et al. Inclusion body myopathy associated with Paget disease of bone and frontotemporal dementia is caused by mutant valosin-containing protein. Nat Genet. 2004;36:377–81.
91. Johnson JO, Mandrioli J, Benatar M, Abramzon Y, Van Deerlin VM, Trojanowski JQ, et al. Exome sequencing reveals VCP mutations as a cause of familial ALS. Neuron. 2010;68:857–64.
92. Ben Hamida M, Hentati F, Ben Hamida C. Hereditary motor system diseases (chronic juvenile amyotrophic lateral sclerosis). Conditions combining a bilateral pyramidal syndrome with limb and bulbar amyotrophy. Brain. 1990;113(Pt 2):347–63.
93. Devon RS, Helm JR, Rouleau GA, Leitner Y, Lerman-Sagie T, Lev D, Hayden MR. The first nonsense mutation in alsin results in a homogeneous phenotype of infantile-onset ascending spastic paralysis with bulbar involvement in two siblings. Clin Genet. 2003;64:210–5.
94. Eymard-Pierre E, Yamanaka K, Haeussler M, Kress W, Gauthier-Barichard F, Combes P, et al. Novel missense mutation in ALS2 gene results in infantile ascending hereditary spastic paralysis. Ann Neurol. 2006;59:976–80.
95. Gros-Louis F, Meijer IA, Hand CK, Dube MP, MacGregor DL, Seni MH, et al. An ALS2 gene mutation causes hereditary spastic paraplegia in a Pakistani kindred. Ann Neurol. 2003;53:144–5.
96. Kress JA, Kuhnlein P, Winter P, Ludolph AC, Kassubek J, Muller U, Sperfeld AD. Novel mutation in the ALS2 gene in juvenile amyotrophic lateral sclerosis. Ann Neurol. 2005;58:800–3.
97. Panzeri C, De Palma C, Martinuzzi A, Daga A, De Polo G, Bresolin N, et al. The first ALS2 missense mutation associated with JPLS reveals new aspects of alsin biological function. Brain. 2006;129:1710–9.
98. Stevanin G, Santorelli FM, Azzedine H, Coutinho P, Chomilier J, Denora PS, et al. Mutations in SPG11, encoding spatacsin, are a major cause of spastic paraplegia with thin corpus callosum. Nat Genet. 2007;39:366–72.
99. Chance PF, Rabin BA, Ryan SG, Ding Y, Scavina M, Crain B, et al. Linkage of the gene for an autosomal dominant form of juvenile amyotrophic lateral sclerosis to chromosome 9q34. Am J Hum Genet. 1998;62:633–40.
100. Myrianthopoulos NC, Lane MH, Silberberg DH, Vincent BL. Nerve conduction and other studies in families with Charcot-Marie-tooth disease. Brain. 1964;87:589–608.
101. Moreira MC, Klur S, Watanabe M, Nemeth AH, Le Ber I, Moniz JC, et al. Senataxin, the ortholog of a yeast RNA helicase, is mutant in ataxia-ocular apraxia 2. Nat Genet. 2004;36:225–7.
102. Chow CY, Zhang Y, Dowling JJ, Jin N, Adamska M, Shiga K, et al. Mutation of FIG4 causes neurodegeneration in the pale tremor mouse and patients with CMT4J. Nature. 2007;448:68–72.
103. Nishimura AL, Mitne-Neto M, Silva HC, Oliveira JR, Vainzof M, Zatz M. A novel locus for late onset amyotrophic lateral sclerosis/motor neurone disease variant at 20q13. J Med Genet. 2004;41:315–20.
104. Landers JE, Leclerc AL, Shi L, Virkud A, Cho T, Maxwell MM, et al. New VAPB deletion variant and exclusion of VAPB mutations in familial ALS. Neurology. 2008;70:1179–85.
105. Wu CH, Fallini C, Ticozzi N, Keagle PJ, Sapp PC, Piotrowska K, et al. Mutations in the profilin 1 gene cause familial amyotrophic lateral sclerosis. Nature. 2012;488:499–503.

Chapter 12
Uncommon Dementias

Camilla Ferrari, Benedetta Nacmias, and Sandro Sorbi

Abstract Dementia is becoming a worldwide phenomenon. Alzheimer's disease represents the first cause of cognitive impairment, followed by vascular and frontotemporal dementia. However, in addition to these well-studied causes, there is a large number of conditions that can cause dementia, such as infections, toxic–metabolic conditions, inflammatory–autoimmune disorders, or metabolic inborn errors. These uncommon causes of dementia, due to their heterogeneous clinical presentation, lack diagnostic criteria and occur rarely, are often misdiagnosed. Prevalence has been only partially estimated in young patients (age at onset <65 years), and management is based only on a few expert suggestions. However, correct diagnosis is of great importance, since some of these dementias are treatable and reversible.

This chapter presents a comprehensive summary of etiologies, clinical presentation, typical features, diagnostic strategies, and treatments of known, uncommon dementias.

Keywords Uncommon dementia • Young-onset dementia • Neurodegenerative disease • Reversible dementia

Introduction

Uncommon dementias indicate a wide heterogeneous group of rare disorders causing cognitive impairment and are generally characterized by an early age at onset. Thus, uncommon dementias greatly overlap the concept of young-onset dementia. Conventionally, young-onset dementia includes conditions that afflict

C. Ferrari, MD (✉) • B. Nacmias, PhD • S. Sorbi, MD
Department of Neuroscience, Psychology, Drug Research and Child Health (NEUROFARBA), University of Florence, Viale Pieraccini 6, 50139 Florence, Italy
e-mail: camilla.ferrari@unifi.it

D. Galimberti, E. Scarpini (eds.), *Neurodegenerative Diseases*,
DOI 10.1007/978-1-4471-6380-0_12, © Springer-Verlag London 2014

patients younger than 65 years of age [1], i.e., early-onset forms of common neurodegenerative dementia, such as familial Alzheimer's disease, dementia associated with other neurological disorders (Huntington's disease, myotonic dystrophies, autosomal dominant cerebellar ataxia, or hereditary spastic paraparesis), or late-onset forms of childhood conditions, such as mitochondrial disorders, lysosomal storage disorders, and leukodystrophies. Potentially reversible etiologies, including inflammatory disorders and infectious or toxic–metabolic abnormalities, can also play a part in the causes of rare dementia. It should be noted that information on the frequency of uncommon dementias among the elderly is not available, while the little epidemiological data available on young-onset dementia comes from restricted geographical settings. Harvey and colleagues estimated, on a population-based study of two London boroughs, a dementia prevalence of 54 per 100,000 people aged 30–65 years and 98 per 100,000 people aged 45–65 years [2].

Alzheimer's disease was the most common single diagnosis (34 %), and the prevalence of metabolic, infective, and inflammatory/autoimmune diseases was generally estimated, cumulatively, to be 19 % [2]. Among others studies [3–5], only one [6], conducted on a population ranging in age between 17 and 45 years old, specifically evaluated the prevalence of all uncommon causes of early-onset dementia, producing the following results: neurodegenerative etiology, 31.1 %; autoimmune and inflammatory, 21.3 %; metabolic, 10.6 %; others, 7.7 %; vascular, 6 %; and infective, 4.7 % [6]. The age-stratified analysis showed a decreasing frequency of a metabolic etiology with aging and the opposite with a neurodegenerative etiology.

The diagnosis of uncommon dementias is challenging, due to the large number of pathologies with heterogeneous clinical presentation and lack of diagnostic criteria. Suggestions for diagnostic procedure, management, and treatment are generally based on small, uncontrolled studies and on expert opinion. To guide clinicians toward differential diagnosis and to avoid misdiagnosis, uncommon dementias have been differently categorized following clinical, pathological, or etiological criteria [1, 6–8]. Rossor and colleagues proposed a flow chart for diagnostic procedure [1]. This chapter presents a comprehensive list of uncommon dementias, grouped in diagnostic categories (Table 12.1), and suggests a two-step diagnostic procedure for the differential diagnosis of rare dementias. A systematic and updated summary of clinical, neuroimaging, and therapeutic aspects of each disease is also included.

Diagnostic Procedure

Uncommon causes of dementia should be suspected in the presence of at least one of:

- Young-onset dementia
- Predominance of psychiatric symptoms
- Association with other neurological signs
- Systemic involvement
- Subacute onset and rapid progression
- Positive family history for dementia or other neurological disturbances

Table 12.1 Diagnostic categories

1. *Neurodegenerative dementias*[a]	5. *Vascular*	9. *Infective dementia*
1.1. Familial Alzheimer's disease	5.1. CADASIL	9.1. HIV-related dementia
1.2. Frontotemporal dementia	5.2. Cerebral amyloid angiopathy	9.2. Neurosyphilis
1.3. Lewy body dementia		9.3. Whipple's disease
1.4. Corticobasal degeneration		9.4. Lyme disease
1.5. Progressive supranuclear palsy		9.5. Tuberculosis meningitis
2. *Dementia-plus*	6. *Mithocondrial*	10. *Inflammatory–auto-immune disorders*
2.1. Huntington's disease	6.1. MELAS	10.1. Limbic Encephalitis
2.2. Myotonic dystrophies	6.2. MERRF	10.2. Hashimoto encephalopathy
2.3. Autosomal dominant cerebellar ataxia	6.3. Kearns–Sayre syndrome	10.3. Neurosarcoidosis
2.4. Hereditary spastic paraparesis		10.4. NeuroLES
2.5. Fragile X-associated tremor/ataxia syndrome		10.5. Bechet
3. *Leukodystrophies (adult onset)*	7. *Basal ganglia pathologies*	11. *Toxic–metabolic*
3.1. Adrenoleukodystrophy	7.1. Neuroacanthocytoses (chorea-acanthocytosis, McLeod syndrome, pantothenate kinase-associated neurodegeneration)	11.1. Alcohol-related dementia
3.2. Krabbe disease	7.2. Fahr disease	11.2. B12 deficiency
3.3. Metachromatic leukodystrophy	7.3. Neuroferritinopathy	11.3. Heavy metal poisoning
3.4. Cerebrotendinous xanthomatosis	7.4. Wilson disease	
3.5. Pelizaeus–Merzbacher disease	*Others*	
3.6. Alexander disease	Nasu–Hakola disease (polycystic lipomembranous osteodysplasia with sclerosing leukoencephalopathy)	
3.7. Adult polyglucosan body disease		
3.8. Vanishing white matter disease		
4. *Lysosomal storage disorders*	8. *Prion disease*	
4.1. Fabry disease	8.1. Creutzfeldt–Jakob disease (CJD) sporadic and variant	
4.2. Gaucher's disease	8.2. Hereditary prion disorders:	
4.3. Niemann–Pick disease, type C	Familial CJD	
4.4. Kuf's disease (neuronal ceroid lipofuscinosis)	Gerstmann–Sträussler–Scheinker	
4.5. Tay–Sachs disease	Familial fatal insomnia	
4.bis Other metabolic inborn errors		
Lesch–Nyhan disease		

CADASIL cerebral autosomal dominant arteriopathy subcortical infarcts and leukoencephalopathy, *MELAS* mitochondrial encephalopathy, lactic acidosis, and stroke, *MERFF* myoclonic epilepsy and ragged red fibers

[a]Discussed elsewhere in this book

All patients should undergo a complete clinical assessment with neurological, neuropsychological, and general examination, basic blood tests, and neuroimaging, preferably magnetic resonance imaging (MRI) [1, 8]. These evaluations represent the first step of the diagnostic procedure. A second diagnostic step, different for each diagnostic category, includes more complex blood or urinary examination, cerebrospinal fluid (CSF) analysis, electroencephalography (EEG), electromyography (EMG), fluorodeoxyglucose positron emission tomography (PET-FDG), tissue biopsy, and genetic testing.

First Diagnostic Step

Clinical history includes family history; specific dementia risk factors, such as alcohol or heavy metal exposure; and description of symptoms using the temporal profile at onset, progression, and degree of impairment. The objective of the neurological examination is to define the pattern of cognitive and behavioral deficits and to investigate the presence of specific neurological signs (pyramidal, extrapyramidal, cerebellar, peripheral). General examination is also important in case of systemic illness and can reveal stigmata of some disorders, such as Achilles tendon xanthomata in cerebrotendinous xanthomatosis. A basic blood test may be useful in detecting toxic–metabolic encephalopathy, infective dementia, or autoimmune illnesses [1, 7, 8].

Once the type of cognitive impairment and its association with specific neurological signs is defined, an MRI is of a great utility. The pattern of cortical atrophy may distinguish between different neurodegenerative dementias, while the presence of leukoencephalopathy suggests leukodystrophies, some lysosomal storage disorders, vascular diseases, or mitochondrial disorders. MRI can be diagnostic with the detection of basal ganglia calcification.

Second Diagnostic Step

After the first orientative step, a variety of investigations are suggested for the different categories: more complex blood and urinary examination are necessary to distinguish between different forms of leukodystrophies, lysosomal storage disorders, or other inborn metabolism errors, such as the dosage of very long-chain fatty acids (VLCFA) or testing the activity of a specific enzyme.

An EEG is mandatory in the presence of subacute onset, as in case of prion disease, or limbic encephalitis or in the case of a disease associated with epileptic disturbances [8, 9]. CSF may facilitate the differential diagnosis of neurodegenerative dementias and is recommended in the diagnosis of infective and transmittable dementia [9]. PET-FDG can be used for the differential diagnosis of neurodegenerative dementias. EMG should be performed in patients with dementia-plus syndrome, lysosomal storage disorders, or toxic–metabolic alterations. Tissue biopsy can be used to confirm diagnosis in the case of mitochondrial disease, metabolic inborn errors, or leukodystrophies [1, 9]. Most of the uncommon dementias are inheritable disorders, and genetic testing represents, in many cases, the gold standard to confirm diagnosis. Genetic investigation is also important to predict susceptibility in family members and is fundamental for some disorders in which presymptomatic treatment can avoid or delay disease onset (i.e., Wilson disease, adrenoleukodystrophy).

Neurodegenerative Dementias

Although rare, these dementias represent a group of very well-studied disorders with specific diagnostic criteria. Dementia is the first and predominant symptom, sometimes associated with extrapyramidal or other neurological signs. Their management is extensively deliberated in the European Guidelines [9].

Dementia Plus

Dementia is always associated with other neurological signs and often occurs later in the disease course. Generally, MRI presents cortical atrophy. All of these diseases are inherited, and genetic testing is necessary to confirm diagnosis.

Huntington's Disease

Autosomal dominant disease caused by the expansion of CAG trinucleotide repeat sequence on chromosome 4, a gene encoding for "*huntingtin*," a protein of unknown function.

Epidemiology: Prevalence in the European population is 0.5–8 in 1,000,000.

Clinical features: Age at onset is 30–50 years of age with behavioral symptoms and rarely occurs before 20 years. The disease is characterized by chorea, psychiatric symptoms, and cognitive decline. Cognitive deficits are mostly in executive function and judgment capacity. Language and semantic memory are generally spared. Disease duration is about 15–20 years.

Neuroimaging: MRI shows atrophy of the caudate and putamen and frontal lobes.

Diagnosis: Genetic testing, CAG expansion (normal, <27 repetitions; incomplete penetrance, 35–39; pathological, >39)

Therapy: Typical neuroleptics, tetrabenazine for motor and psychiatric symptoms.

A Cochrane Library review found no data on the treatment of cognitive impairment [9, 10]. Intrastriatal transplantation of striatal neuroblasts taken from human fetus is currently being explored as potential treatment. Data from initial pilot clinical studies seems to show a delay in disease progression and transient clinical improvement [11].

Myotonic Dystrophies

Myotonic Dystrophy is a CTG trinucleotide repeat expansion disorder transmitted with autosomal dominant inheritance.

Clinical features: Symptoms are progressive myopathy, myotonia, cataracts, dementia, and multiorgan involvement. It presents a wide range of age at onset, from congenital form to adulthood. Dementia is generally observed later in the disease course of the adult-onset form, 32–67 years of age.

Neuroimaging: MRI shows cortical atrophy, especially in the frontal, temporal, and parietal lobes.

Diagnosis: EMG indicates dystrophic and myotonic phenomena. Genetic testing [12].

Autosomal Dominant Cerebellar Ataxia

It encompasses a group of neurodegenerative disorders characterized by ataxia and different combination of pyramidal, extrapyramidal signs, and peripheral neuropathy. Dementia occurs only in some forms of spinocerebellar ataxia (SCA), developing in the latest stage of the disease: SCA1, SCA2, SCA3, and SCA12. Regardless, dementia is a constant feature of dentatorubral-pallidoluysian atrophy (DRPLA) and SCA17, with behavioral disorders and frontal-type dementia preceding ataxia [13].

Hereditary Spastic Paraparesis (SPG)

Hereditary Spastic Paraparesis (SPG) are a group of heterogeneous neurodegenerative inherited disorders. The main clinical features are slowly progressive spasticity and leg weakness. Dementia has been reported only in some families with the SPG4 form, and it is specifically associated with the deletion of exon 17 of the spastin gene. The degree of cognitive impairment is not correlated with the severity of spastic symptoms but seems to be related to aging [14].

Fragile X-Associated Tremor/Ataxia Syndrome (FXTAS)

Fragile X-Associated Tremor/Ataxia Syndrome (FXTAS) is a trinucleotide repeat disorder caused by the expansion of CGG on the fragile X mental retardation 1 gene (FMR1) on the X chromosome. Normal alleles present 5–44 repeats, while more than 200 CGG repeats determine the most common inherited form of intellectual disability and autism.

Premutation alleles 55–200 can be associated with FXTAS. Premutation expansion prevalence is 1 in 113–259 females and 1 in 260–800 males [15]. FXTAS affects nearly 40 % of premutation males and 8 % of premutation females.

Clinical features: Age at onset is over 50 years old. Clinical features are kinetic, intentional or postural tremor, cerebellar gait and limb ataxia, parkinsonism, and dementia. Patients may have autonomic dysfunction and peripheral neuropathy. Dementia in FXTAS presents memory loss associated with both frontal lobe features (disinhibition, poor executive functioning, perseveration, mood disturbance) and subcortical features (psychomotor slowing, bradyphrenia, attention and concentration difficulties). The onset of cognitive symptoms often follows the onset of movement disorders.

Neuroimaging: MRI shows diffuse cerebellar and cerebral atrophy and hyperintensity of middle cerebellar peduncles and subcortical regions on T2 sequences.

Diagnosis: Genetic testing. It is important to recognize the disorder among family members in order to identify premutant alleles [15].

Therapy: Treatment of cognitive impairment is based on off-label application of dementia therapies. In early phases of memory impairment, the use of cholinesterase inhibitors can results in short-term improvement. Expert opinion reported some benefits from memantine use [16].

Leukodystrophies (Adult Onset)

These are childhood neurodegenerative disorders, involving myelin development, that can be divided into (1) dysmyelinating (abnormally formed myelin), (2) hypomyelinating (decreased myelin production), and (3) spongiform (cystic degeneration of myelin) [17].

Clinical onset is normally in infancy; however, an adult-onset form has been described. Information on epidemiology and clinical data are based on single-case reports or specific clinical settings. The hallmark of this group of disorders is MRI leukoencephalopathy. Generally, age at onset is earlier than 45 years old and is predominantly characterized by psychiatric symptoms.

Some disorders present typical physical features in the general examination that can help in the diagnosis [18].

Adrenoleukodystrophy

Adrenoleukodystrophy is a hereditary, X-linked disease caused by mutations of the gene encoding a protein necessary for metabolization of the very long-chain fatty acids (VLCFA).

Epidemiology: Adrenoleukodystrophy occurs usually in childhood, only 1–3 % of cases present adult onset.

Clinical features: Age at onset of the adult form is around 20–30 years old. Cognitive disorders are characterized by psychotic symptoms, character changes, hyperorality, tendency to wander, and stereotypical vocal expression and by subcortical signs, such as bradyphrenia and concentration deficits. Impaired vision and hearing are characteristic. Paraparesis, sphincteric disturbance, and sexual dysfunction can be present [17–19].

Neuroimaging: MRI T2 sequences show white matter hyperintensity especially in parieto-occipital regions.

Diagnosis: High plasma level of VLCFA (C24–C30 chain length). Genetic testing.

Therapy: Dietary treatment with the use of glycerol trioleate and glycerol trierucate that lower VCLFA in culture. Unfortunately, the use of these oils does not improve clinical course in symptomatic patients; however, genetically detected asymptomatic cases have a less-severe illness course [17].

Krabbe Disease (Globoid Cell Leukodystrophy)

Krabbe Disease is an autosomal recessive lysosomal storage disease caused by a deficiency of the lysosomal enzyme galactocerebrosidase (GALC). The GALC deficiency impairs the degradation of galactocerebroside, a major myelin lipid, the excess of which elicits formation of multinucleated macrophages, the globoid cells.

Epidemiology: Adult onset is rare. A recent review of published cases reported 28 adult-onset cases [20].

Clinical features: Age at onset in the adult form is between 25 and 72 years. Symptom progression is slow and disease duration can be more than 10 years. Patients present pyramidal signs (96 % of cases), dysarthria (31 %), cerebellar ataxia (27 %), deep sensory signs (23 %), tongue atrophy (15 %), and optic neuropathy (12 %). Cognitive decline is described in 12 % of cases [20].

Neuroimaging: MRI shows T2 hyperintensity of optic radiations, the posterior part of the corpus callosum, and the corticospinal tracts.

Diagnosis: Deficient GALC activity in leukocytes or fibroblasts.

Therapy: Hematopoietic stem cell transplantation.

Metachromatic Leukodystrophy (MLD)

Autosomal recessive lysosomal sphingolipid storage disorder [17, 21] is caused by a deficiency in the enzyme arylsulfatase A resulting in the accumulation of no degradated sulfatide in oligodendrocytes, Schwann cells, and some neurons. Sulfatide accumulation is the trigger to demyelination.

Epidemiology: Incidence in Europe is 1 per 100,000 live births. Even if more than 100 mutations have been described, among Caucasians only three are frequent (splice donor site mutation of exon 2/intron 2; missense mutations causing Pro-426Leu substitution, missense mutation causing Ile-179Ser substitution).

Clinical features: There is genotype–phenotype correlation with disease severity based on the amount of residual enzyme activity. The adult form, with onset beyond the age of 16 years, corresponds to 18–20 % of MLD cases. The adult form shows two possible distinct phenotypes, one with a predominant cerebello-pyramidal presentation and the other with predominantly psychiatric features. The psychiatric presentation is often associated with a specific mutation in Caucasians (Ile170S). Neurological signs appear later with seizures, chorea, or dystonia. Cognitive impairment is characterized by attentional disturbances, reduced speed of processing, and executive functions impairment. In some cases, patients present frontotemporal-like dementia symptoms.

Neuroimaging: Symmetric confluent T2 MRI hyperintensity in the periventricular regions and corpus callosum. Within the abnormal white matter, low-density tigroid stripes are present. The tigroid stripes are typical of MLD, but not specific.

Diagnosis: There is a reduced arylsulfatase A enzyme activity in blood leukocytes and an increased sulfatide excretion in 24-h urine sample.

Therapy: No available treatment. Hematopoietic stem cell transplantation is beneficial only in late juvenile or adult patients in the early stages of the disease.

Cerebrotendinous Xanthomatosis

Cerebrotendinous Xanthomatosis is an autosomal recessive disorder due to mutations on the gene for the mitochondrial enzyme sterol 27-hydroxylase that is responsible for the production of bile acids [17, 22]. The deficiency in this mitochondrial enzyme determines increased plasma levels of cholestanol and deposition in different body tissues, such as the Achilles tendon, nervous system, and lungs.

Epidemiology: Few cases are described in the literature [22].

Clinical features: The characteristic triad of the disorder is tendon xanthomata (especially of the Achilles tendons), juvenile ocular cataracts, and nervous system dysfunction. Nervous system symptoms consist of behavioral problems, dementia, psychiatric disorders, pyramidal weakness, cerebellar ataxia, and seizures.

Neuroimaging: MRI shows white matter hyperintensity above and especially below the tentorium and in some cases focal lesions and diffuse brain and spine atrophy.

Diagnosis: Increased plasma level of cholestanol associated with low or normal levels of cholesterol.

Therapy: Since the disease results from a defect in bile acid synthesis, treatment consists of the assumption of chenodeoxycholic acid. Chenodeoxycholic acid can reverse encephalopathy during the early stages. Long-term treatment with chenodeoxycholic acid (750 mg/day) suppresses abnormal bile acid synthesis.

Pelizaeus–Merzbacher Disease

Pelizaeus-Merzbacher Disease is an X-linked leukoencephalopathy due to mutations in the proteolipid (PLP) gene. The PLP gene encodes for two PLP proteins, PLP, the prominent protein in CNS myelin, and DM 20, involved in oligodendrocyte differentiation.

Epidemiology: Only rare, single cases are reported.

Clinical features: Classic presentation occurs before 5 years of age and consists in nystagmus, stridor, hypotonia, spasticity, ataxia, and choreoathetosis. Rare adult-onset cases present spastic paraparesis and sometimes tremor ataxia and dementia. Interestingly, PLP gene mutations are also associated with type 2 spastic paraplegia (SPG2) [17].

Neuroimaging: Central white matter is reduced in volume and presents diffuse hyperintensity (cerebral hemispheres, cerebellum, and brainstem) with a thin corpus callosum. Some preserved myelin islands are present and have a "tigroid" appearance.

Diagnosis: Genetic testing.

Therapy: No treatment.

Alexander Disease

Alexander Disease is an autosomal dominant disorder caused by mutations on the glial fibrillary acidic protein gene (GFAP). GFAP gene mutations cause an overexpression of abnormal protein.

Epidemiology: Few cases are reported in the literature [23].

Clinical features: Age at onset is between 13 and 62 years. Disease duration can be a few years or decades. Symptoms are dysarthria, dysphonia, dysphagia (bulbar and pseudobulbar signs), pyramidal signs, ataxia, and palatal myoclonus. Cognitive decline occurs late in the disease course.

Neuroimaging: Paucity of myelin especially in the frontal lobe, and cystic degeneration and cavitation of white matter are frequently present. Basal ganglia and thalami are also affected as well as the medulla oblongata and the cervical spinal cord.

Diagnosis: Genetic testing. Characteristic histological finding: Rosenthal fibers, which are eosinophilic inclusion localized in astrocyte cytoplasm.

Therapy: No treatment.

Adult Polyglucosan Body Disease

Adult Polyglucosan Body Disease is an autosomal recessive polyglucosan storage disorder caused by mutations of the glycogen branching enzyme (GBE1) gene. This disorder is often observed among Ashkenazi Jewish families.

Clinical features: Onset is in the fifth or sixth decades of life with myelopathy signs, peripheral axonal sensorimotor neuropathy, and neurogenic bladder. Weakness and sensory loss typically starts in the lower limbs. Around 2/3 of patients have cognitive impairment at onset, with cortical and subcortical deficits.

Neuroimaging: MRI shows periventricular, subcortical, and deep white matter changes that extend to the cervicomedullary junction. Brain and spinal cord atrophy.

Diagnosis: Decreased GBE1 activity on skin fibroblasts or muscle. Intra-axonal polyglucosan bodies on sural nerve biopsy. Genetic testing [8, 24].

Therapy: No treatment.

Vanishing White Matter Disease

Vanishing White Matter Disease is an autosomal recessive disorder caused by mutations on one of the five genes encoding subunits of eukaryotic translation initiation factor 2B (EIF2B). Mutations in the EIF2B gene disrupt the normal stress-elicited compensatory mechanisms (synthesis of new protein and signals promoting both cellular survival and apoptosis).

Epidemiology: Case reports [25].

Clinical features: Based on a recent review (16 cases), the mean age at onset is 31 years (range, 16–62). Characteristics of the disease are episodes of acute deterioration with hypotonia, irritability, vomiting, seizures, unconsciousness after minor head trauma, febrile infections, sun exposure, or fear. Extracranial involvement often includes ovarian dysgenesis. Cognitive decline is described in 62 % of cases.

Neuroimaging: MRI shows diffuse abnormalities in white matter that can have signal intensity near the signal intensity of CSF, diffuse disappearance of cerebral white matter. Relative sparing of temporal lobes.

Diagnosis: Genetic testing.

Therapy: No treatment [17, 25].

Lysosomal Storage Disorders

This is a group of metabolic inborn errors with clinical onset usually during infancy and childhood. Psychiatric disorders are predominant, as well as clinical signs of diffuse nervous system involvement (pyramidal, extrapyramidal, cerebellar). Rare, late-onset cases can present cognitive decline and less devastating neurological deficits. Some disorders are characteristically associated with leukoencephalopathy, while others with gray matter alterations. Diagnosis is based on the demonstration of decreased activity of specific metabolic enzymes. Some of these disorders can be treated with enzyme replacement therapy [8].

Fabry Disease

Fabry Disease is an X-linked lysosomal storage disorder which results from the deficiency of the enzyme alpha-galactosidase A that determines accumulation of glycosphingolipids in tissues (renal, cardiac, ocular, skin, nervous system).

Epidemiology: Estimated incidence is between 1 in 40,000 and 1 in 117,000 live male births. Heterozygous females can present some symptoms.

Clinical features: Typically, subjects are affected by small fiber neuropathy characterized by acute pain, including acroparesthesia and "Fabry crises," which are episodes of severe pain in the extremities in response to physical exercise, fever, or temperature changes. Fabry disease represents one of the most frequent causes of stroke in young subjects. Depression is a common symptom with prevalence ranging between 15 and 62 % [26]. Cognitive impairment can be also present and is characterized by deficits in executive functioning, information processing speed and attention, with relative spared memory and naming. Systemic involvement includes renal and cardiac disorders, angiokeratoma, hypohidrosis, and corneal and lenticular opacities.

Neuroimaging: It shows periventricular and deep white matter hyperintensity and lacunar strokes.

Diagnosis: Decreased alpha-galactosidase A enzyme activity in blood leukocytes.

Therapy: Enzyme replacement therapy [27].

Gaucher's Disease (GD)

Gaucher's disease is an autosomal recessive disease caused by mutations on the gene encoding the lysosomal enzyme glucocerebrosidase (GBA) [28, 29]. A deficiency in GBA activity determines the accumulation of glucosylceramide in several organs (liver, spleen, kidneys, lungs, bone marrow, nervous system).

Clinical features: GD type 1 presents hepatosplenomegaly, anemia, or thrombocytopenia without neurological signs. Type 2 is characterized by neurological signs, predominantly bulbar signs, and juvenile presentation. GD type 3 has an adult onset (17–55 years of age) and includes different neurological symptoms: akinetic-rigid syndrome that responds poorly to dopatherapy, supranuclear palsy, seizures, or cerebellar ataxia. Cognitive and psychotic disturbances are frequent features. Interestingly, it was recently reported that 5–10 % of Parkinson disease patients are heterozygous for GBA mutations [30].

Neuroimaging: Atrophy of basal ganglia.

Diagnosis: Decreased glucocerebrosidase enzyme activity in blood leukocytes. Genetic testing.

Therapy: Enzyme replacement therapy (however, it does not cross the blood–brain barrier; no efficacy on neurological signs).

Niemann–Pick Disease, Type C

Niemann-Pick Disease is an autosomal recessive disease due to mutations on the NPC1 and NPC2 genes that are involved in intracellular transport of endocytosed cholesterol. These mutations determine sequestration of unesterified cholesterol in lysosomes.

Epidemiology: Reports from European countries describe an incidence of 1 in 120,000 live births.

Clinical features: Neurological disturbances consist mainly of cerebellar ataxia, dysarthria, dysphagia, supranuclear gaze palsy, and progressive dementia. Cataplexy, seizures, and dystonia are other common features. Cognitive impairment has been described as the initial manifestation in 25–40 % of adult-onset cases, 20–30 years of age. Often psychiatric symptoms, such as psychosis, are present alone, and patients are treated with neuroleptics for many years before gait impairment and cognitive decline appear. Splenomegaly is present in more than half of adult patients [31].

Neuroimaging: It shows severe atrophy of white matter tracts, huge neuronal loss in the corpus callosum, and atrophy of the cerebellum (loss of Purkinje neurons), striatum, thalamus, and hippocampus.

Diagnosis: Accumulation of unesterified cholesterol in perinuclear vesicles (lysosomes) of skin fibroblasts. Genetic testing.

Therapy: Miglustat, an inhibitor of glucosylceramide synthase [31].

Kuf's Disease (Neuronal Ceroid Lipofuscinosis)

Kuf's Disease is characterized by accumulation of lipofuscin-like material in lysosomes in neuronal and extraneuronal tissue. Adult onset is rare. Recently, mutations on the ceroid lipofuscinosis neuronal 6 (CLN6) gene were identified as associated with recessive Kuf's disease, presenting with progressive myoclonus epilepsy (type A). The causes of the type B form, characterized by dementia and motor symptoms, are still unknown [32].

Clinical features: Adult age at onset cases have been described between 13 and 41 years. Predominant neurological features are seizure, dementia, ataxia, and extrapyramidal signs. In older patients, clinical presentation may mimic frontotemporal dementia [33]. Neuropsychological deficits affect psychomotor speed, new learning, executive function, and attention.

Neuroimaging: It shows cerebral (hippocampus) and cerebellar atrophy, callosal thinning, and altered signal in basal ganglia.

Diagnosis: Specific findings on peripheral lymphocytes or skin biopsy (fingerprint and curvilinear profiles, granular osmiophilic deposits, and rectilinear complex).

Therapy: Symptomatic; carbamazepine, phenytoin, and lamotrigine may increase seizure activity and myoclonus and result in clinical deterioration.

Tay–Sachs Disease

Autosomal recessive gangliosidosis GM2 caused by a deficiency in the beta (β)-hexosaminidase A (HEXA) enzyme resulting in accumulation of complex glycosphingolipids in the nervous system and other tissue. About 130 mutations have been reported in the HEXA gene.

Epidemiology: It was common in Ashkenazi Jews until carrier screening was introduced in the 1970s, and disease incidence was reduced by 90 %.

Clinical features: Tay–Sachs disease in early infancy presents hypotonia, early vegetative state, and death. Late-onset forms of Tay–Sachs are infrequent. Some cases have been reported with an age at onset ranging between 14 and 47 years [34, 35], presenting spinocerebellar ataxia, psychosis, or progressive muscular atrophy. Cognitive impairment is common in this group [35].

Neuroimaging: It shows cerebellar atrophy.

Diagnosis: Decreased or absent hexosaminidase A enzyme activity on serum or leukocytes. Genetic testing.

Therapy: No treatment.

Other Metabolic Inborn Errors

Lesch–Nyhan Disease

Lesch-Nyhan Disease is an X-linked metabolic defect causing overproduction of uric acid. It is caused by mutations on the hypoxanthine-guanine phosphoribosyltransferase (HPRT) enzyme gene; more than 400 mutations have been described.

Clinical features: Clinical phenotype is heterogeneous and can vary from isolated gout to severe motor handicap and mental retardation with recurrent self-injurious behavior.

Diagnosis: Decreased or absent HPRT enzyme activity measured on erythrocytes or fibroblasts.

Therapy: Allopurinol for gout; no effect on neurobehavioral problems [36].

Vascular Diseases

Vascular dementia is a heterogeneous category of disorders causing dementia with a subcortical deficit cognitive profile and typical neuroimaging findings. CADASIL and genetically determined cerebral angiopathy are causes of young-onset vascular dementia. For review: [37–40].

Mitochondrial Diseases

It is a group of progressive neurodegenerative disorders associated with polygenetic, maternally inherited, mitochondrial DNA mutations. Prevalence studies report that mitochondrial disease affects 9 in 100,000 adults aged less than 65 years. The clinical presentation of mitochondrial disease is varied and can occur in almost any stage of life. Dementia is generally of a subcortical type, and neuroimaging shows characteristic involvement of white matter, especially in MELAS [41]. For a comprehensive review: [42].

Basal Ganglia Pathologies

Neuroacanthocytoses (NA): [43]

It is a group of disorders characterized by peripheral blood acanthocytes and central nervous system and neuromuscular symptoms. All the disorders present pathology of the basal ganglia, which accounts for the neurological features: chorea or dystonia, dysexecutive syndrome, obsessive–compulsive disorder, depression, and schizophrenia-like psychosis.

NA can be divided into four groups: (1) basal ganglia degeneration, choreiform movement disorders, and acanthocytosis; (2) neurodegeneration with only occasional acanthocytosis; (3) acanthocytosis with paroxysmal exertion-induced dyskinesia; and (4) low lipoprotein blood level, ataxia, and no movement disorder. Groups 1 and 2 present neuropsychiatric symptoms.

Chorea-Acanthocytosis (Group 1)

Chorea-Acanthocytosis is an autosomal recessive disease associated with mutations on the vacuolar protein sorting 13 homolog A (VPS13A) gene which encodes for the membrane protein chorein. Loss of chorein particularly affects basal ganglia, especially the caudate nucleus.

Clinical features: Neurological disturbance onset is commonly in the third and fifth decades with limb and orobuccal chorea. Psychiatric symptoms may precede movement disturbances by up to a decade. Cognitive impairment is also present as a dysexecutive syndrome (impaired judgment, disinhibition).

Neuroimaging: MRI shows dramatic caudate atrophy and increased T2 signals in basal ganglia.

Diagnosis: Peripheral blood acanthocytes. Absent or reduced chorein on red blood cells. Genetic testing.

McLeod Syndrome (Group 1)

McLeod Syndrome is an X-linked multisystemic disease due to mutations of the XK gene.

Clinical features: Illness onset is usually between 25 and 60 years of age. More than 80 % of patients have neuropsychiatric disturbance with schizophrenia-like psychosis or obsessive–compulsive disorders. Executive impairment is common and can be present even in unaffected female carriers. One patient has been described with frontotemporal-like syndrome onset [44].

Neuroimaging: MRI can show striatal atrophy.

Diagnosis: Peripheral blood acanthocytosis and elevated creatine kinase. Genetic testing.

Pantothenate Kinase-Associated Neurodegeneration (PKAN) or Hallervorden–Spatz Syndrome (Group 2)

Pantothenate Kinase-Associated Neurodegeneration is an autosomal recessive disease characterized by iron accumulation in the basal ganglia due to mutations in the pantothenate kinase 2 (PANK2) gene, necessary for coenzyme A synthesis.

Clinical features: Atypical late-onset form occurs in the second or third decade of life and has a disease duration of up to 15 years. Symptoms include hyperkinetic movements, dystonia, rigidity, and dysarthria. Psychiatric symptoms are present in more than half of patients. Cognitive decline is universal in PKAN patients and may precede motor signs. The pattern of cognitive decline implicates deficits in executive function, attention, and sometimes in memory functions.

Neuroimaging: The hallmark is the "eye of the tiger" sign, bilateral areas of hyperintensity (iron deposition) within a region of hypointensity in the globus pallidus on T2 MRI.

Diagnosis: Genetic testing [45].

Fahr Disease

Fahr Disease is an autosomal dominant disease characterized by abnormal calcium deposition in the basal ganglia, thalamus, dentate nucleus, cerebral cortex, cerebellum, or hippocampus. It is associated with mutations on SLC20A2 gene (sodium-dependent phosphate transporter).

Epidemiology: Prevalence <1/1,000,000.

Clinical data: Onset ranges between 20 and 50 years of age with extrapyramidal symptoms. Additionally, cerebellar dysfunction, dementia, and neuropsychiatric disturbances may be present.

Neuroimaging: It shows bilateral calcification of the basal ganglia.

Diagnosis: Diagnostic criteria according to Manyam (2005) [46]. Secondary causes of altered calcium metabolism should be investigated (i.e., hypo- or hyper-parathyroidism) [47].

Neuroferritinopathy

Neuroferritinopathy is characterized by iron storage or cystic degeneration in the putamen. Clinical symptoms are adult-onset chorea, dystonia, and cognitive impairment. Only a few cases have been described in the literature [47].

Wilson Disease [48, 49]

It is an autosomal recessive copper metabolism deficit due to mutations in the ATP7B gene (copper-transporting ATPase) on chromosome 13. Mutations determine the loss of function of a transmembrane copper-binding protein. Excessive copper deposition is seen primarily in the liver and brain.

Clinical features: Late-onset presentation (second–third decades of life) is associated with the most common European mutation (H1069Q). This is a multisystemic disease characterized by hepatic, neurological, and osteomuscular involvement. Copper accumulation starts after birth. Initial symptoms can be asymptomatic elevation of liver enzyme, irritability, or recurrent joint pain and swelling. Neurological extrapyramidal signs are the most frequent features: involuntary movements and dystonia. Psychiatric symptoms, such as antisocial behaviors, can start during childhood, as well as cognitive problems, such as changes in handwriting or a drop in school grades [48].

Neuroimaging: It shows basal ganglia symmetrical hyperintensity on T2 MRI.

Diagnosis: Kayser–Fleischer rings on slit-lamp examination. Low serum ceruloplasmin, increased 24-h urinary copper excretion. Genetic testing.

Therapy: Penicillamine (decoppering treatment). Initial dose of 125–250 mg per day, gradually increasing up to 1–3 g/day. Maintenance phase 250–750 mg per day. Clinical improvement is seen after a few months, even in patients with severe neurological disability. Decoppering ensures that presymptomatic individuals remain symptom-free [49].

Others

Nasu–Hakola Disease (Polycystic Lipomembranous Osteodysplasia with Sclerosing Leukoencephalopathy)

This is an autosomal recessive, inherited disorder characterized by progressive dementia and repeated fractures during adolescence. Recently, mutations on the DAP12 gene (DNAX-activating protein 12) and the TREM2 gene (triggering receptors expressed on myeloid cell 2) have been described [50].

Epidemiology: 200 cases described worldwide, especially in Japan and Finland.

Clinical features: Onset ranges from 10 to 46 years of age. Average duration 16 years. Disease starts with foot and knee pains and repeated pathological fractures. Dementia is characterized by personality changes, memory disorder, apraxia, agnosia, acalculia, and disorientation. Some patients can have urinary incontinence, seizures, and pyramidal signs. Bone X-rays show cystic lesion in epiphyses of long bones [51]. Interestingly, mutations in the TREM2 gene have been recently described in a family with frontotemporal dementia-like presentation without bone involvement [52].

Neuroimaging: It shows general diffuse atrophy of white matter (sclerosing leukoencephalopathy) and gray matter. Sometimes basal ganglia calcification can be seen [51].

Prion Disease

This is a group of diseases characterized by spongiform degeneration of the whole brain due to the deposition of misfolded prion proteins, a normal component of neurons cells [53, 54]. The most common form of prion disorder is Creutzfeldt–Jakob disease, which is sporadic and occurs with a frequency of 1 per million inhabitants. It is a devastating subacute dementia with ataxia and myoclonus.

Diagnosis is based on characteristic MRI changes, an EEG pattern with periodic sharp waves, CSF containing a 14-3-3 protein elevation [53, 55, 56]. Rarely prion disorders are inherited in an autosomal dominant manner, and they can have different and heterogeneous presentations. For management, see the European Guidelines [9].

Infective Dementia, Inflammatory–Autoimmune Disorders, and Toxic–Metabolic Disorders

These are acquired and often treatable causes of dementia and should be suspected in early-onset dementia, rapid-progressive course, or in the presence of systemic involvement. Generally, a correct clinical assessment and blood tests, performed in the first diagnostic step, can orientate the diagnosis.

Infective Dementia [8, 57–61]

Cognitive decline is associated with other systemic symptoms: mood disorders, frequent infectious, and systemic illnesses in HIV [57]; meningitis and tabes dorsalis in neurosyphilis [58]; lymphocytic meningitis, arthralgia, peripheral, or facial neuropathies in Lyme disease [59, 60]; and arthralgia, gastrointestinal symptoms, and ataxia in Whipple's disease [61].

Inflammatory–Autoimmune Disorders

Limbic encephalitis [1, 9, 62, 63]. It is a subacute disorder characterized by cognitive, mood, and behavior alterations and temporal seizures. Generally, the etiology is paraneoplastic based n autoantibodies or on the inflammatory response to tumors (often lung, testicular, or ovarian cancer). In addition to clinical history and blood markers, diagnosis is made by MRI, CSF, EEG, and searching specifically for cancer. Therapy consists in antiepileptic drugs, immunosuppression with steroids, and tumor-specific treatment, when possible. Limbic encephalitis can precede tumor presentation by up to 5 years.

Hashimoto encephalopathy presents seizures, stroke-like episodes, cognitive decline, neuropsychiatric symptoms, and myoclonus. Clinical course can also be relapsing-remitting. Diagnosis is made, after excluding other subacute causes of dementia, based on high titers of antithyroid peroxidase antibodies. Treatment with corticosteroids is almost always successful [9, 64].

Toxic–Metabolic Disorders [1, 7, 8, 65–67]

Alcohol-related dementia represents one of the most frequent causes of dementia in the young population; Rossor reported a prevalence of 10 % [1]. Dementia is associated with cerebellar signs and, in the case of thiamine deficiency, with confusion and ophthalmoplegia (Wernicke–Korsakoff encephalopathy). Treatment is supplementation and nutritional support [1, 7, 8, 65].

Conclusion

Uncommon causes of dementia comprise a wide number of very rare and often misdiagnosed disorders, including late-onset forms of childhood metabolic inborn errors, inflammatory disorders, infectious diseases, and toxic–metabolic abnormalities.

Clinical data on most of them are based only on a single-case report, and often diagnosis is challenging due to the clinical heterogeneity among and within the various disorders. Thus, a complete list of uncommon dementia is not possible.

The creation of diagnostic categories, even if arbitrary, can help clinicians make differential diagnoses and may reduce diagnostic errors, which is of great importance since disease-modifying therapies are available in some cases.

Creation of a regional or national registry may be useful to make a real estimate of the prevalence of uncommon dementia and to improve our clinical knowledge overall.

References

1. Rossor MN, Fox NC, Mummery CJ, et al. The diagnosis of young-onset dementia. Lancet Neurol. 2010;9:793–806.
2. Harvey RJ, Skelton-Robinson M, Rossor MN, et al. The prevalence and causes of dementia in people under the age of 65 years. J Neurol Neurosurg Psychiatry. 2003;74:1206–9.
3. Fujihara S, Brucki SM, Rocha MS, et al. Prevalence of presenile dementia in a tertiary outpatient clinic. Arq Neuropsiquiatr. 2004;62(3):592–5.
4. Panegyres PK, French K. Course and causes of suspected dementia in young adults: a longitudinal study. Am J Alzheimers Dis Other Demen. 2007;22(1):48–56.
5. Shinagawa S, Ikeda M, Toyota Y, et al. Frequency and clinical characteristics of early-onset dementia in consecutive patients in a memory clinic. Dement Geriatr Cogn Disord. 2007;24(1):42–7.
6. Kelley BJ, Boeve BF, Josephs KA. Young-onset dementia: demographic and etiologic characteristics of 235 patients. Arch Neurol. 2008;65(11):1502–8.
7. Sampson EL, Warren JD, Rossor MN. Young onset dementia. Postgrad Med J. 2004;80:125–39.
8. Kuruppu DK, Matthews BR. Young-onset dementia. Semin Neurol. 2013;33:365–85.
9. Sorbi S, Hort J, Erkinjuntti T, et al. EFNS-ENS guidelines on the diagnosis and management of disorders associated with dementia. Eur J Neurol. 2012;19:1159–79.
10. Roos RA. Huntington's disease: a clinical review. Orphanet J Rare Dis. 2010;5:40–8.
11. Gallina P, Paganini M, Lombardini L, et al. Progress in restorative neurosurgery: human fetal striatal transplantation in Huntington's disease. Review. J Neurosurg Sci. 2011;55(4): 371–81.
12. Kumar A, Agarwal S, Agarwal D, et al. Myotonic dystrophy type 1 (DM1): a triplet repeat expansion disorder. Gene. 2013;522:226–30.
13. Bruni AC, Takahashi-Fujigasaki J, Maltecca F, et al. Behavioral disorder, dementia, ataxia, and rigidity in a large family with TATA box-binding protein mutation. Arch Neurol. 2004;61(8):1314–20.
14. Murphy S, Gorman G, Beetz C, et al. Dementia in SPG4 hereditary spastic paraplegia: clinical, genetic, and neuropathologic evidence. Neurology. 2009;73(5):378–84.
15. Bourgeois J, Coffey S, Rivera SM, et al. Fragile X premutation disorders – expanding the psychiatric perspective. J Clin Psychiatry. 2009;70(6):852–62.
16. Hagerman RJ, Hall DA, Coffey S, et al. Treatment of fragile X-associated tremor ataxia syndrome (FXTAS) and related neurological problems. Clin Interv Aging. 2008;3(2):251–62.
17. Kaye EM. Update on genetic disorders affecting white matter. Pediatr Neurol. 2001;24:11–24.
18. Köhler W. Leukodystrophies with late disease onset: an update. Curr Opin Neurol. 2010; 23(3):234–41.
19. Luda E, Barisone MG. Adult-onset adrenoleukodystrophy: a clinical and neuropsychological study. Neurol Sci. 2001;22:21–5.
20. Debs R, Froissart R, Aubourg P, et al. Krabbe disease in adults: phenotypic and genotypic update from a series of 11 cases and a review. J Inherit Metab Dis. 2013;36:859–68.
21. Gieselmann V, Krägeloh-Mann I. Metachromatic leukodystrophy – an update. Neuropediatrics. 2010;41:1–6.
22. Fraidakis MJ. Psychiatric manifestations in cerebrotendinous xanthomatosis. Translational Psychiatry 2013:3,e302. doi:10.1038/tp.2013.76.
23. Pareyson D, Fancellu R, Mariotti C, et al. Adult-onset alexander disease: a series of eleven unrelated cases with review of the literature. Brain. 2008;131:2321–31.

24. Klein CJ, Boes CJ, Chapin JE, et al. Adult polyglucosan body disease: case description of an expanding genetic and clinical syndrome. Muscle Nerve. 2004;29(2):323–8.
25. Labauge P, Horzinski L, Ayrignac X, et al. Natural history of adult-onset eIF2B-related disorders: a multi-centric survey of 16 cases. Brain. 2009;132:2161–9.
26. Bolsover FE, Murphy E, Cipolotti L, et al. Cognitive dysfunction and depression in Fabry disease: a systematic review. J Inherit Metab Dis. 2013. doi:10.1007/s10545-013-9643-x
27. El Dib RP, Nascimento P, Pastores GM. Enzyme replacement therapy for Anderson-Fabry disease. Cochrane Database Syst Rev. 2013;2:CD006663. doi:10.1002/14651858.CD006663.pub3.
28. Guimarãesa J, Amaralb O, SáMiranda MC. Adult-onset neuronopathic form of Gaucher's disease: a case report. Parkinsonism Relat Disord. 2003;9:261–4.
29. Mistry PK, Cappellini MD, Lukina E, et al. A reappraisal of Gaucher disease-diagnosis and disease management algorithms. Am J Hematol. 2011;86(1):110–5.
30. Beavan MS, Schapira AHV. Glucocerebrosidase mutations and the pathogenesis of Parkinson disease. Ann Med. 2013;45:511–21.
31. Vanier MT. Niemann-Pick disease type C. Orphanet J Rare Dis. 2010;5:16.
32. Babacan-Yildiz G, Hanagasi H, Gurvit H, et al. A rare dementing disease: adult neuronal ceroid lipofuscinoses. J Neuropsychiatry Clin Neurosci. 2012;24(4):493–8.
33. Zini A, Cenacchi G, Nichelli P, et al. Early-onset dementia with prolonged occipital seizures: an atypical case of Kufs disease. Neurology. 2008;71(21):1709–12.
34. Neudorfer O, Pastores GM, Zeng BJ, et al. Late-onset Tay-Sachs disease: phenotypic characterization and genotypic correlations in 21 affected patients. Genet Med. 2005;7(2):119–23.
35. Patterson MC. Gangliosidoses. Handb Clin Neurol. 2013;113:1707–8. doi:10.1016/B978-0-444-59565-2.00039-3.
36. Torres RJ, Puig JG, Jinnah HA. Update on the phenotypic spectrum of Lesch-Nyhan disease and its attenuated variants. Curr Rheumatol Rep. 2012;14:189–94.
37. Pantoni L. Cerebral small vessel disease: from pathogenesis and clinical characteristics to therapeutic challenges. Lancet Neurol. 2010;9(7):689–701.
38. Chouldhary S, Mcleod M, Torchia D, et al. Cerebral Autosomal Dominant Arteriopathy with Subcortical Infarcts and Leukoencephalopathy (CADASIL). J Clin Aesthet Dermatol. 2013;6(3):29–32.
39. Pescini F, Nannucci S, Bertaccini B, et al. The Cerebral Autosomal-Dominant Arteriopathy with Subcortical Infarcts and Leukoencephalopathy (CADASIL) scale. A screening tool to select patients for NOTCH3 gene analysis. Stroke. 2012;43:2871–6.
40. Dichgans M, Markus HS, Salloway S, et al. Donepezil in patients with subcortical vascular cognitive impairment: a randomised double-blind trial in CADASIL. Lancet Neurol. 2008;7:310–8.
41. Kaufman KR, Zuber N, Rueda-Lara MA, et al. MELAS with recurrent complex partial seizures, nonconvulsive status epilepticus, psychosis, and behavioral disturbances: case analysis with literature review. Epilepsy Behav. 2010;18(4):494–7.
42. McFarland R, Taylor RT, Turnbull DM. A neurological perspective on mitochondrial disease. Lancet Neurol. 2010;9:829–40.
43. Walterfanga M, Evansa A, Leong Looid JC, et al. The neuropsychiatry of neuroacanthocytosis syndrome. Neurosci Biobehav Rev. 2011;35:1275–83.
44. Zeman A, Daniels G, Tilley L, et al. McLeod syndrome: life-long neuropsychiatric disorder due to a novel mutation of the XK gene. Psychiatr Genet. 2005;15(4):291–3.
45. Hayflick J, Westaway SK, Levinson B, et al. Genetic, clinical, and radiographic delineation of Hallervorden–Spatz syndrome. N Engl J Med. 2003;348:33–40.
46. Manyam BV. What is and what is not 'Fahr's disease'. Parkinsonism Relat Disord. 2005;11(2):73–80.
47. Saleem S, Aslam HM, Anwar M. Fahr's syndrome: literature review of current evidence. Orphanet J Rare Dis. 2013;8:156.
48. Taly AB, Meenakshi-Sundaram S, Sinha S, Swamy HS, Arunodaya GR. Wilson disease: description of 282 patients evaluated over 3 decades. Medicine (Baltimore). 2007;86(2):112–21.

49. Aggarwal A, Bhatt M. Recovery from severe neurological Wilson's disease with copper chelation. Int Rev Neurobiol. 2013;110:313–48.
50. Paloneva J, Manninen T, Christman G, et al. Mutations in two genes encoding different subunits of a receptor signaling complex result in an identical disease phenotype. Am J Hum Genet. 2002;71:656–62.
51. Kaneko M, Sano K, Nakayama J, et al. Nasu-Hakola disease: the first case reported by Nasu and review. Neuropathology. 2010;30:463–70.
52. Guerreiro RJ, Lohmann E, Brás JM, et al. Using exome sequencing to reveal mutations in TREM2 presenting as a frontotemporal dementia–like syndrome without bone involvement. JAMA Neurol. 2013;70(1):78–84.
53. Sikorska B, Knight R, Ironside JW, Liberski PP. Creutzfeldt-Jakob disease. Adv Exp Med Biol. 2012;724:76–90.
54. Cali I, Castellani R, Yuan J, et al. Classification of sporadic Creutzfeldt–Jakob disease revisited. Brain. 2006;129:2266–77.
55. Zerr I, Kallenberg K, Summers DM, et al. Updated clinical diagnostic criteria for sporadic Creutzfeldt–Jakob disease. Brain. 2009;132:2659–68.
56. Rosenbloom MH, Atri A. The evaluation of rapidly progressive dementia. Neurologist. 2011;17(2):67–74.
57. Valcour V, Paul R, Chiao S, Wendelken LA, Miller B. Screening for cognitive impairment in human immunodeficiency virus. Clin Infect Dis. 2011;53(8):836–42.
58. Read PJ, Donovan B. Clinical aspects of adult syphilis. Intern Med J. 2012;42(6):614–20.
59. Halperin JJ. Nervous system lyme disease: diagnosis and treatment. Curr Treat Options Neurol. 2013;15(4):454–64.
60. Halperin JJ. Lyme disease: a multisystem infection that affects the nervous system. Continuum (Minneap Minn). 2012;18(6 Infectious Disease):1338–50.
61. Puéchal X. Whipple's disease. Ann Rheum Dis. 2013;72(6):797–803.
62. Graus F, Saiz A. Limbic encephalitis: an expanding concept. Neurology. 2008;70:500–1.
63. Serratrice G, Pellissier JF, Serratrice J, De Paula A. Limbic encephalitis – evolving concepts. Bull Acad Natl Med. 2008;192(8):1531–41.
64. Mocellin R, Walterfang M. Velakoulis D Hashimoto's encephalopathy: epidemiology, pathogenesis and management. CNS Drugs. 2007;21(10):799–811.
65. Victor M. Alcoholic dementia. Can J Neurol Sci. 1994;21:88–99.
66. Chalouhi C, Faesch S, Anthoine-Milhomme MC, et al. Neurological consequences of vitamin B12 deficiency and its treatment. Pediatr Emerg Care. 2008;24(8):538–41.
67. Ibrahim D, Froberg B, Wolf A, Rusyniak DE. Heavy metal poisoning: clinical presentations and pathophysiology. Clin Lab Med. 2006;26(1):67–97.

Chapter 13
Neurodevelopment Alterations, Neurodegeneration, and Immunoinflammatory Patterns in the Pathophysiology of Schizophrenia

Bernardo Dell'Osso, Gregorio Spagnolin, Neva E. Suardi, and A. Carlo Altamura

Abstract Schizophrenia is a highly disabling syndrome, with frequent onset in the first half of adult life. As for other major psychoses, the etiology of schizophrenia is supposed to involve a gene–environment interaction. In terms of pathophysiology, however, immune alterations have been repeatedly reported in schizophrenic patients, involving both the unspecific and specific pathways of the immune system and suggesting that infectious/autoimmune processes play an important role in the development of the disorder. In such perspective, it seems that schizophrenia may be associated with an imbalance in inflammatory cytokines. Alterations in the inflammatory and immune systems, moreover, seem to be already present in the early stages of schizophrenia, likely connected to specific neurodevelopmental abnormalities, which identify the roots of the disorder during brain development with consequences that do not manifest themselves until adolescence or early adulthood. At the same time, neuropathological studies and longitudinal observation in schizophrenia, showing progressive losses of gray matter in the frontal and temporal lobes of the brain, also support a neurodegenerative hypothesis, and, more recently, a novel mixed hypothesis, integrating the aforementioned models, has been put forward.

Keywords Cytokines • Neurodegeneration • Neurodevelopmental model Schizophrenia

B. Dell'Osso, MD (✉) • G. Spagnolin, MD • N.E. Suardi, MD • A.C. Altamura, MD
Department of Psychiatry, University of Milan, Fondazione IRCCS Cà Granda, Ospedale Maggiore Policlinico, Via Francesco Sforza 35, 20122 Milan, Italy
e-mail: bernardo.dellosso@policlinico.mi.it

D. Galimberti, E. Scarpini (eds.), *Neurodegenerative Diseases*,
DOI 10.1007/978-1-4471-6380-0_13,

Introduction

Schizophrenia is a chronic, severe, and disabling brain disorder which frequently onset in the first half of adult life. Kraepelin and Bleuler already recognized that a significant part of schizophrenic subjects had previously shown behavioral abnormalities over childhood [1]. Subsequent genetic studies reported differences in neurological development in high-risk children [2–4]. Indeed, neurodevelopmental abnormalities, occurring throughout childhood, have been reported in up to 50 % of high-risk children, born from schizophrenic mothers [4], comprising hypoactivity, hypotonia, soft neurological signs – poor motor coordination, in particular – and deficits in attention and information processing in late childhood.

Taken as a whole, converging evidence supports the hypothesis that at least part of the genetic vulnerability to schizophrenia involves abnormal neurodevelopment [1]. Actually, many environmental risk factors seem to operate before, around, or immediately after birth, including pregnancy and birth complications, perinatal and early childhood brain damages, altered fetal development, season of birth, and heavy cannabis intake [1]. Therefore, up to one third of the variance in liability to schizophrenia may be attributable to nongenetic factors. Despite consistent evidence supporting the presence of neurodevelopmental alterations in schizophrenia, many authors have put more emphasis on the neurodegenerative processes that occur over the course of the illness [5]. Currently, however, the traditional neurodegenerative hypothesis has been largely questioned and, at least to some extent, revisited [6]. As a matter of fact, the debate, as to whether there is an abnormal developmental or degenerative process, likely stems from a spurious dichotomy and depends on the stage at which its observation begins [1].

Over the last two decades, moreover, within the pathophysiological alterations present in schizophrenia – either of neurodevelopmental and/or neurodegenerative nature – a dysregulation of the inflammatory response system has been largely documented as well [7, 8]. For instance, evidence of immune activation was observed from the detection of abnormal levels of proinflammatory cytokines and their receptors in peripheral blood and cerebrospinal fluid from schizophrenic patients [9]. Cytokines, in particular, are involved in normal central nervous system (CNS) development, as well as in the pathogenesis of many neuropsychiatric disorders, acting directly on neural cells or modulating neurotransmitter and peptidergic pathways [9]. In such perspective, neurobiological hypotheses linking the neurodevelopmental alterations occurring in schizophrenia with the inflammatory processes, largely documented over the course of the illness, have been put forward [10].

Neurodevelopmental Hypothesis of Schizophrenia

Several lines of evidence strongly indicate that schizophrenia may be a neurodevelopmental disorder [10]. The "neurodevelopmental model" of schizophrenia, in its simplest form, postulates that the disorder could be the correlate of an aberrant neurodevelopmental process starting much earlier than the onset of clinical symptoms, caused by a combination of genetic and environmental factors [11, 12].

The cerebral multiple damage, which is characteristic of the disorder, that becomes more evident in the long-tem run, in fact, might have, actually, begun early in life [13, 14]. In particular, several investigators believe that the damage occurs during brain development, over the intrauterine period, and the first few years after birth [15]. Main neurodevelopmental abnormalities in schizophrenia consist of changes in the normal expression of proteins involved in early migration of neurons and glia, cell proliferation, axonal outgrowth, synaptogenesis, and apoptosis [16].

The "neurodevelopmental model" seems to be based on reports of an excess of adverse events occurring during the pre- and perinatal periods, which would lead to the presence of cognitive and behavioral signs, particularly in adolescence and childhood. Another element supporting this hypothesis is represented by the lack of clear-cut neurodegenerative patterns in many schizophrenics [17]. Furthermore, multiple markers of congenital anomalies, indicative of neurodevelopmental insults, have been indicated as supportive for the neurodevelopmental model of schizophrenia [18, 19] including agenesis of corpus callosum, stenosis of Sylvian aqueduct, cerebral hematomas, and cavum septum pellucidum. Presence of low-set ears, epicanthal eye folds, wide spaces between the first and second toes, and abnormal dermatoglyphics are, in turn, suggestive of both first and second trimester abnormalities [20]. Multiple records, moreover, indicate the presence of premorbid neurological soft signs in children, who later develop schizophrenia [2, 21]. Additionally, children at high risk for schizophrenia were found to show a broad range of abnormalities, the most prominent of which seemed to occur in attention, motor function, coordination, sensory integration, mood, and social behaviors [22]. Indeed, such abnormalities may have predictive value in determining which children will later keep on showing overt signs of either schizophrenia spectrum disorders or schizophrenia itself [2].

Several studies have established an important role for genetic factors in the pathophysiology of schizophrenia, with polygenic model acting additively or multiplicatively, likely providing the best explanation for the disorder. Linkage and association studies [23, 24] have shown 12 chromosomal regions, containing 2,181 known genes [24] and 9 specific genes, involved in the possible etiology of the disorder [24]. On the other hand, environmental factors play an active role in the pathogenesis of schizophrenia, including pre- and perinatal complications, as well as maternal infections occurring during pregnancy. A meta-analysis of population-based data found significant estimates for three main categories of pre- and perinatal complications: (1) complications of pregnancy (e.g., bleeding, preeclampsia, diabetes), (2) abnormal fetal growth and development (e.g., low birth weight, congenital malformations, small head circumference), and (3) complications of delivery (e.g., asphyxia, uterine atony, emergency caesarean section) [25].

Obstetric complications, in turn, are supposed to increase the risk of developing schizophrenia in two main ways: acting alone and/or interacting with genetic risk factors [25, 26]. In fact, it has been suggested that specific susceptibility genes for schizophrenia may be regulated by hypoxia/ischemia [27] occurring during birth.

Other environmental factors, potentially causing abnormal neurodevelopment, include the infective processes occurring during pregnancy. A large body of evidence, in fact, indicates that environmental factors, such as maternal infections, can increase the risk for the offspring to develop schizophrenia during adulthood [28, 29].

The available body of research in the field suggests that pre-/perinatal infections and other environmental insults, that adversely affect infant brain development, may increase the likelihood to develop schizophrenia in later life, particularly in genetically susceptible individuals [30]. Such increased risk has been associated with maternal infections with viruses, including influenza [31], measles [32], polio [33], herpes simplex type 2 [34], as well as specific bacterial infections, such as diphtheria and pneumonia [35]. Association studies regarding the influenza A virus showed that the maximum risk for the embryonic brain to be hit is represented by the exposure to the infective agent during the fourth and seventh months of gestation [36]. Subsequent studies have shown that other viruses, such as rubella, may increase the risk for development of schizophrenia in the progeny of exposed mothers [37]. Prenatal exposure to rubella, in particular, was found to increase 10–20-fold the risk of developing schizophrenia [38], prenatal exposure to influenza in the first trimester increased risk sevenfold, and infection in early to mid-gestation increased risk threefold. Also presence of maternal antibodies against *Toxoplasma gondii* leads to 2.5-fold increased risk.

Alterations of Inflammatory Pathways in Schizophrenia

Dysregulation of the inflammatory response system appears to be a major piece of evidence in the pathophysiology of schizophrenia, along with genetic and environmental factors, ultimately affecting the neurodevelopmental process [39, 40]. Recently, there has been growing interest on the interface between immunology and chronic mental illness, including areas such as stress, neuroplasticity, genetics, and cytokines [41]. The latter ones, in particular, playing a pivotal role in infectious and inflammatory processes and mediating of the cross talk between the brain and the immune system, are supposed to be the main actors of the immune and inflammatory abnormalities, documented in schizophrenia [42].

Because cytokines are large hydrophilic polypeptides, their ability to cross the Blood-brain barrier is reduced, at least under physiologic conditions. The presence of abnormal circulating levels of proinflammatory cytokines and their receptors is well established in peripheral blood and cerebrospinal fluid of schizophrenic patients [43–45] and their first-degree relatives [44, 46], thus confirming the presence of immune abnormalities in schizophrenia [47–49].

In the last two decades, different hypotheses in relation to the cytokine-mediated development of schizophrenia have been proposed.

As a matter of fact, cytokines play an important role during neurodevelopment and in CNS functions at all stages, starting with the induction of neuroepithelium

[50]. Later on, cytokines monitor the renewal of neuroepithelial cells, which act as precursors for all neurons, microglia, and adult progenitors, as well as framework for radially migrating neurons [51]. Such processes are orchestrated by cytokines and related responses of their target cells [52]. As general rule, there is an overproduction of neurons and glia, and cytokines are crucial either to promote survival of cells, properly connected in neural network, or to induce apoptosis of cells with impaired connections [53]. Therefore, even minimal variation on cytokine levels could result in subsequent functional impairment [54].

An increase of cytokines, following maternal infection, may alter the immune status of the brain, causing abnormal cells development with subsequent brain damage [55]. It is clear that maternal immune activation (MIA) induces increase of cytokines in the placenta (IL-1beta, IL-6, TNF-alpha) and amniotic fluid (IL-6, TNF-alpha) [56]. The action of cytokines on the placenta might alter transfer of cells, nutrient, oxygen, growth factors, and maternal antibodies, each of which with potential crucial effect on fetal development [56].

Besides affecting neurodevelopment, some cytokines (i.e., IL-2 and IL-6) appear to have a role in the progression of schizophrenic illness.

For instance, IL-2 stimulates the proliferation of T lymphocytes and its inhibition contributes to humoral immunity enhancement [57]. Kim et al. found lower IL-2 serum levels in schizophrenics with long duration of illness [58]. Such findings suggest that IL-2 may be a key modulator of dopaminergic metabolism and psychotic symptoms in schizophrenia [59].

Another contribution to the progression of the illness might be due to an hyperactivation of humoral immunity, which stimulates the tryptophan 2,3-dioxygenase enzyme, with an increased transformation of the amino acid tryptophan in kynurenic acid that acts as a NMDA antagonist [59].

Among cytokines, IL-6 potentiates B lymphocyte proliferation, and it seems to play a key role in the immunological abnormalities observed in schizophrenic patients [48]. It is also worthwhile to highlight that several studies showed that a long duration of illness in schizophrenia is associated with higher serum levels of IL-6 [48].

Moreover, elevated IL-6 serum concentrations have been proposed as key factors, responsible for cerebral atrophy observed in schizophrenics with long duration of illness [60, 61].

Neurodegeneration in Schizophrenia

Neuroanatomical abnormalities are frequent in schizophrenic patients. Such anomalies are thought to represent the structural substrate for the disorder and may originate from a neurodevelopmental defect [15], even though there is growing evidence that the magnitude and pattern of such abnormalities could progress over time [62], involving a proper neurodegenerative process. The combination of neurodevelopmental and neurodegenerative processes in the disorder's pathogenesis is a challenging but plausible possibility [5]. Tissue losses in brain can involve different areas: for example, decreases in the volume of the temporal lobe [63], in the hippocampal volume [64], and in the volume of parahippocampal gyrus [65] were reported. Similarly, several studies have shown reductions in the gray matter of volume of cortical structures in schizophrenic patients.

The molecular basis of gray matter volume losses in schizophrenic subjects is still poorly understood, even though such anomalies seem to be more likely connected to the loss or disorganization of neuronal processes than to the loss of neuronal cell bodies. In fact, postmortem studies in schizophrenic brains showed abnormal neuronal organization within corticolimbic structures [66, 67]. For instance, a magnetic resonance (MR) imaging study reported that schizophrenic patients showed vertical sulcal patterns more frequently than healthy controls [68], while other studies also demonstrated distortions of normal patterns of cortical asymmetries in schizophrenia and hippocampal volume reductions only on the left side [69, 70]. Indeed, postmortem studies also reported larger abnormalities in the left temporal lobes of schizophrenics, i.e., temporal horn enlargement [71] and neuronal heterotopia [67].

Even though some studies report the progression of neuroanatomical abnormalities in schizophrenic patients, the point of whether such alterations are static or dynamic is still open to argument. Some studies report that ventricular enlargement and gray matter volume losses are progressive over periods of 1–5 years in schizophrenic subjects [72, 73], while other studies describe that such structural measures are highly stable over time [74, 75]. On the other hand, recent studies of individuals with "prodromal" schizophrenia showed that we can find relatively rapid changes in neuroanatomical structure early in the course of illness [62]. Some investigators reported the presence of cortical thickness reductions in schizophrenic patients; in particular, the absence of widespread cortical thinning before disease onset implies that the cortical thinning is unlikely to simply reflect genetic liability to schizophrenia but is predominantly driven by disease-associated factors [76].

Several different mechanisms of neuronal injury are now being investigated in relation to the pathogenesis of schizophrenia. Some investigators suggested that a developmental deficit of *N*-methyl-D-aspartate (NMDA) receptor-bearing gamma-aminobutyric acid (GABA)ergic interneurons would place an individual at increased risk for excitotoxic neuronal injury later in life [77].

Excitotoxicity (i.e., neurodegeneration via the overactivity of excitatory neurotransmission) represents an interesting mechanism to explain neuronal injury in schizophrenia, because it could be initiated and maintained through the action of neurotransmitter systems, such as the monoamines, that have long been implicated in schizophrenia [78]. Another intriguing theory to explain neuronal injury in schizophrenia is the dysregulation of apoptosis [79], a process normally associated with the elimination of redundant neurons during development [80]. Also, glucocorticoid hormones [81], triggered by environmental stressors, including those (e.g., famine) associated with an increased risk for schizophrenia [82], have been implicated as factors contributing to neurodegenerative impairment.

Conclusions

The present chapter tried to summarize the most intriguing patterns linking abnormalities in the neurodevelopment with altered immune/inflammatory mechanisms in schizophrenic patients. However, such perspective does not exclude the possibility to consider also the presence of progressive neurodegeneration as prominent biological feature of the disorder. In fact, it seems likely that what we currently diagnose as a unitary disorder includes, actually, highly heterogeneous schizophrenic entities, in terms of pathophysiology [83]. These would include forms predominantly characterized by neurodevelopmental alterations (e.g., inflammatory features), as well as others with minor or absent neurodevelopmental aspects, but marked and progressive neurodegeneration, starting from the early adolescence, as main biological feature. Therefore, the attempt to solve the question whether schizophrenia is or is not a neurodevelopmental disorder or a progressive neurodegenerative seems to be outdated [1]. Differences in the genetic background could give account of these two different timing and patterns of presentation.

In conclusion, all the reviewed inflammatory and neurodevelopmental data represent the most robust evidence confuting the conceptualization of schizophrenia as a "functional psychosis." Indeed, they encourage to consider it as a pure "brain disease," as Kraepelin had correctly defined it, as a dementing process occurring in the frontal part of the brain, regardless of how and when such deep dysregulations of neural mechanisms, leading to neuronal death in specific brain areas, occur. Neuronal loss seems, likely, to be due to the neurodevelopmental/inflammatory abnormalities reported in this chapter.

References

1. Jones PB, Buckley PF. Schizophrenia. Philadelphia: Elsevier; 2006.
2. Fish B, Marcus J, Hans S, Auerbach JG, Perdue S. Infants at risk for schizophrenia: sequelae of a genetic neurointegrative defect. Arch Gen Psychiatry. 1992;49:221–35.
3. Altamura AC, Pozzoli S, Fiorentini A, Dell'Osso B. Neurodevelopment and inflammatory patterns in schizophrenia in relation to pathophysiology. Prog Neuropsychopharmacol Biol Psychiatry. 2013;42:63–70.
4. Walker E, Lewine RJ. Prediction of adult-onset schizophrenia from childhood home movies of the patients. Am J Psychiatry. 1990;147(8):1052–6.
5. Csernansky JG. Neurodegeneration in schizophrenia: evidence from in vivo neuroimaging studies. ScientificWorldJournal. 2007;7:135–43.
6. Rund BR. Is schizophrenia a neurodegenerative disorder? Nord J Psychiatry. 2009;63(3):196–201.
7. Boin F, Zanardini R, Pioli R, Altamura CA, Maes M, Gennarelli M. Association between –G308A tumor necrosis factor alpha gene polymorphism and schizophrenia. Mol Psychiatry. 2001; 6(1):79–82.
8. Lin A, Kenis G, Bignotti S, Tura GJ, De Jong R, Bosmans E, et al. The inflammatory response system in treatment-resistant schizophrenia: increased serum interleukin-6. Schizophr Res. 1998;32(1):9–15.
9. Maes M, Bocchio Chiavetto L, Bignotti S, Battisa Tura GJ, et al. Increased serum interleukin-8 and interleukin-10 in schizophrenic patients resistant to treatment with neuroleptics and the

stimulatory effects of clozapine on serum leukemia inhibitory factor receptor. Schizophr Res. 2002;54(3):281–91.
10. Gourion D, Gourevitch R, Leprovost JB, Olié H lôo JP, Krebs MO. Neurodevelopmental hypothesis in schizophrenia. Encéphale. 2004;30:109–18.
11. Rapoport JL, Addington AM, Frangou S, Psych MR. The neurodevelopmental model of schizophrenia: update 2005. Mol Psychiatry. 2005;10:434–49.
12. Singh SM, McDonald P, Murphy B, O'Reilly R. Incidental neurodevelopmental episodes in the etiology of schizophrenia: an expanded model involving epigenetics and development. Clin Genet. 2004;65:435–40.
13. Buckley P. The clinical stigmata of aberrant neurodevelopment in schizophrenia. J Nerv Ment Dis. 1998;186(2):79–86.
14. Keshavan MS, Murray RM. Neurodevelopment and adult psychopathology. Cambridge: Cambridge University Press; 1997.
15. Weinberger DR. Implications of normal brain development for the pathogenesis of schizophrenia. Arch Gen Psychiatry. 1987;44:660–9.
16. Fatemi SH, Folsom TD. The neurodevelopmental hypothesis of schizophrenia, revisited. Schizophr Bull. 2009;35(3):528–48.
17. Lewis DA, Levitt P. Schizophrenia as a disorder of neurodevelopment. Annu Rev Neurosci. 2002;25:409–32.
18. Lloyd T, Dazzan P, Dean K, Park SB, Fearon P, Doody GA, et al. Minor physical anomalies in patients with first-episode psychosis: their frequency and diagnostic specificity. Psychol Med. 2008;38:71–7.
19. Meltzer HY, Fatemi SH. Schizophrenia and other psychotic disorders. In: Ebert MH, Loosen PT, Nurcombe B, editors. Current diagnosis and treatment in psychiatry. Norwalk: Appleton and Lange; 2000. p. 260–77.
20. Fatemi SH, Folsom TD. The neurodevelopmental hypothesis of schizophrenia. Revisited. Schizophr Bull. 2009;35(3):528–54.
21. Barkus E, Stirling J, Hopkins R, Lewis S. The presence of neurological soft signs along the psychosis proneness continuum. Schizophr Bull. 2006;32:573–7.
22. Niemi LT, Suvisaari JM, Tuulio-Henriksson A, Lonnqvist JK. Childhood developmental abnormalities in schizophrenia: evidence from high-risk studies. Schizophr Res. 2003;60: 239–58.
23. Lewis CM, Levinson DF, Wise LH, DeLisi LE, Straub RE, Hovatta I, et al. Genome scan meta-analysis of schizophrenia and bipolar disorder, part II: schizophrenia. Am J Hum Genet. 2003;73:34–48.
24. Sullivan PF, Eaves LJ, Kendler KS, Neale MC. Genetic case–control association studies in neuropsychiatry. Arch Gen Psychiatry. 2001;58:1015–24.
25. Cannon TD, van Erp TG, Rosso IM, Huttunen M, Lönnqvist J, Pirkola T, et al. Fetal hypoxia and structural brain abnormalities in schizophrenic patients, their siblings, and controls. Arch Gen Psychiatry. 2002;59:35–41.
26. Boog G. Obstetrical complications and subsequent schizophrenia in adolescent and young adult offsprings: is there a relationship? Eur J Obstet Gynecol Reprod Biol. 2004;114:130–6.
27. Schmidt-Kastner R, van Os J, W M Steinbusch H, Schmitz C. Gene regulation by hypoxia and the neurodevelopmental origin of schizophrenia. Schizophr Res. 2006;84:253–71.
28. Karlsson H, Bachmann S, Schroder J, McArthur J, Torrey EF, Yolken RH. Retroviral RNA identified in the cerebrospinal fluids and brains of individuals with schizophrenia. Proc Natl Acad Sci U S A. 2001;98:4634–9.
29. Lewis DA. Retroviruses and the pathogenesis of schizophrenia. Proc Natl Acad Sci U S A. 2001;94:4293–4.
30. Jones P, Cannon M. The new epidemiology of schizophrenia. Psychiatr Clin North Am. 1998;21(1):1–25.
31. Mednick SA, Machon RA, Huttunen MO, Bonett D. Adult schizophrenia following prenatal exposure to an influenza epidemic. Arch Gen Psychiatry. 1988;45:189–92.

32. Torrey EF, Rawlings R, Waldman IN. Schizophrenic births and viral diseases in two states. Schizophr Res. 1988;1:73–7.
33. Suvisaari J, Haukka J, Tanskanen A, Hovi T, Lönnqvist J. Association between prenatal exposure to poliovirus infection and adult schizophrenia. Am J Psychiatry. 1999;156:1100–2.
34. Buka SL, Tsuang MT, Torrey EF, Klebanoff MA, Bernstein D, Yolken RH. Maternal infections and subsequent psychosis among offspring. Arch Gen Psychiatry. 2001;58:1032–7.
35. Watson CG, Kucala T, Tilleskjor C, Jacobs L. Schizophrenic birth seasonality in relation to the incidence of infectious diseases and temperature extremes. Arch Gen Psychiatry. 1984;41:85–90.
36. Brown AS, Begg MD, Gravenstein S, Schaefer CA, Wyatt RJ, Bresnahan M, et al. Serologic evidence of prenatal influenza in the etiology of schizophrenia. Arch Gen Psychiatry. 2004;61:774–80.
37. Brown AS, Schaefer CA, Wyatt RJ, Goetz R, Begg MD, Gorman JM, Susser ES. Maternal exposure to respiratory infections and adult schizophrenia spectrum disorders: a prospective birth cohort study. Schizophr Bull. 2000;26:287–95.
38. Brown AS. Prenatal infection as a risk factor for schizophrenia. Schizophr Bull. 2006;32:200–2.
39. Altamura AC, Boin F, Maes M. HPA axis and cytokines dysregulation in schizophrenia: potential implications for the antipsychotics treatment. Eur Neuropsychopharmacol. 1999;10:1–4.
40. Jablensky A. Epidemiology of schizophrenia: the global burden of disease and disability. Eur Arch Psychiatry Clin Neurosci. 2000;250(6):274–85.
41. Mundo E, Altamura AC, Vismara S, Zanardini R, Bignotti S, Randazzo R, et al. MCP-1 gene (SCYA2) and schizophrenia: a case–control association study. Am J Med Genet B Neuropsychiatr Genet. 2005;132B:1–4.
42. Altamura AC, Pozzoli S, Fiorentini A, Dell'Osso B. Neurodevelopment and inflammatory patterns in schizophrenia in relation to pathophysiology. Prog Neuropsychopharmacol Biol Psychiatry. 2013;5(42):63–70.
43. Garver DL, Tamas RL, Holcomb JA. Elevated interleukin-6 in the cerebrospinal fluid of a previously delineated schizophrenia subtype. Neuropsychopharmacology. 2003;28:1515–20.
44. Miller BJ, Buckley P, Seabolt W, Mellor A, Kirkpatrick B. Meta-analysis of cytokine alterations in schizophrenia: clinical status and antipsychotic effects. Biol Psychiatry. 2011;70: 663–71.
45. Potvin S, Stip E, Sepehry AA, Gendron A, Bah R, Kouassi E. Inflammatory cytokines alterations in schizophrenia: a systematic quantitative review. Biol Psychiatry. 2008;63:801–8.
46. Nunes SO, Matsuo T, Kaminami MS, Watanabe MA, Reiche EM, Itano EN. An autoimmune or an inflammatory process in patients with schizophrenia, schizoaffective disorder, and their biological relatives. Schizophr Res. 2006;84:180–2.
47. Ganguli R, Yang Z, Shurin G, Chengappa KN, Brar JS, Gubbi AV, Rabin BS. Serum interleukin-6 concentration in schizophrenia: elevation associated with duration of illness. Psychiatry Res. 1994;51:1–10.
48. Naudin J, Mege JL, Azorin JM, Dassa D. Elevated circulating levels of IL-6 in schizophrenia: an overview. Eur Arch Psychiatry Clin Neurosci. 1996;20(3):269–73.
49. Müller N, Riedel M, Scheppach C, Brandstätter B, Sokullu S, Krampe K, et al. Beneficial antipsychotic effects of celecoxib add-on therapy compared to risperidone alone in schizophrenia. Am J Psychiatry. 2002;159:1029–34.
50. Gulden J, Reiter JF. Nuer-ons and neur-offs: regulators of neural induction in vertebrate embryos and embryonic cells. Hum Mol Genet. 2008;17:R60–6.
51. Pinto L, Gotz M. Radial glial cell heterogeneity—the source of diverse progeny in the CNS. Prog Neurobiol. 2007;7:797–805.
52. Lee RH, Mills EA, Schwartz N, Bell MR, Deeg KE, Ruthazer ES, et al. Neurodevelopmental effects of chronic exposure to elevated levels of pro-inflammatory cytokines in a developing visual system. Neural Dev. 2010;5:2.
53. Deverman BE, Patterson P. Cytokines and CNS development. Neuron. 2009;64:61–77.

54. Chklovskii DB. Exact solution for the optimal neuronal layout problem. Neural Comput. 2004;16(10):2067–78.
55. Braun AS, Derkits EJ. Prenatal infection and schizophrenia: a review of epidemiologic and translational studies. Am J Psychiatry. 2010;167:261–80.
56. Patterson PH. Immune involvement in schizophrenia and autism. Etiology, pathology and animal models. Behav Brain Res. 2009;204:313–21.
57. Bresee C, Rapaport MH. Persistently increased serum soluble interleukin-2 receptors in continuously ill patients with schizophrenia. Int J Neuropsychopharmacol. 2009;12(6):861–5.
58. Kim YK, Kim L, Lee MS. Relationships between interleukins, neurotransmitters and psychopathology in drug-free male schizophrenics. Schizophr Res. 2000;44:165–75.
59. Muller N, Schwarz M. Schizophrenia as an inflammation-mediated dysbalance of glutamatergic neurotransmission. Neurotox Res. 2006;10(2):131–48.
60. Akiyama K. Serum levels of soluble IL-2 receptor alpha, IL-6 and IL-1 receptor antagonist in schizophrenia before and during neuroleptic administration. Schizophr Res. 1999;37:97–106.
61. Waddington JL. Neurodynamics of abnormalities in cerebral metabolism and structure in schizophrenia. Schizophr Bull. 1993;19:55–69.
62. Keshavan MS, Berger G, Zipursky RB, Wood SJ, Pantelis C. Neurobiology of early psychosis. Br J Psychiatry Suppl. 2005;48:s8–18.
63. Suddath RL, Casanova MF, Goldberg TE, et al. Temporal lobe pathology in schizophrenia: a quantitative MRI study. Am J Psychiatry. 1989;146:464–72.
64. Breier A, Buchanan RW, Elkashef A, Munson RC, Kirkpatrick B, Gellad F. Brain morphology and schizophrenia. A magnetic resonance imaging study of limbic, prefrontal cortex, and caudate structures. Arch Gen Psychiatry. 1992;49:921–6.
65. McCarley RW, Shenton ME, O'Donnell BF, Faux SF, Kikinis R, Nestor PG, Jolesz FA. Auditory P300 abnormalities and left posterior superior temporal gyrus volume reduction in schizophrenia. Arch Gen Psychiatry. 1993;50:190–7.
66. Falkai P, Bogerts B. Cell loss in the hippocampus of schizophrenics. Eur Arch Psychiatry Neurol Sci. 1986;236:154–61.
67. Jakob H, Beckmann H. Prenatal developmental disturbances in the limbic allocortex in schizophrenics. J Neural Transm. 1986;65:303–26.
68. Kikinis R, Shenton ME, Gerig G, Hokama H, Haimson J, O'Donnell BF, et al. Temporal lobe sulco-gyral pattern anomalies in schizophrenia: an in vivo MR three-dimensional surface rendering study. Neurosci Lett. 1994;182:7–12.
69. Luchins DJ, Weinberger DR, Wyatt RJ. Schizophrenia and cerebral asymmetry detected by computed tomography. Am J Psychiatry. 1982;139:753–7.
70. Johnstone EC, Crow TJ, Frith CD, Husband J, Kreel L. Cerebral ventricular size and cognitive impairment in chronic schizophrenia. Lancet. 1976;2(7992):924–6.
71. Heckers S, Heinsen H, Heinsen Y, Beckmann H. Morphometry of the parahippocampal gyrus in schizophrenics and controls. Some anatomical considerations. J Neural Transm. 1990;80:151–5.
72. Davis KL, Buchsbaum MS, Shihabuddin L, Spiegel-Cohen J, Metzger M, Frecska E, et al. Ventricular enlargement in poor-outcome schizophrenia. Biol Psychiatry. 1998;43:783–93.
73. Nair TR, Christensen JD, Kingsbury SJ, Kumar NG, Terry WM, Garver DL. Progression of cerebroventricular enlargement and the subtyping of schizophrenia. Psychiatry Res. 1997;74(3):141–50.
74. Illowsky BP, Juliano DM, Bigelow LB, Weinberger DR. Stability of CT scan findings in schizophrenia: results of an 8 year follow-up study. J Neurol Neurosurg Psychiatry. 1998;51:209–13.
75. Nasrallah HA, Olson SC, McCalley-Whitters M, Chapman S, Jacoby CG. Cerebral ventricular enlargement in schizophrenia: a preliminary follow-up study. Arch Gen Psychiatry. 1986;43:157–9.
76. Sprooten E, Papmeyer M, Smyth AM, Vincenz D, Honold S, Conlon GA, et al. Cortical thickness in first-episode schizophrenia patients and individuals at high familial risk: a cross-sectional comparison. Schizophr Res. 2013;151:259–64. doi:10.1016/j.schres.2013.09.024. pii: S0920-9964(13)00523-9.

77. Olney JW, Newcomer JW, Farber NB. NMDA receptor hypofunction model of schizophrenia. J Psychiatr Res. 1999;33:523–33.
78. Farber NB, Hanslick J, Kirby C, McWilliams L, Olney JW. Serotonergic agents that activate 5HT2A receptors prevent NMDA antagonist neurotoxicity. Neuropsychopharmacology. 1998;18:57–62.
79. Glantz LA, Gilmore JH, Lieberman JA, Jarskog LF. Apoptotic mechanisms and the synaptic pathology of schizophrenia. Schizophr Res. 2006;81:47–63.
80. De Zio D, Giunta L, Corvaro M, Ferraro E, Cecconi F. Expanding roles of programmed cell death in mammalian neurodevelopment. Semin Cell Dev Biol. 2005;16:281–94.
81. Sapolsky RM. Glucocorticoids and hippocampal atrophy in neuropsychiatric disorders. Arch Gen Psychiatry. 2000;57:925–35.
82. Susser E, Hoek HW, Brown A. Neurodevelopmental disorders after prenatal famine: the story of the Dutch Famine Study. Am J Epidemiol. 1998;147:213–6.
83. Altamura AC. A multidimensional (pharmacokinetic and clinical–biological) approach to neuroleptic response in schizophrenia. With particular reference to drug resistance. Schizophr Res. 1993;8(3):187–98.

Chapter 14
Neurodegeneration and Multiple Sclerosis

Axel Petzold

Abstract Neurodegeneration causes inexorable loss of neurons and function in both diseases and aging. Neurodegeneration damage produces a range of progressive disabilities from cognitive decline, behavioral, and mood disorders to problems with movement, coordination, and sensory dysfunction. Neurodegeneration is a major and growing public health issue which in its broadest sense embraces classical neurodegenerative disorders such as Alzheimer's disease and Parkinson's disease, as well as multiple sclerosis (MS), diabetes, acute brain injury among many other conditions. This chapter discusses the clinical and pathophysiological features of neurodegeneration in MS.

Keywords Demyelinating disease • Multiple sclerosis • Neurodegeneration • Transsynaptic axonal degeneration • Protein biomarker • Cerebrospinal fluid • Retina • Optical coherence tomography

Introduction

Neurodegeneration causes inexorable loss of neurons and function in both diseases and aging. Neurodegeneration damage produces a range of progressive disabilities from cognitive decline, behavioral, and mood disorders to problems with movement, coordination, and sensory dysfunction. Neurodegeneration is a major and growing public health issue which in its broadest sense embraces classical neurodegenerative disorders such as Alzheimer's disease and Parkinson's disease, as well as multiple sclerosis (MS), diabetes, acute brain injury, among many other conditions. This chapter discusses the clinical and pathophysiological features of neurodegeneration in MS.

A. Petzold, MD, PhD
Department of Neurology, VU University Medical Center,
Ce boelelaan, Amsterdam NL 1081 hv, The Netherlands
e-mail: a.petzold@vumc.nl

D. Galimberti, E. Scarpini (eds.), *Neurodegenerative Diseases*,
DOI 10.1007/978-1-4471-6380-0_14, © Springer-Verlag London 2014

The historical context will be discussed first, because our understanding of MS pathology has been much influenced by demyelination and a concept of dissemination in time and space. Next, the classical pathological features of neurodegeneration in MS are reviewed in more detail. Axonal loss will be placed centrally because of the important link to irreversible loss of function. The resulting disability has a major impact on an individual patient's life. Here limitations will be reviewed of those clinical and paraclinical assessments which were predominantly focused on demyelination and/or evidence for dissemination in time and space. It is against this backdrop that biomarkers for neurodegeneration will be presented. The chapter closes with an outlook on how this knowledge may be applied to future treatment trials targeted at halting neurodegeneration in MS.

Historical Context

Most of the credited clinicopathological descriptions of MS date back to the mid-nineteenth century [1–4]. The classical pathological features embrace inflammation, demyelination, and gliosis [5, 6].

Jean Marin Charcot, who pioneered the pathophysiological explanation of the symptoms observed in patients, distinguished three steps in the pathology of MS, which he called "la sclérose en plaques disseminé, la sclerose generalizé et la sclerose multiloculaire." First, astrocytic and microglial activation: "la multiplication des noyaux et l'hypertroplasie concomitante des fibres réticulé es de la névroglie sont le fait initial." Second, neuroaxonal degeneration: "l'atrophie dé gé nerative des é lé ments nerveux est secondaire." The interested reader is referred to a wonderful historical account on axonal pathology for more details [7]. And third, astrogliosis: "la né vroglie fait place au tissu fibrillaire" [2]. Ultimately, it was demyelination ("dépouillés de leur myé line" [8]) which became the key pathological feature of the disease, here depicted in a frequently cited sketch (Fig. 14.1).

The cause for these features has remained enigmatic ever since James Dawson's dichotomization into "inflammatory" and "developmental" concepts [9]. (This dichotomization remains a persistent intellectual concept with changing names such as "exogenous versus endogenous," "outside-in versus inside-out.")

While pathologically succinct, the difficulty for the treating physician remains to recognize and communicate a diagnosis of MS to the patient. Historically, MS was recognized in the preantibiotic area where inflammatory diseases such as syphilis presented major public health issues. Separating one from the other was not always straightforward. Not surprisingly, given the multitude of symptoms and signs mimicking other diseases, MS was also considered a chameleon. In absence of a diagnostic test, the clinical judgment cannot be substituted for. This notion is reflected in a series of diagnostic criteria, all more or less stating that the patient's symptoms and signs ought to be compatible with the characteristics of MS [10–13]. The careful and systematic, evidence-based approach on which these criteria rest distilled a conceptual framework which may be phrased as "dissemination in time and space."

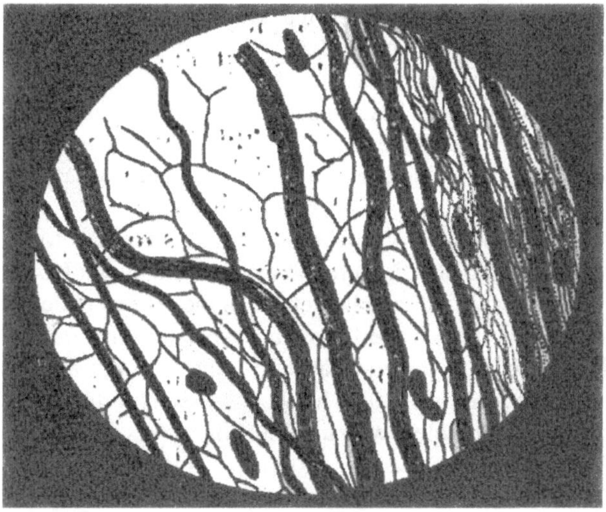

Fig. 14.1 The figure shows the original sketch of an MS lesion from the landmark paper of Charcot [8]. The image depicts a fresh MS plaque colored with carmine. Charcot's text implies presence of axonal pathology based on morphological observations of diameter and continuity. His interpretation is careful as he does not exclude possible preparation-related artifacts. The original text reads as "Elle représente une préparation frâche, provenant du centre d'une plaque scléreuse, colorié par le carmin et traité e par delacération. Au centre, vaisseau capillaire portant plusieurs noyaux. A droite et à gauche, cylindres d'axe, les uns volumineux, les autres d'un très–petit diamètre, tous dé pouillés de leur myéline. Le vaisseau capillaire et les cylindres d'axe étaient fortement colorés par le carmin. Les cylindres d'axe ont des bords parfaitement lisses, ne presentant aucune ramification. Dans l'intervalle des cylindres d'axe, membranes fibrilles de formation récente, à peu près parallèles les unes aux autres dans la partie droite de la préparation, formant à gauche et au centre, une sorte de réseau résultant, soit de l'enchevêment, soit de l'anastomose des fibrilles. Celles–ci se distinguent des cylindres d'axe, 1 par leur diamètre qui est beaucoup moindre; 2 par les ramifications qu'elles offrent dans leur trajet; 3 parce qu'elles ne se colorent pas par le carmin. — C¸ á et et là , noyaux disséminés. Quelques–uns paraissant en connexion avec les fibrilles conjonctives; d'autres ayant pris une forme irré gulière, due à l'action de la solution ammoniacale du carmin" [8]

Dissemination in time (DIT) and dissemination in space (DIS) are well suited to describe the occurrence of radiologically recognizable MS lesions in the brain and spinal cord [14].

It was precisely the absence of clear evidence for these characteristic features which made it so challenging to develop diagnostic criteria for primary progressive multiple sclerosis (PPMS) [15]. Later, Thompson and colleagues phrased this as "Neither set of criteria is appropriate to PPMS, since the basic requirement of two discrete episodes of neurological dysfunction cannot by definition be fulfilled" [16]. The clinical cornerstone of what emerged in International Panel diagnostic criteria was the documented clinical progression for more than 1 year [12].

To give one example, an Irish study on 111 PPMS patients showed that application of the Poser criteria would result in reclassification of the diagnosis to clinical probable MS (46 %), laboratory-supported MS (37 %), and unclassifiable (17 %) [17]. Therefore, an international panel, led by Thompson, phrased the core problem as "Neither set of criteria is appropriate to PPMS, since the basic requirement of two discrete episodes of neurological dysfunction cannot by definition be fulfilled" [16]. The clinical cornerstone of what emerged in International Panel diagnostic criteria was the documented clinical progression for more than 1 year [12]. Disease progression in PPMS is driven by neurodegeneration [18].

Paradoxically, the first in vivo observation of axonal loss in MS was difficult to publish at all, according to anecdotal reports from the authors. Hoyt and colleagues had observed retinal nerve fiber bundle defects in the eyes of patients with MS [19]. Much more frequently cited is the follow-up paper on this observation by Frisen et al. stating the presence of "insidious atrophy" of retinal nerve fibers in the eyes of patients with multiple sclerosis [20]. The second case reported by Frisen and Hoyt was a 15-year-old student athlete with a clinical diagnosis of "multifocal demyelinating disease," but without any history of optic neuritis. One may speculate that one argument for rejection at the time might have been that multiple sclerosis was a demyelinating disease, and the question was raised: why should there be at all atrophy of the nonmyelinated axons in the eye of a patient who did not even suffer from optic neuritis?

Axonal loss was, only some 24 years later, firmly put on the MS research agenda by the American cell biologist Bruce Trapp and the Norwegian pathologist Lars Bo [21]. The conceptional change this influential pathological study had will be discussed in the next section.

Pathological Features

Axonal Loss in Multiple Sclerosis

In order to put the observation by Trapp et al. into context, one needs to recall that axonal pathology may not be the most striking feature in the MS brain, but certainly is the one with a high impact for the patient [21–26]. Historically, axonal loss in MS has been associated with the "burnt-out" phase of the disease [27, 28]. Only with the wide availability of immunohistological techniques it was possible to demonstrate axonal pathology in active MS lesions [29]. There was extensive staining for amyloid precursor protein (APP) and the APP-positive structures resembled transected axons. It was, however, the three-dimensional reconstruction of these axonal ovoids, using confocal microscopy, which conclusively demonstrated axonal transections within acute MS lesions [21]. Interestingly, an accumulation of neurofilament protein was observed in the so-called end-bulbs.

In other words, the important new insight from this work was that a high number of transected axons were already present in acute lesions [21, 29] and in patients with a short clinical course [21]. This data changed the earlier perception of axonal loss in MS [1, 3, 4, 21, 29–36].

The data from Trapp et al. is consistent with the concept that MS lesions are an important trigger for axonal loss. But because disability continued to progress even after successful suppression of the inflammatory part of the disease, other aspects of axonal pathology were discussed [37]. Axons might be driven into a fatal energy deficit [38, 39]. There is good evidence that mitochondrial pathology and sodium channel redistribution contribute to an "ATP penalty" [40–45]. Axonal transport might be impaired [46–48]. Next, there might be loss of trophic support or increase of inhibitory substances such as Nogo. A barrier may result from astrogliosis. A low-grade inflammatory process might persist. There is the problem of failure to remyelinate. There may be acceleration of physiological processes of aging-related neurodegeneration. Endogenous capacities of repair might have their limits. In sum, those factors causing axonal degeneration might eventually outnumber those which were protective.

It is worthwhile to remember some limitations, axonal injury remains a dynamic process, and quantification of axonal loss in histological material might be complicated by tissue edema, the presence of inflammatory cells, and the problem of establishing a relationship with the number of healthy axons. There is a crucial dependence on well-preserved tissue with limited capacities of the existing brain banks. Most postmortem studies were biased to tissue from patients with long-standing disease duration, and there is a lack of representative tissue from the clinically and therapeutically relevant early disease phase. Some early tissue might be available through biopsy, but again questions might be asked how representative such tissue really is if taken because the presentation was very atypical. Finally, there are shortcomings to the analytical methods, dyes, and antibodies used.

Concepts of Axonal Degeneration

Like axonal injury, axonal degeneration is also a dynamic process. Most recent insights come from experimental studies in mice on fluorescently labeled axons [49]. It may be opportune to go back in time and revisit the first systematic description of axonal injury by Waller which gave rise to the eponym "Wallerian degeneration" [50].

In brief, Wallerian degeneration is a complex process which describes the degeneration of the distal axonal stump after axonal transection from the neuron. Wallerian degeneration begins with the enzymatic proteolysis of the axonal cytoskeleton [51]. Additionally, Wallerian degeneration affects also the sheathing glial cells, causes alterations in the adjacent blood–tissue barriers, and stimulates cells of macrophage lineage. From a mechanistic point of view, Wallerian degeneration is of anterograde direction.

Wallerian degeneration has to be distinguished from dying-back neuropathy, defined as the slow proximal spread of nerve fiber breakdown and ultimate apoptosis of the neuron [52]. The term dying back was introduced to describe the spatiotemporal pattern of central and peripheral nerve fiber pathology in degenerative diseases. Particular experiments with 2,5-hexanedione (2,5-HD) and acrylamide showed that the initial changes in the ultrastructure of the axon consisted of neurofilament (Nf) accumulation which was accompanied by clearly visible focal fiber swelling [53, 54]. This observation can also be seen in experimental autoimmune encephalomyelitis (EAE) [55].

An important, mechanistic question to be asked is: how the process of neurodegeneration can spread from a sick to a healthy neuron/axon? One attractive concept is transsynpatic axonal degeneration [56, 57]. These authors used a noninvasive, ultrarapid imaging technique, readily tolerated by patients, retinal optical coherence tomography (OCT) [58]. The study design was elegant and simple by focusing on neurodegeneration in the visual pathways. Following a stroke in the posterior visual pathways, dying-back neuropathy spreads (transsynaptic) from the second-order neuron located in the lateral geniculate nucleus (LGN) to the axons (retinal nerve fiber layer, RNFL) of the first-order neuron (retinal ganglion cell, RGC) [56, 57]. Likewise, there is evidence for anterograde transsynaptic axonal degeneration from a postmortem study of the visual system of patients with multiple sclerosis [59].

Taken together, this data suggests a concept of bidirectional (transsynaptic) axonal degeneration (Fig. 14.2).

The attraction of this unified concept of bidirectional (transsynaptic) axonal degeneration is that not only it is convenient to explaining how neurodegeneration spreads in MS, but more importantly it may contribute to opening a therapeutic window for future neuroprotective strategies in MS. The aim here will be to prevent the transsynaptic part of the degenerative process and thereby at least limit the ultimate impairment for the patient.

The Patient

The use and definition of terms to describe a patient's impairment, disability, and handicap in this section were based on the recommendations of the system adopted by the World Health Organization (WHO) [60].

Impairment describes the "loss or abnormality… of structure of function." Disability describes "a restriction or lack… of ability to perform an activity in the manner of within the range considered normal for a human being." Handicap describes "the disadvantage for an individual… that prevents or limits the performance of a role that is normal… for that individual." To be more specific, handicap represents the effects of impairments or disabilities in a wide social context and may be substantially influenced by the cultural background.

By definition (DIS and DIT), a patient will suffer from MS-related symptoms causing potentially reversible impairment in different parts of his or her body. From

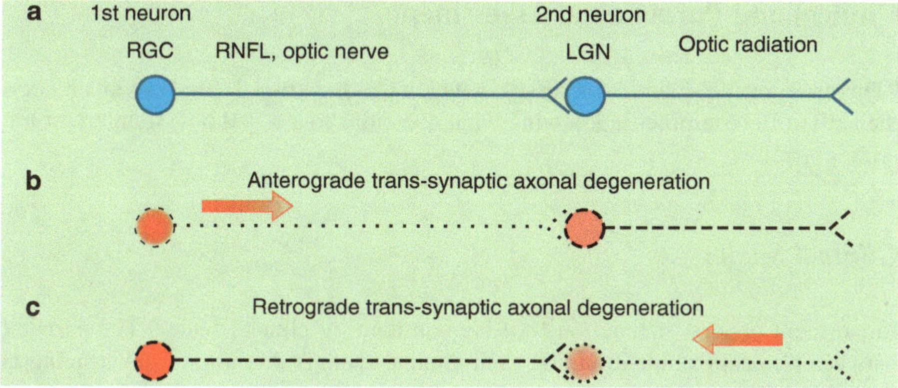

Fig. 14.2 A simplified and uniform mechanistic concept of axonal degeneration. (**a**) The normal situation is here shown for the visual system. The first-order neuron is represented by the retinal ganglion cell (*RGC*). The first axon is represented by the retinal nerve fiber layer (*RNFL*) which is named optic nerve after the axons passed through the lamina cribrosa. Here an axon is shown to synapse in the lateral geniculate nucleus (*LGN*) with the second-order neuron. Next, the second neuron sends its axon through the optic radiations to the occipital cortex. (**b**) Anterograde axonal degeneration starts at the RGC/RNFL/optic nerve (e.g., with optic neuritis). Once anterograde axonal degeneration reaches the LGN, it continues as transsynaptic anterograde axonal degeneration. (**c**) Retrograde axonal degeneration starts with axonal transections in the optic radiations (e.g., with eloquently placed white matter lesions). Once retrograde axonal degeneration reaches the LGN, the process continues as transsynaptic retrograde axonal degeneration. Ultimately, this leads to loss of retinal nerve fibers and apoptosis of the RGC. Longitudinally, the transsynaptic part of this concept of bidirectional axonal degeneration will always have to occur with a time lag. Understanding this time lag may potentially open a new therapeutic window for future neuroprotective strategies in MS

a patient's perception, gait and vision are the two most valuable functions [61]. Both gait and vision topped a list of 13 bodily functions during the early (<5 years) and late (>15 years) disease course. Importantly, early in the disease where patients were still ambulatory, gait was rated more valuable compared to visual function, but there was a crossover with long disease duration. With the ever-increasing use of visual communication channels (e.g., smartphones, tablets, social media), it can be anticipated that from a patient's point of view, the value and dependence on the visual system will continue to increase in the near future. This may be particularly true for those handicapped patients who crucially depend on the visual system for social interaction.

Two questions are frequently asked by patients: "Will this happen again?" (relapse) and "Will I end up in a wheelchair?" (neurodegeneration). The first one may, with caution, be answered based on the momentary clinical and radiological disease activity. Addressing the second question is more challenging because of a relative lack of longitudinal data from well-validated outcome measures for neurodegeneration.

Clinical and Paraclinical Assessments

"There are few neurological diseases in which the diagnosis depends so much upon the skill of the examiner in knowing what questions to ask and how to interpret the replies" [62].

Clinical Scales

Impairment or loss of function [60] is quantified by clinical scales. The paradox between clinical examination and each clinical scale is that normal functioning is tested, but loss of function is quantified. Because of the potential of CNS regeneration and plasticity, the clinical appearance of disability is a dynamic process. This has been illustrated in Fig. 14.1 and forms the basis on which MS patients are classified.

A range of validated clinical scales is now in use. For MS the most widely applied scale is the extended disability status scale (EDSS) for multiple sclerosis developed by Kurtzke in 1983 [63]. The EDSS combines a disability status scale [64] with functional systems [65–68]. A simple assessment of lower limb function is provided by the ambulation index (AI) or the timed walk test (TWT) [69]. Motor function of the upper limbs is quantified by the 9 hole peg test (9HPT) [69].

Psychometry is tested by the Paced Auditory Serial Addition Test (PASAT) [70]. The National Adult Reading Test (NART) is used to give an estimate of the premorbid IQ [71]. Current intellectual function is assessed by the Advance Progressive Matrices, Set 1 (Ravens). Memory is assessed by recognition of words and faces [72]. The paired associated learning test estimates learning abilities. Attention is readily quantified by the speed of letter counting [73]. Tests of executive function include the Wisconsin Card Sorting Test (Nelson) and the Cambridge Neuropsychological Test Automated Battery (CANTAB) [71, 74]. Fatigue is commonly estimated by Krupp's Fatigue Rating Scale [75]. Anxiety and depression have been measured using the National Hospital Anxiety and Depression Scale (HAD) measuring quality of life and measures for outcome of neurorehabilitation [76].

The TWT, 9HPT, and PASAT have been combined mathematically to give the Multiple Sclerosis Functional Composite (MSFC) [77, 78]. The MSFC has the potential to provide a more reliable measure of changes of function in MS than the EDSS, which is nonlinear and biased toward locomotion [79]. In addition, the MSFC may be perceived as a "melting pot" which permits to embrace other relevant clinical measures within a statistically valid concept. One potential extension of the MSFC may be contrast letter acuity [80].

A cross-sectional measure of disease severity in individual patients is provided by the global Multiple Sclerosis Severity Score (MSSS) [81]. The global MSSS is taken from a statistically constructed "lookup table." This table provides normally distributed disease severity scores for patients with an EDSS between 0 and 9.5 and a disease duration between 1 and 30 years.

Finally, there are patient-based outcome measures such as the MSIS-29 [82].

The advantages of clinical scales (and questionnaires) are that they may provide a more holistic view of an individual patient's disability compared to paraclinical tests. But there are also limitations to be considered:

1. Psychophysiological testing heavily depends on the patient's cooperation and motivation.
2. Biased to data from the system tested. This has been a frequently discussed limitation of the EDSS which is biased to the pyramidal system.
3. Learning effects. This is particularly challenging for testing cognition longitudinally.
4. Challenges of validation across cultural and language barriers. This may impact on the use as an outcome measure in multicenter studies.
5. Multiple biological causes for poor performance. In MS this includes:
    (a) Conduction block
    (b) Demyelination
    (c) Axonal loss

Paraclinical Tests

"The technological advances that have contributed to a better understanding of the pathophysiology and pathogenesis of MS have resulted in a disturbing increase in the number of false diagnoses of MS based exclusively on the results of test procedures." [62].

Paraclinical tests are a double-edged sword, but do have their merits in experienced hands if used as an extension of the clinical reasoning. The four most frequently used paraclinical tests over the past 50 years comprise in alphabetical order: CSF, CT, MRI, and visual evoked potentials (VEP), acknowledging that MRI has become the sole paraclinical test of the 2010 revision of the McDonald criteria for RRMS [12]. A historical head-to-head comparison based on the earlier Poser criteria is presented in Table 14.1.

Of note, none of these studies investigated the relevance of any of these tests for axonal loss, which as pointed out earlier was not the main focus of MS research at the time.

While sensitive for diagnostic purposes, the limitations of MRI to predict development disability were elegantly summarized by Kappos and colleagues in a thoroughly conducted meta-analysis: "Neither the initial scan nor monthly scans over 6 months were predictive of change in the EDSS in the subsequent 12 months or 24 months. The mean of gadolinium-enhancing lesion counts in the first six monthly scans was weakly predictive of EDSS change after 1 year (odds ratio = 1.34, $P = 0.082$) and 2 years (odds ratio = 1.65, $P = 0.049$)" [86].

This meta-analysis demonstrates the difficulties in predicting accumulation of irreversible disability, which is related to neurodegeneration, based on a paraclinical test focused on inflammatory disease activity. In contrast, MRI data on CNS atrophy

Table 14.1 Paraclinical tests used in multiple sclerosis

Reference	Test	Sensitivity, %	Conclusion
Polman et al. [83]	CSF	72.2	Diagnostic classification
	CT*	17.0	Differential diagnosis
	VEP	62.0	Diagnostic classification
Beer et al. [84]	CSF	77	Best reclassification specificity
	MRI	84	Highly sensitive, demonstrates DIS
	VEP	37	Useful if MRI and CSF not diagnostic
Filipini et al. [85]	CSF		
	MRI	70	Most sensitive test
	VEP		

*This study also included a very small, n=3, number of MRI scans, BAER and SSER

are much better correlated to sustained disability [87–89]. There is data on perfusion, functional MRI, high-field MRI, new sequences specifically addressing iron storage, double inversion recovery (DIR), and MR spectroscopy (MRS). For in-depth review of these and other MRI techniques, the reader is referred to recent reviews on the issue [90–94].

Likewise, for the CSF there is conflicting evidence on the relationship of CSF oligoclonal bands (OCBs) and disability. There are some reports suggesting that the absence of OCBs in the CSF of patient with MS may be a good prognostic sign [95–100]. Others did not find any prognostic value of either presence or absence of CSF OCBs [101, 102].

There may, however, be leverage using VEPs (and other evoked potentials) as a paraclinical test for neurodegeneration in MS [103, 104].

It may be suggested to separate those paraclinical tests which permit detection of axonal loss (and neurodegeneration) in the acute phase from those which are superior for documenting axon loss after some time has elapsed. Tentatively, retinal OCT was added to this list as an emerging paraclinical test for retinal layer atrophy:

1. Early phase of ensuing axonal injury and loss:
   - Biomarkers for acute axonal damage [105–107]
   - Imaging markers for neuronal dysfunction and apoptosis [108–110]
2. Late phase of axonal loss having resulted in manifest atrophy:
   - MRI atrophy markers [90, 111]
   - OCT [112, 113]
   - VEP and motor evoked potentials (MEP) [103, 104]

Acute Neurodegeneration in MS: Body Fluid Biomarkers

In MS disintegration of the axonal membrane causes release of biomarkers from injured axons and neurons in the surrounding extracellular fluid (ECF) [114]. These biomarkers diffuse from the brain ECF into the CSF and blood. Sampling from each of these body fluid compartments is possible with related advantages and disadvantages.

Table 14.2 Blood biomarkers in MS and their cellular sources

Blood biomarker	Neuron and axon	Astrocyte	Microglia	Oligodendrocyte	Other cells
14-3-3γ	+	+	+	+	+
Amyloid β42	+				
ApoE	+	+	+		
FABPs	+	+	+	+	+
FFA	+	+	+	+	+
Ferritin			+		+
GAP-43	+				
Gelsolin	+				+
GFAP		+			
HNE	+	+	+	+	+
NSE	+				+
Neurofilaments	+				
S100B		+		+	+
Tau	+	+	+	+	+
UCHL-1	+				

A review of the biomarker literature in MS shows that most studies were cross sectional and frequently of limited sample size [105, 107, 115–119]. Because of the essentially correlative nature of clinical biomarker investigations, only a snapshot in time is provided by cross-sectional studies. Not surprisingly, some studies find a clinical relevant correlation for a particular biomarker, while others do not. Some of these issues can be addressed by a meta-analysis. It will, however, be much more important to obtain high-quality long-term data. Therefore, Table 14.2 summarizes blood biomarkers categorized to their cell-type specificity.

For an extended biomarker table and in-depth review on CSF biomarkers for neurodegeneration, see Dujmovic [118] and Petzold [114].

The measurement of cell-type-specific biomarkers indirectly permits to estimate the degree of damage to the respective cellular source. For example, an increase of blood Nf levels gives indirect evidence for neuroaxonal damage. Neurofilaments have consistently found to be of prognostic value in MS [120–129].

Importantly, there has been convincing analytical and experimental work to substantiate the hypothesis that Nf levels are related to neurodegeneration [129–139]. Several commercial sandwich enzyme-linked immunosorbent assays (ELISAs) have been made available for both the neurofilament heavy (NfH) and light (NfL) chains.

Emerging Atrophy-Related Imaging Biomarkers for Neurodegeneration: Optical Coherence Tomography

An emerging imaging technology for neurodegeneration in MS is retinal OCT [113]. The results from a recent meta-analysis suggest that OCT provides an elegant, noninvasive, and rapid outcome measure for neurodegeneration in MS.

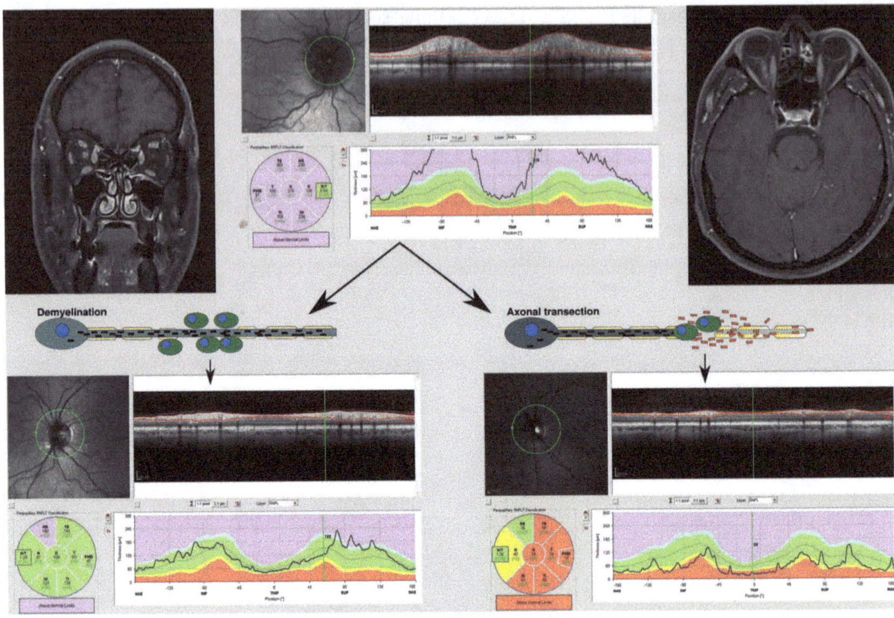

Fig. 14.3 A holistic model combining the strength of biomarkers suited for diagnosis (whole brain and spinal cord MRI) of the acute phase of neurodegeneration (e.g., body fluid neurofilament levels) with those more reliable during the later phase of neurodegeneration-related atrophy measures (retinal OCT)

While it is well known that optic neuritis causes loss of the retinal nerve fiber layer [20], it only recently emerged that such atrophy can also be present in nonoptic neuritic eyes [113, 140–152]. Because RNFL thickness also correlated with clinical scales and MRI measures, there is a need to test the reliability and validity of OCT in a multicenter setting.

Outlook

Taken together, neurodegeneration is an important feature of MS pathology because it is responsible for irreversible disability in patients. The dynamic nature of neurodegeneration poses challenges to the techniques used for monitoring. Some methods have their strengths in the acute phase; others only become reliable once neurodegeneration becomes manifest as atrophy [153]. A holistic model combining the respective strength and weaknesses is presented in Fig. 14.3.

Acknowledgment The MS Center VUmc is partially funded by a program grant of the Dutch MS Research Foundation.

References

1. Carswell R. Pathological anatomy: illustrations on elementary forms of disease. London: Longman; 1838.
2. Charcot M. Lecons sur les maladies chroniques du systéme nerveux: I – Des scléroses de la moelle épinière. Gaz des Hop. 1868;14:405–6, 554–5, 557–8, 566.
3. Curveilhier J, editor. Anatomie pathologique du crops humain. Paris: Bailliere; 1841.
4. Rindfleisch E. Histologisches Detail zur grauen Degeneration von Gehirn und Rückenmark Zugleich ein Beitrag zu der Lehre von der Entstehung und Verwandlung der Zelle. Arch Pathol Anat Physiol Klin Med. 1863;26:474–83.
5. McDonald WI, Miller DH, Barnes D. The pathological evolution of multiple sclerosis. Neuropathol Appl Neurobiol. 1992;18:319–34.
6. Raine CS. The neuropathology of multiple sclerosis. In: Raine CS, McFarland HF, Tourtellotte WW, editors. Multiple sclerosis, vol. 1. London: Chapman & Hall Medical; 1997. p. 151–70.
7. Kornek B, Lassmann H. Axonal pathology in multiple sclerosis. A historical note. Brain Pathol. 1999;9:651–6.
8. Charcot M. Histologie de la sclérose en plaques (II). Gaz des Hop. 1868;14:557–8.
9. Dawson J. The histology of disseminated sclerosis. Trans R Soc Edinb. 1916;50:517–740.
10. McDonald WI, Compston A, Edan G, et al. Recommended diagnostic criteria for multiple sclerosis: guidelines from the International Panel on the diagnosis of multiple sclerosis. Ann Neurol. 2001;50:121–7.
11. Polman CH, Reingold SC, Edan G, Filippi M, Hartung HP, Kappos L, et al. Diagnostic criteria for multiple sclerosis: 2005 revisions to the "McDonald Criteria". Ann Neurol. 2005;58: 840–6.
12. Polman CH, Reingold SC, Banwell B, Clanet M, Cohen JA, Filippi M, et al. Diagnostic criteria for multiple sclerosis: 2010 revisions to the McDonald criteria. Ann Neurol. 2011;69:292–302.
13. Poser CM, Donald WP, Scheinberg L, McDonald WI, et al. New diagnostic criteria for multiple sclerosis: guidelines for research protocols. Ann Neurol. 1983;13:227–31.
14. Swanton JK, Rovira A, Tintore M, Altmann DR, Barkhof F, Filippi M, Huerga E, et al. MRI criteria for multiple sclerosis in patients presenting with clinically isolated syndromes: a multicentre retrospective study. Lancet Neurol. 2007;6:677–86.
15. Thompson AJ, Polman CH, Miller DH, McDonald WI, Brochet B, Filippi M, Montalban M, De Sa J. Primary progressive multiple sclerosis. Brain. 1997;120:1085–96.
16. Thompson AJ, Montalban X, Barkhof F, Brochet B, Filippi M, Miller DH, et al. Diagnostic criteria for primary progressive multiple sclerosis: a position paper. Ann Neurol. 2000;47:831–5.
17. McDonnell GV, Hawkins SA. Application of the Poser criteria in primary progressive multiple sclerosis. Ann Neurol. 1997;42:982–3.
18. Lassmann H, van Horssen J, Mahad D. Progressive multiple sclerosis: pathology and pathogenesis. Nat Rev Neurol. 2012;8:647–56.
19. Hoyt WF, Schlicke B, Eckelhoff RJ. Fundoscopic appearance of a nerve-fibre-bundle defect. Br J Ophthalmol. 1972;56:577–83.
20. Frisen L, Hoyt WF. Insidious atrophy of retinal nerve fibers in multiple sclerosis. Funduscopic identification in patients with and without visual complaints. Arch Ophthalmol. 1974;92:91–7.
21. Trapp BD, Peterson JP, et al. Axonal transection in the lesions of multiple sclerosis. N Engl J Med. 1998;338:278–85.
22. Bjartmar C, Trapp BD. Axonal and neuronal degeneration in multiple sclerosis: mechanisms and functional consequences. Curr Opin Neurol. 2001;14:271–8.
23. McDonald WI. Relapse, remission, and progression in multiple sclerosis. N Engl J Med. 2000;343:1486–7.
24. Trapp BD, Ransohoff RM, Fisher E, Rudick RA. Neurodegeneration in multiple sclerosis, relationship to neurological disability. Neuroscientist. 1999;5:48–57.

25. Waxman SG. Demyelinating diseases – new pathological insights, new therapeutic targets. N Engl J Med. 1998;338:323–5.
26. Wujek JR, Bjartmar C, Richer E, Ransohoff RM, Yu M, Tuohy VK, Trapp BD. Axon loss in the spinal cord determines permanent neurological disability in an animal model of multiple sclerosis. J Neuropathol Exp Neurol. 2002;61:23–32.
27. Greenfield JG, King LS. Observations on the histopathology of the cerebral lesions in disseminated sclerosis. Brain. 1936;59:445–58.
28. Putnam TJ. Studies in multiple sclerosis VII Similarities between some forms of "encephalomyelitis" and multiple sclerosis. Arch Neurol Psychiatry. 1935. Read before the American College of Physicians, Philadelphia, April 29, 1935, pp. 1289–1308.
29. Ferguson B, Matyszak MK, Esiri MM, Perry VH. Axonal damage in acute multiple sclerosis lesions. Brain. 1997;120:393–9.
30. Arnold DL, de Stefano N, Matthews PM, Trapp BD. N-acetylaspartate: usefulness as an indicator of viable neuronal tissue. Ann Neurol. 2001;50:823–5.
31. Charcot M. Histologie de la sclérose en plaques (I). Gaz des Hop. 1868;14:554–5.
32. Davies SE, Newcombe J, Williams SR, McDonald WI, Clark JB. High resolution proton NMR spectroscopy of multiple sclerosis lesions. J Neurochem. 1995;64:742–8.
33. Matthews PM, Pioro E, Narayanan S, et al. Assessment of lesion pathology in multiple sclerosis using quantitative MRI morphometry and magnetic resonance spectroscopy. Brain. 1996;119:715–22.
34. McDonald WI. The pathological and clinical dynamics of multiple sclerosis. J Neuropathol Exp Neurol. 1994;53:338–43.
35. McFarland HF, Frank JA, Albert PS, Smith ME, Martin R, Harris JO, et al. Using gadolinium-enhanced magnetic resonance imaging lesions to monitor disease activity in multiple sclerosis. Ann Neurol. 1992;32:758–66.
36. Narayanan S, Fu L, Pioro E, De Stefano N, Collins DL, Francis GS, et al. Imaging of axonal damage in multiple sclerosis: spatial distribution of magnetic resonance imaging lesions. Ann Neurol. 1997;41:385–91.
37. Stadelmann C. Multiple sclerosis as a neurodegenerative disease: pathology, mechanisms and therapeutic implications. Curr Opin Neurol. 2011;24:224–9.
38. Su KG, Banker G, Bourdette D, Forte M. Axonal degeneration in multiple sclerosis: the mitochondrial hypothesis. Curr Neurol Neurosci Rep. 2009;9:411–7.
39. Trapp BD, Stys PK. Virtual hypoxia and chronic necrosis of demyelinated axons in multiple sclerosis. Lancet Neurol. 2009;8:280–91.
40. Black JA, Newcombe J, Trapp BD, Waxman SG. Sodium channel expression within chronic multiple sclerosis plaques. J Neuropathol Exp Neurol. 2007;66:828–37.
41. Cambron M, D'Haeseleer M, Laureys G, Clinckers R, Debruyne J, De Keyser J. White-matter astrocytes, axonal energy metabolism, and axonal degeneration in multiple sclerosis. J Cereb Blood Flow Metab. 2012;32:413–24.
42. Herrero-Herranz E, Pardo LA, Gold R, Linker RA. Pattern of axonal injury in murine myelin oligodendrocyte glycoprotein induced experimental autoimmune encephalomyelitis: implications for multiple sclerosis. Neurobiol Dis. 2008;30:162–73.
43. Lazzarino G, Amorini AM, Eikelenboom MJ, Killestein J, Belli A, Di Pietro V, et al. Cerebrospinal fluid ATP metabolites in multiple sclerosis. Mult Scler. 2010;16:549–54.
44. Mahad DJ, Ziabreva I, Campbell G, Lax N, White K, Hanson PS, Lassmann H, Turnbull DM. Mitochondrial changes within axons in multiple sclerosis. Brain. 2009;132:1161–74.
45. Witte ME, Bø L, Rodenburg RJ, Belien JA, Musters R, Hazes T, Wintjes LT, Smeitink JA, Geurts JJ, De Vries HE, van der Valk P, van Horssen J. Enhanced number and activity of mitochondria in multiple sclerosis lesions. J Pathol. 2009;219:193–204.
46. Kreutzer M, Seehusen F, Kreutzer R, Pringproa K, Kummerfeld M, Claus P, et al. Axonopathy is associated with complex axonal transport defects in a model of multiple sclerosis. Brain Pathol. 2011;22:454–71.

47. Petzold A, Gveric D, Groves M, Schmierer K, Grant D, Chapman M, Keir G, Cuzner L, Thompson EJ. Phosphorylation and compactness of neurofilaments in multiple sclerosis: indicators of axonal pathology. Exp Neurol. 2008;213:326–35.
48. Schirmer L, Merkler D, König FB, Brück W, Stadelmann C. Neuroaxonal regeneration is more pronounced in early multiple sclerosis than in traumatic brain injury lesions. Brain Pathol. 2013;23:2–12.
49. Kerschensteiner M, Schwab ME, Lichtman JW, Misgeld T. In vivo imaging of axonal degeneration and regeneration in the injured spinal cord. Nat Med. 2005;11:572–7.
50. Waller A. Experiments on the section of glossopharyngeal and hypoglossal nerves of the frog and observations of the alternatives produced thereby in the structure of their primitive fibres. Philos Trans R Soc Lond. 1850;140:423–9.
51. George R, Griffin JW. Delayed macrophage responses and myelin clearance during Wallerian degeneration in the central nervous system: the dorsal radiculotomy model. Exp Neurol. 1994;129:225–36.
52. Spencer PS, Schaumburg HH. Ultrastructural studies of the dying-back process IV Differential vulnerability of PNS and CNS fibers in experimental central-peripheral distal axonopathies. J Neuropathol Exp Neurol. 1977;36:300–20.
53. Politis MJ, Pellegrino RG, Spencer PS. Ultrastructural studies of the dying-back process. V. Axonal neurofilaments accumulate at sites of 2,5-hexanedione application: evidence for nerve fibre dysfunction in experimental hexacarbon neuropathy. J Neurocytol. 1980;9:505–16.
54. Schaumburg HH, Spencer PS. Degeneration in central and peripheral nervous systems produced by pure n-hexane: an experimental study. Brain. 1976;99:183–92.
55. Nikić I, Merkler D, Sorbara C, Brinkoetter M, Kreutzfeldt M, Bareyre FM, Brück W, Bishop D, Misgeld T, Kerschensteiner M. A reversible form of axon damage in experimental autoimmune encephalomyelitis and multiple sclerosis. Nat Med. 2011;17:495–9.
56. Jindahra P, Petrie A, Plant GT. Retrograde trans-synaptic retinal ganglion cell loss identified by optical coherence tomography. Brain. 2009;132:628–34.
57. Jindahra P, Petrie A, Plant GT. The time course of retrograde trans-synaptic degeneration following occipital lobe damage in humans. Brain. 2012;135:534–41.
58. Jindahra P, Hedges TR, Mendoza-Santiesteban CE, Plant GT. Optical coherence tomography of the retina: applications in neurology. Curr Opin Neurol. 2010;23:16–23.
59. Evangelou N, Konz D, Esiri MM, Smith S, Palace J, Matthews PM. Size-selective neuronal changes in the anterior optic pathways suggest a differential susceptibility to injury in multiple sclerosis. Brain. 2001;124:1813–20.
60. World Health Organization. International classification of impairments, disabilities and handicaps. Geneva. 1980. Available at: http://www.cdc.gov/nchs/icd/icf.htm. Accessed 10 Jan 2014.
61. Heesen C, Böhm J, Reich C, Kasper J, Goebel M, Gold SM. Patient perception of bodily functions in multiple sclerosis: gait and visual function are the most valuable. Mult Scler. 2008;14:988–91.
62. Poser CM. The unfortunate triumph of mechanodiagnosis in multiple sclerosis: a clinician's lament. Clin Neurol Neurosurg. 1992;94(Suppl):S139–42.
63. Kurtzke JF. Rating neurological impairment in multiple sclerosis: an expanded disability status scale (EDSS). Neurology. 1983;33:1444–52.
64. Kurtzke JF. A new scale for evaluating disability in multiple sclerosis. Neurology. 1955;5:580–3.
65. Kurtzke JF. Further notes on disability evaluation in multiple sclerosis. Neurology. 1965;15: 654–61.
66. Kurtzke JF. Natural history and clinical outcome measures for multiple sclerosis studies. Why at the present time does EDSS scale remain a preferred outcome measure to evaluate disease evolution? Neurol Sci. 2000;21:339–41.
67. Kurtzke JF. On the evaluation of disability in multiple sclerosis. Neurology. 1961;11:686–94.
68. Kurtzke JF. The disability Status Scale for multiple sclerosis: aplogia pro DSS sua. Neurology. 1989;39:291–302.

69. Kalkers NF, Bergers E, Castelijns JA, et al. Optimizing the association between disability and biological markers in MS. Neurology. 2001;57:1253–8.
70. Gromwall DMA. Paced auditory serial addition task: a measure of recovery from concussion. Percept Mot Skills. 1977;44:367–73.
71. Nelson HE, editor. National adult reading test: manual. Windsor: NFER-Nelson; 1982.
72. Warrington EK, editor. Recognition memory tests. Windsor: NFER Nelson; 1984.
73. Willison JR, Thomas DJ, du Boulay GH, et al. Effect of high haematocrit on alertness. Lancet. 1980;19:846–8.
74. Sahakian BJ, Owen MA. Computerized assessment in neuropsychiatry using CANTAB: discussion paper. J R Soc Med. 1992;85:399–402.
75. Krupp LB, LaRocca NG, Muir-Nash J, Steinberg AD. The fatigue severity scale. Application to patients with multiple sclerosis and systemic lupus erythematosus. Arch Neurol. 1989;46:1121–3.
76. Thompson AJ. Multiple sclerosis: rehabilitation measures. Semin Neurol. 1998;18:397–403.
77. Cutter GR, Baier ML, Rudick RA, Cookfair DL, Fischer JS, Petkau J, et al. Development of a multiple sclerosis functional composite as a clinical trial outcome measure. Brain. 1999;122:871–82.
78. Hoogervorst EL, Kalkers NF, Uitdehaag BM, Polman CH. A study validating changes in the multiple sclerosis functional composite. Arch Neurol. 2002;59:113–6.
79. Barkhof F. The clinico–radiological paradox in multiple sclerosis. Curr Opin Neurol. 2002;15:239–45.
80. Balcer LJ, Baier ML, Cohen JA, Kooijmans MF, Sandrock AW, Nano-Schiavi ML, Pfohl DC, et al. Contrast letter acuity as a visual component for the Multiple Sclerosis Functional Composite. Neurology. 2003;61:1367–73.
81. Roxburgh RHSR, Seaman SR, Masterman T, Hensiek AE, Sawcer SJ, Vukusic S, et al. Multiple sclerosis severity score. Using disability and disease duration to rate disease severity. Neurology. 2005;64:1144–51.
82. Hawton A, Green C, Telford C, Zajicek J, Wright D. Using the Multiple Sclerosis Impact Scale to estimate health state utility values: mapping from the MSIS-29, version 2, to the EQ-5D and the SF-6D. Value Health. 2012;15:1084–91.
83. Polman CH, Koetsier JC, Wolters EC. Multiple sclerosis: incorporation of results of laboratory techniques in the diagnosis. Clin Neurol Neurosurg. 1985;87:187–92.
84. Beer S, Rösler KM, Hess CW. Diagnostic value of paraclinical tests in multiple sclerosis: relative sensitivities and specificities for reclassification according to the Poser committee criteria. J Neurol Neurosurg Psychiatry. 1995;59:152–9.
85. Filippini G, Comi GC, Cosi V, Bevilacqua L, Ferrarini M, Martinelli V, et al. Sensitivities and predictive values of paraclinical tests for diagnosing multiple sclerosis. J Neurol. 1994;241:132–7.
86. Kappos L, Moeri D, Radue EW, Schoetzau A, Schweikert K, Barkhof F, et al. Predictive value of gadolinium–enhanced magnetic resonance imaging for relapse rate and changes in disability or impairment in multiple sclerosis: a meta–analysis. Lancet. 1999;353:964–9.
87. Bonati U, Fisniku LK, Altmann DR, Yiannakas MC, Furby J, Thompson AJ, et al. Cervical cord and brain grey matter atrophy independently associate with long-term MS disability. J Neurol Neurosurg Psychiatry. 2011;82:471–2.
88. Fisniku LK, Brex PA, Altmann DR, Miszkiel KA, Benton CE, Lanyon R, et al. Disability and T2 MRI lesions: a 20-year follow-up of patients with relapse onset of multiple sclerosis. Brain. 2008;131:808–17.
89. Fisniku LK, Chard DT, Jackson JS, Anderson VM, Altmann DR, Miszkiel KA, et al. Gray matter atrophy is related to long-term disability in multiple sclerosis. Ann Neurol. 2008;64:247–54.
90. Barkhof F, Calabresi PA, Miller DH, Reingold SC. Imaging outcomes for neuroprotection and repair in multiple sclerosis trials. Nat Rev Neurol. 2009;5:256–66.
91. Ropele S, de Graaf W, Khalil M, Wattjes MP, Langkammer C, Rocca MA, et al. MRI assessment of iron deposition in multiple sclerosis. J Magn Reson Imaging. 2011;34:13–21.

92. Rovaris M, Gass A, Bammer R, Hickman SJ, Ciccarelli O, Miller DH, Filippi M. Diffusion MRI in multiple sclerosis. Neurology. 2005;65:1526–32.
93. De Stefano N, Filippi M. MR spectroscopy in multiple sclerosis. J Neuroimaging. 2007;17 Suppl 1:31S–5.
94. Wattjes MP, Barkhof F. High field MRI in the diagnosis of multiple sclerosis: high field-high yield? Neuroradiology. 2009;51:279–92.
95. Amato MP, Ponziani G. A prospective study on the prognosis of multiple sclerosis. Neurol Sci. 2000;21:S831–8.
96. Avasarala JR, Cross AH, Trotter JL. Oligoclonal band number as a marker for prognosis in multiple sclerosis. Arch Neurol. 2001;58:2044–5.
97. Joseph FG, Hirst CL, Pickersgill TP, Ben-Shlomo Y, Robertson NP, Scolding NJ. CSF oligoclonal band status in- forms prognosis in multiple sclerosis: a case control study of 100 patients. J Neurol Neurosurg Psychiatry. 2009;80:292–6.
98. Lechner-Scott J, Spencer B, de Malmanche T, Attia J, Fitzgerald M, Trojano M, et al. The frequency of CSF oligoclonal banding in multiple sclerosis increases with latitude. Mult Scler. 2012;18:974–82.
99. Moulin D, Paty DW, Ebers GC. The predictive value of cerebrospinal fluid electrophoresis in 'possible' multiple sclerosis. Brain. 1983;106(Pt 4):809–16.
100. Zeman AZ, Kidd D, McLean BN, Kelly MA, Francis DA, Miller DH, et al. A study of oligoclonal band negative multiple sclerosis. J Neurol Neurosurg Psychiatry. 1996;60: 27–30.
101. Imrell K, Landtblom A-M, Hillert J, Masterman T. Multiple sclerosis with and without CSF bands: clinically indistinguishable but immunogenetically distinct. Neurology. 2006;67:1062–4.
102. Koch M, Heersema D, Mostert J, Teelken A, De Keyser J. Cerebrospinal fluid oligoclonal bands and progression of disability in multiple sclerosis. Eur J Neurol. 2007;14:797–800.
103. Fuhr P, Borggrefe-Chappuis A, Schindler C, Kappos L. Visual and motor evoked potentials in the course of multiple sclerosis. Brain. 2001;124:2162–8.
104. Klistorner A, Garrick R, Barnett MH, Graham SL, Arvind H, Sriram P, Yiannikas C. Axonal loss in non-optic neuritis eyes of patients with multiple sclerosis linked to delayed visual evoked potential. Neurology. 2013;80:242–5.
105. Petzold A. Neurofilament phosphoforms: surrogate markers for axonal injury, degeneration and loss. J Neurol Sci. 2005;233:183–98.
106. Stangel M, Fredrikson S, Meinl E, Petzold A, Stüve O, Tumani H. The utility of cerebrospinal fluid analysis in patients with multiple sclerosis. Nat Rev Neurol. 2013;9:267–76.
107. Ziemann U, Wahl M, Hattingen E, Tumani H. Development of biomarkers for multiple sclerosis as a neurodegenerative disorder. Prog Neurobiol. 2011;95:670–85.
108. Cordeiro MF, Guo L, Luong V, Harding G, Wang W, Jones HE, et al. Real-time imaging of single nerve cell apoptosis in retinal neurodegeneration. Proc Natl Acad Sci U S A. 2004;101:13352–6.
109. Narayana PA, Doyle TJ, Lai D, Wolinsky JS. Serial proton magnetic resonance spectroscopic imaging, contrast-enhanced magnetic resonance imaging, and quantitative lesion volumetry in multiple sclerosis. Ann Neurol. 1998;43:56–71.
110. Wattjes MP, Harzheim M, Lutterbey GG, Bogdanow M, Schmidt S, Schild HH, Träber F. Prognostic value of high-field proton magnetic resonance spectroscopy in patients presenting with clinically isolated syndromes suggestive of multiple sclerosis. Neuroradiology. 2008;50:123–9.
111. Lukas C, Sombekke MH, Bellenberg B, Hahn HK, Popescu V, Bendfeldt K, et al. Relevance of spinal cord abnormalities to clinical disability in multiple sclerosis: MR imaging findings in a large cohort of patients. Radiology. 2013;269:542–52.
112. Kupersmith MJ, Anderson S, Durbin MK, Kardon RH. Scanning laser polarimetry, but not optical coherence tomography predicts permanent visual field loss in acute nonarteritic anterior ischemic optic neuropathy. Invest Ophthalmol Vis Sci. 2013;54:5514–9.

113. Petzold A, de Boer JF, Schippling S, Vermersch P, Kardon R, Green A, Calabresi PA, Polman C. Optical coherence tomography in multiple sclerosis: a systematic review and meta-analysis. Lancet Neurol. 2010;9:921–32.
114. Petzold A. Biomarkers of disease progression. In: Wilkins A, editor. Progressive multiple sclerosis. London: Springer; 2013. p. 115–46.
115. Awad A, Hemmer B, Hartung H-P, Kieseier B, Bennett JL, Stuve O. Analyses of cerebrospinal fluid in the diagnosis and monitoring of multiple sclerosis. J Neuroimmunol. 2010;219:1–7.
116. Berger T, Reindl M. Multiple sclerosis: disease biomarkers as indicated by pathophysiology. J Neurol Sci. 2007;259:21–6.
117. Bielekova B, Martin R. Development of biomarkers in multiple sclerosis. Brain. 2004;127: 1463–78.
118. Dujmovic I. Cerebrospinal fluid and blood biomarkers of neuroaxonal damage in multiple sclerosis. Mult Scler Int. 2011;2011:767–83.
119. Kuhle J, Petzold A. What makes a prognostic biomarker in CNS diseases: strategies for targeted biomarker discovery? Part 2: chronic progressive and relapsing diseases. Expert Opin Med Diagn. 2011;5:393–410.
120. Brettschneider J, Petzold A, Junker A, Tumani H. Axonal damage markers in cerebrospinal fluid of patients with clinically isolated syndrome improve predicting conversion to definite MS. Mult Scler. 2006;12:143–8.
121. Gunnarsson M, Malmeström C, Axelsson M, Sundström P, Dahle C, Vrethem M, Olsson T, Piehl F, Norgren N, Rosengren L, Svenningsson A, Lycke J. Axonal damage in relapsing multiple sclerosis is markedly reduced by natalizumab. Ann Neurol. 2011;69:83–9.
122. Kuhle J, Leppert D, Petzold A, Regeniter A, Schindler C, Mehling M, et al. Neurofilament heavy chain in CSF correlates with relapses and disability in multiple sclerosis. Neurology. 2011;76:1206–13.
123. Lycke J, Andersen O, Rosengren L. Neurofilament in cerebrospinal fluid: a potential marker of activity in multiple sclerosis. Eur J Neurol. 1996;3:100.
124. Lycke JN, Karlsson JE, Andersen O, Rosengren LE. Neurofilament protein in cerebrospinal fluid: a potential marker of activity in multiple sclerosis. J Neurol Neurosurg Psychiatry. 1998;64:402–4.
125. Malmeström C, Haghighi S, Rosengren L, Andersen O, Lycke J. Neurofilament light protein and glial fibrillary acidic protein as biological markers in MS. Neurology. 2003;61:1720–5.
126. Petzold A, Rejdak K, Plant GT. Axonal degeneration and inflammation in acute optic neuritis. J Neurol Neurosurg Psychiatry. 2004;75:1178–80.
127. Petzold A, Mondria T, Kuhle J, Rocca MA, Cornelissen J, Te Boekhorst P, et al. Evidence for acute neurotoxicity after chemotherapy. Ann Neurol. 2010;68:806–15.
128. Petzold A, Eikelenboom MJ, Keir G, Grant D, Lazeron RH, Polman CH, et al. Axonal damage accumulates in the progressive phase of multiple sclerosis: three year follow up study. J Neurol Neurosurg Psychiatry. 2005;76:206–11.
129. Teunissen CE, Iacobaeus E, Khademi M, Brundin L, Norgren N, Koel-Simmelink MJA, et al. Combination of CSF N-acetylaspartate and neurofilaments in multiple sclerosis. Neurology. 2009;72:1322–9.
130. Abdo WF, van de Warrenburg BP, Munneke M, van Geel WJ, Bloem BR, Kremer HP, Verbeek MM. CSF analysis differentiates multiple-system atrophy from idiopathic late-onset cerebellar ataxia. Neurology. 2006;67:474–9.
131. Jackson SJ, Baker D, Cuzner ML, Diemel LT. Cannabinoid-mediated neuroprotection following interferon- gamma treatment in a three-dimensional mouse brain aggregate cell culture. Eur J Neurosci. 2004;20:2267–75.
132. Jackson SJ, Diemel LT, Pryce G, Baker D. Cannabinoids and neuroprotection in CNS inflammatory disease. J Neurol Sci. 2005;233:21–5.
133. Jackson SJ, Pryce G, Diemel LT, Cuzner ML, Baker D. Cannabinoid-receptor 1 null mice are susceptible to neurofilament damage and caspase 3 activation. Neuroscience. 2005;134:261–8.
134. Kuhle J, Regeniter A, Leppert D, Mehling M, Kappos L, Lindberg RLP, Petzold A. A highly sensitive electrochemiluminescence immunoassay for the neurofilament heavy chain protein. J Neuroimmonol. 2010;220:114–9.

135. Lu C-H, Kalmar B, Malaspina A, Greensmith L, Petzold A. A method to solubilise protein aggregates for immunoassay quantification which overcomes the neurofilament "hook" effect. J Neurosci Methods. 2011;195:143–50.
136. Petzold A, Keir G, Green AJE, Giovannoni G, Thompson EJ. A specific ELISA for measuring neurofilament heavy chain phosphoforms. J Immunol Methods. 2003;278:179–90.
137. Petzold A, Baker D, Pryce G, Keir G, Thompson EJ, Giovannoni G. Quantification of neurodegeneration by measurement of brain-specific proteins. J Neuroimmunol. 2003;138:45–8.
138. Pryce G, Ahmed Z, Hankey DJ, et al. Cannabinoids inhibit neurodegeneration in models of multiple sclerosis. Brain. 2003;127:2191–202.
139. Shaw G, Yang C, Ellis R, Anderson K, et al. Hyperphosphorylated neurofilament NF-H is a serum biomarker for axonal injury. Biochem Biophys Res Commun. 2005;336:1268–77.
140. Albrecht P, Fröhlich R, Hartung H-P, Kieseier BC, Methner A. Optical coherence tomography measures axonal loss in multiple sclerosis independently of optic neuritis. J Neurol. 2007;254:595–6.
141. Bock M, Brandt AU, Dörr J, Kraft H, Weinges-Evers N, Gaede G, et al. Patterns of retinal nerve fiber layer loss in multiple sclerosis patients with or without optic neuritis and glaucoma patients. Clin Neurol Neurosurg. 2010;112:647–52.
142. Fisher JB, Jacobs DA, Markowitz CE, Galetta SL, Volpe NJ, Nano-Schiavi ML, et al. Relation of visual function to retinal nerve fiber layer thickness in multiple sclerosis. Ophthalmology. 2006;113:324–32.
143. Frohman EM, Dwyer MG, Frohman T, Cox JL, Salter A, Greenberg BM, et al. Relationship of optic nerve and brain conventional and non-conventional MRI measures and retinal nerve fiber layer thickness, as assessed by OCT and GDx: a pilot study. J Neurol Sci. 2009;282:96–105.
144. Gundogan FC, Demirkaya S, Sobaci G. Is optical coherence tomography really a new biomarker candidate in multiple sclerosis? A structural and functional evaluation. Invest Ophthalmol Vis Sci. 2007;48:5773–81.
145. Jeanjean L, Castelnovo G, Carlander B, Villain M, Mura F, Dupeyron G, Labauge P. Retinal atrophy using optical coherence tomography (OCT) in 15 patients with multiple sclerosis and comparison with healthy subjects. Rev Neurol (Paris). 2008;164:927–34.
146. Klistorner A, Arvind H, Nguyen T, Garrick R, Paine M, Graham S, et al. Axonal loss and myelin in early ON loss in postacute optic neuritis. Ann Neurol. 2008;64:325–31.
147. Pulicken M, Gordon-Lipkin E, Balcer LJ, Frohman E, Cutter G, Calabresi PA. Optical coherence tomography and disease subtype in multiple sclerosis. Neurology. 2007;69:2085–92.
148. Quelly A, Cheng H, Laron M, Schiffman S, Tang RA. Comparison of optical coherence tomography and scanning laser polarimetry measurements in patients with multiple sclerosis. Optom Vis Sci. 2010;87:576–84.
149. Ratchford JN, Quigg ME, Conger A, Frohman T, Frohman E, Balcer LJ, et al. Optical coherence tomography helps differentiate neuromyelitis optica and MS optic neuropathies. Neurology. 2009;73:302–8.
150. Siger M, Dziegielewski K, Jasek L, Bieniek M, Nicpan A, Nawrocki J, Selmaj K. Optical coherence tomography in multiple sclerosis: thickness of the retinal nerve fiber layer as a potential measure of axonal loss and brain atrophy. J Neurol. 2008;255:1555–60.
151. Trip SA, Schlottmann PG, Jones SJ, Altmann DR, Garway-Heath DF, Thompson AJ, Plant GT, Miller DH. Retinal nerve fiber layer axonal loss and visual dysfunction in optic neuritis. Ann Neurol. 2005;58:383–91.
152. Zaveri MS, Conger A, Salter A, Frohman TC, Galetta SL, Markowitz CE, et al. Retinal imaging by laser polarimetry and optical coherence tomography evidence of axonal degeneration in multiple sclerosis. Arch Neurol. 2008;65:924–8.
153. Petzold A, Plant GT. Diagnosis and classification of autoimmune optic neuropathy. Autoimmun Rev. 2014;13(4–5):539–45.

Index

D. Galimberti, E. Scarpini (eds.), *Neurodegenerative Diseases*,
DOI 10.1007/978-1-4471-6380-0, © Springer-Verlag London 2014

M

N

O

P

FSC
www.fsc.org
MIX
Papier | Fördert gute Waldnutzung
FSC® C083411

Kolibri 360 GmbH
Ferdinand-Jühlke-Straße 7
99095 Erfurt, Deutschland
produktsicherheit@kolibri360.de